Vienna Circle Institute Library

Volume 10

Editorial Work/Production
Zarah Weiss

Editorial Address
Wiener Kreis Gesellschaft
Universitätscampus, Hof 1, Eingang 1.2
Spitalgasse 2-4, A–1090 Wien, Austria
Tel.: +431/4277 46504 (international) or 01/4277 46504 (national)
Email: vcs@univie.ac.at
Homepage: https://vcs.univie.ac.at/

This peer-reviewed series includes monographs and edited volumes, which complement the format of the related series "Vienna Circle Institute Yearbook" and "Veröffentlichungen des Institut Wiener Kreis", both published with Springer Nature. The books mainly deal with individual members of the Vienna Circle, the entire Schlick-Circle as a collective, its adherents and critics, as well as with related topics of Logical Empiricism and its periphery in historical and philosophical perspective. Specifically, they are based on so far unpublished primary sources and feature also forgotten and marginalized issues as significant contributions to the most recent research in these fields.

Matteo Collodel • Eric Oberheim
Editors

Feyerabend's Formative Years. Volume 2. Feyerabend on Logical Empiricism, Bohm & Kuhn

Correspondence and Unpublished Papers

Springer

Editors
Matteo Collodel
Ca' Foscari University of Venice
Venice, Italy

Eric Oberheim
Humboldt-Universität zu Berlin
Berlin, Germany

ISSN 1571-3083
Vienna Circle Institute Library
ISBN 978-3-031-57518-1 ISBN 978-3-031-57519-8 (eBook)
https://doi.org/10.1007/978-3-031-57519-8

This Springer imprint is published by the registered company Springer Nature Switzerland AG
The registered company address is: Gewerbestrasse 11, 6330 Cham, Switzerland

Contents

Abbreviations

AAAS	American Association for the Advancement of Science
CASBS	Center for Advanced Studies in the Behavioral Sciences
c	Copy or carbon copy
CH	Hempel's annotation
d	Draft
E	Editorial note
e	Envelope
EFA	*Europäisches Forum Alpbach* [European Forum Alpbach; known before 1949 as IHA]
ERE	"Explanation, Reduction and Empiricism", see FEYERABEND 1962b
FFY1	*Feyerabend's Formative Years. Volume 1. Feyerabend and Popper. Correspondence and Unpublished Papers*, see COLLODEL–OBERHEIM (eds.) 2020
FFY3	*Feyerabend's Formative Years. Volume 3. Feyerabend Among Popperians. Correspondence and Unpublished Papers*, see COLLODEL–OBERHEIM (eds.) forthcoming
H	Handwritten (or mostly handwritten) document
HF	Feigl's annotation
HF	Herbert Feigl Papers, Archives and Special Collections, Elmer L. Andersen Library, University of Minnesota
IEUS	International Encyclopedia of Unified Science
IHA	*Internationale Hochschulwochen Alpbach des Österreichischen College* [Alpbach International Summer Seminar of the Austrian College Society; known after 1949 as EFA]
IL	Imre Lakatos Collection, *LSE Library Archives and Special Collections*
JW	John W. N. Watkins Collection, *LSE Library Archives and Special Collections*
KP	Karl Popper Papers, Hoover Institution Library & Archives, Stanford University; and *Karl Popper-Sammlung, Universitätsbibliothek der Alpen-Adria-Universität Klagenfurt*
MCPS	Minnesota Center for Philosophy of Science

MSPS	Minnesota Studies in the Philosophy of Science (book series)
NSU	*Nordisk Sommeruniversitet*
ÖC	*Österreichisches College* [Austrian College Society]
PM	Postmark
PC	Postcard
PF	*Sammlung Paul Feyerabend, Philosophisches Archiv der Universität Konstanz*
SHQP	Sources for History of Quantum Physics
SSR	*The Structure of Scientific Revolutions,* see Kuhn 1962
R	*Folium rectum* (front side of a sheet of paper)
RC	Rudolf Carnap Papers, Archive of Scientific Philosophy
T	Typed (or mostly typed) document
T/H	Typed document with substantial handwritten annotations
TK	Kuhn's annotation

Chapter 1
Introduction

Matteo Collodel and Eric Oberheim

One of the most fascinating features of Paul Feyerabend's formative years was his direct involvement, both intellectually and personally, in several major movements in the history of twentieth century philosophy of science; most notably the rise of Karl Popper's critical rationalism, the demise of Rudolf Carnap's logical empiricism, and the 'historical turn' primarily attributed to Thomas Kuhn. The materials collected in this volume contain detailed discussions and analyses of specific arguments and assumptions underpinning these movements. They provide fresh insights into Feyerabend's personal relations to many prominent individuals and their evolving institutional roles,[1] as well as into Feyerabend's philosophical breakthrough at the turn of the 1960s.[2] The intellectual exchanges to which these materials testify jointly contributed to shaping Feyerabend's methodological argument for theoretical pluralism, initially developed in his controversial "Explanation, Reduction and Empiricism" (FEYERABEND 1962b, *abbr. ERE*). In this landmark essay, Feyerabend drew together his earlier work on Wittgenstein and formal logic, the foundations of quantum theory, and normative methodology, developing his idea of

[1] Indeed, as we shall see, a striking feature of Feyerabend's formative years is the personal interactions Feyerabend had with prominent individuals whose ideas directly shaped specific aspects of his developing views, such as Popper, Bohr, Wittgenstein, Carnap, Hempel, Bohm, and Kuhn.

[2] Understanding Feyerabend's ideas and the different roles that he played in these (and other) movements has been severely hindered by a notorious lack of reliable resources. For example, Carnap and Popper made almost no mention of their acquaintance with Feyerabend in their published philosophical or autobiographical papers (cf. CARNAP 1963; POPPER 1976). Scholars attempting to clarify and compare their mutual influences and to untangle and evaluate their views have been left for the most part with Feyerabend's side of these stories. Moreover, for a variety of reasons, Feyerabend's autobiographical reflections do not always provide an ideal basis for interpreting and assessing how his work developed and for understanding his various roles in these movements. See COLLODEL 2016.

M. Collodel, E. Oberheim (eds.), *Feyerabend's Formative Years. Volume 2. Feyerabend on Logical Empiricism, Bohm & Kuhn*, Vienna Circle Institute Library 10,
https://doi.org/10.1007/978-3-031-57519-8_1

'incommensurability', which laid the foundation for his views on science, knowledge and pluralism that gradually morphed into *Against Method* (FEYERABEND 1975; 1988; 1993).

The first part of this volume contains unpublished papers, discussions and correspondences directly pertaining to the development of Feyerabend's criticism of logical empiricism, including previously unpublished papers written by Feyerabend and detailed discussions of them by several prominent logical empiricists, such as Carnap and Carl Hempel. The second part contains Feyerabend's autobiographical reflections about David Bohm, and a previously unpublished lengthy response by Bohm to Feyerabend's review of Bohm's *Causality and Chance in Modern Physics* (BOHM 1957a).[3] The third part collects Feyerabend's correspondence with Thomas Kuhn, including detailed comments and criticisms that he made on a draft of Kuhn's *The Structure of Scientific Revolutions* (KUHN 1962, *abbr. SSR*). Bohm and Kuhn both made significant contributions to the development of Feyerabend's 1962 criticism of contemporary empiricism.

Taken together and against the background of Feyerabend's relationship to Popper presented in the first volume, the materials collected in this volume provide first-hand depictions of significant developments within these movements as well as succinct summaries of some of Feyerabend's specific criticisms of them. Because Feyerabend often expressed himself much more clearly and directly in his correspondence than in his sometimes rather notoriously opaque academic publications, the autobiographical remarks they contain contribute to a much clearer picture of his own development than the rather unclear image that has emerged based solely on third-person reconstructions that have had to rely on his published papers. By offering crisper and more direct formulations of some of Feyerabend's complex ideas and how they developed, these materials are of great help for understanding and situating those ideas in their proper context, allowing for a deeper and more accurate view, not only of the intellectual and personal sides of Feyerabend's relations, but also of the history of twentieth century philosophy of science more generally.

To facilitate access to, and appreciation of, the materials collected, and how they shed new light on Feyerabend's formative years, the next three sub-sections introduce Feyerabend's developing relationships to logical empiricism (1.1), Bohm (1.2), and then Kuhn (1.3) emulating the structure of the collection. Section (1.4) describes the collection providing specific information about the materials, their origins, and how they have been ordered. Section (1.5) specifies the editorial policies and conventions used to edit and present the material, which is followed by our acknowledgments to the many people and institutions that helped make this project possible.

[3] See FEYERABEND 1960b.

1.1 Feyerabend and Logical Empiricism

Feyerabend's critical interest in logical empiricism developed in the context of post-war Vienna, under the direct supervision of Viktor Kraft, who was Feyerabend's doctoral advisor and the last remaining member of the Vienna circle in Vienna after the war. Feyerabend's 1951 doctoral dissertation "On the Theory of Basic Sentences"[4] as well as his 1951 public lecture "The Dogmas of Logical Empiricism"[5] (Ch. 2) illustrate just how deeply Feyerabend was immersed in the Vienna circle tradition in the early 1950s. They also show that Feyerabend had already been influenced by both Popper and Willard Quine's contemporaneous criticisms of empiricism.[6] In the 1950s, Feyerabend developed ideas on meaning and methodology that would eventually become central to his criticism of contemporary empiricism in *ERE*, which sets out his methodological argument for theoretical pluralism based on the idea that theories separated by a scientific revolution are incommensurables, and concludes that a formal, 'objective' account of explanation cannot be given, because the meaning of a sentences is always relative to framework used to interpret it.[7]

The materials collected in Part I show how Feyerabend developed his infamous criticism of contemporary empiricism in *ERE* 1962 through a series of critical engagements with logical empiricism in the 1950s. This illustrates how Feyerabend developed some of the specific targets of his criticism, and to whom he was trying to explain 'how to be a good empiricist' (FEYERABEND 1963a). As we shall see, initially, in the 1950s, Feyerabend's focus was on Carnap, who had resumed his earlier position as logical empiricism's leading representative in the post-war period. By the early 1960's, Feyerabend's targets also included Hempel and Ernst Nagel. As he developed intellectually, Feyerabend also personally came to know many philosophers gravitating around logical empiricism, such as Béla von Juhos, Philipp Frank, Arthur Pap (for whom he worked as an assistant from 1953–54), and through him, Herbert Feigl. Feigl was director of the Minnesota Center for Philosophy of Science (*abbr.* MCPS), which was the institutional headquarters of logical empiricism in North America. The MCSP supported the work of leading logical empiricists, such as Carnap, Hempel and Nagel. Feyerabend kept in regular contact with Feigl, who was trying to help Feyerabend find an academic position. In 1955, Feigl was looking forward to hosting an extended visit by Feyerabend at the MCPS:

---

[4] Feyerabend's 1951 dissertation, *Zur Theorie der Basisätze,* has not been translated into English. Cf. FEYERABEND 1951.

[5] Cf. "*Die Dogmen des logischen Empirismus*", in STADLER (ed.) 2010. English translation this volume, Part 1, Ch. 2.

[6] Popper tried to develop his views explicitly in contrast to logical empiricism and had coined the term '*Basissätze*' (translated into English as 'basic statements') in contrast to the logical empiricist idea of 'observation statements' because Popper "needed a term not burdened with the connotation of a perception statement" (POPPER 1959a: 12fn*2). Popper's influence on Feyerabend is clearly already discernible in the first title 'On the Theory of Basic Sentences', whereas the second title is clearly a play on Quine's "Two Dogmas of Empiricism" (QUINE 1951a).

[7] See FEYERABEND 1962b: 95.

"I have been reading some of Feyerabend's papers and find most of them very good indeed. It will be a pleasure to have him here for some period next year. I shall write him soon".[8] Feyerabend must have made a very good first impression (which also seems to have had been the case with both Popper and Wittgenstein) as just six weeks later, Feyerabend reported to Popper that "Prof. Feigl told me that there may be a possibility for me staying at his institute for another two years. Here you have all those exciting possibilities of travelling, learning, etc. etc.".[9] Feyerabend even entertained the possibility of settling down in the U.S. for the long term, explaining to Popper that Feigl "is a realist and this is the most important thing in our 'idealistic' time. I am certain that I could stay in the U.S. for more than a year and that there is even a possibility of settling down there, if I am not a complete fool, which happens sometimes (and you know of it)".[10]

Throughout this period, Feyerabend displayed a polemical attitude towards the views of the leaders of logical empiricism, both in his published papers and in his contributions to the epistolary discussion groups organized by the MCPS under Feigl's management. Feyerabend's personal ties with the logical empiricist network further strengthened and consolidated in the summer of 1957, after Feyerabend's participation in a six-week conference on the philosophical foundations of physics at Feigl's institute. This paved the way to the offer of a visiting appointment for 1958–59, which was extended to Feyerabend by the University of California, Berkeley in early 1958, and which he eagerly accepted. In October of 1959, on a leave of absence from Berkeley, Feyerabend returned to the MCPS to visit Feigl. Feyerabend found his stay very stimulating in comparison to his lack of colleagues in Berkely (he had not yet met Kuhn): "Whereas in Berkeley there was nobody at all with whom I could discuss here there are at least two people, Herbert Feigl and Grover Maxwell."[11] During these early years of his American experience, Feyerabend's frequent visits to Minneapolis continued to be a fundamental part of boosting his philosophical breakthrough.[12]

From the very start, Carnap's deep and lasting positive influence on Feyerabend resulted from his understanding of Carnap's 1930's distinction between the occasion of the production of the sentences and its subsequent theoretical interpretation as a statement according to some theory, which is the main idea behind 'objective' theories of observation sentences.[13] Feyerabend adopted this early Carnapian distinction to try to criticize Carnap's later account of scientific language, and to explain his own view, which he initially called a "theory of reactors" (FEYERABEND 1951). This initial presentation was gradually refined into Feyerabend's 'causal' or

[8] Cf. Feigl to Pap, 15 February 1955, HF 03-134-33.

[9] COLLODEL-OBERHEIM (eds.) 2020 (*abbr. FFY1*), Feyerabend to Popper, 19 March 1955-8: 165.

[10] *FFY1*, *Feyerabend* to Popper, 19 March 1955-8: 166. As Feyerabend took up his first academic post as lecturer in the philosophy of science at Bristol in the fall of 1955, he did not visit the MCPS until 1957.

[11] Cf. *FFY1*, Feyerabend to Popper, 26 October 1959-12: 328.

[12] Cf. FEYERABEND 1966a; 1978; 1995.

[13] Cf. CARNAP 1929; 1936; 1937.

'pragmatic' theory of observation sentences complemented by his contextual theory of meaning in *ERE* 1962. This account allows one observation sentence to make two or more different statements depending on which theory is used to interpret it. This is how incommensurable theories, according to Feyerabend, give incompatible meanings to the same observation sentences. For example, according to Feyerabend, 'The ball fell' was understood to make at least two incompatible statements that imply two incompatible processes: that it was pushed by its inner impetus or pulled by an external force.[14]

After his initial engagement with Carnap in his 1951 Ph.D. thesis, a second engagement with Carnap was made possible thanks to Herbert Feigl. By circulating drafts and collecting responses, Feigl facilitated Feyerabend's participation in lively discussions with members of the newly founded MCPS, who were engaging in internal, epistolary debates on Carnap's 1956 criterion of empirical meaningfulness, and on Carnap's explication of the structure of scientific theories more generally. This resulted in two distinct controversies (see *below*, Part I, Ch. 3). Feyerabend explicitly stressed the similarity between his own views and Carnap's physicalism in the 1930s (and Popper and Neurath's views).[15] In this way, Feyerabend tried to use ideas from Carnap's earlier pre-war work to criticize the logical empiricist tradition that had grown out of it in the US after the war, which had largely abandoned realism as metaphysical speculation. As the materials collected in Part I are directly relevant to these developments, the following subsections will focus more closely on each of them in turn.

The Kraft circle and Feyerabend's theory of basic sentences (1951). As mentioned, Feyerabend's first visible engagement with logical empiricism and Carnap's work occurred during Feyerabend's university studies in post-war Vienna, which can be seen in his 1951 dissertation, 'On the Theory of Basic Sentences'. Feyerabend's investigation reviews the pre-war Vienna Circle's protocol sentence debate. It was developed through thorough discussions at meetings of the Kraft Circle[16], also known today as "the Third Vienna Circle" (Stadler 2006). Feyerabend was the leader of this small circle of STEM students (science, technology, engineering, and mathematics), which included Rudolf Goldberger de Buda (1924–1988), Peter Schiske (1924–2012), Heinrich K. Eichhorn (1927–1999), Hans Sagan (1928–2000), and Erich Jantsch (1929–1980). They were trying to revive the spirit of the Vienna Circle, but this time under the patronage of Feyerabend's supervisor, Victor Kraft, who was a former senior but peripheral member of the Vienna Circle, and the successor of Mach, Boltzmann, and Schlick's Chair at the University of Vienna.[17] Most of the members of this study group went on to have brilliant academic careers in

---

[14] See Feyerabend 1962b: 56–57.

[15] See Feyerabend 1962b: 34f.

[16] See Feyerabend 1995a: p. 85.

[17] Against the backdrop of the forced migration of the adherents of the scientific worldview, an unlikely, transient revival of scientific philosophy occurred during the allied occupation from 1946 to 1955. Cf. Kuby 2010; Schorner 2010; Stadler (ed.) 2010.

North America in astronomy, physics or mathematics. Feyerabend also managed a brilliant academic career, but along a very different path.

During this period, on various occasions in Vienna and Alpbach, in events promoted by para-academic institutions such the *Österreichisches College* [Austrian College Society] or the *Institut für Wissenschaft und Kunst* [Institute for Science and Fine Arts], Feyerabend had the chance to meet and interact intensively and repeatedly with acclaimed physicists and distinguished philosophers of science such as Erwin Schrödinger (1949, 1953, 1955),[18] Léon Rosenfeld (1949),[19] Maurice Pryce (1949),[20] Philipp Frank (1950, 1955),[21] and Karl Popper (1948).[22] This afforded him the opportunity to establish himself as a philosopher in the foundations of quantum theory within academic circles on both sides of the Atlantic, virtually overnight in the second half of the 1950s.

In his doctoral thesis, Feyerabend examined what came to be known as the protocol sentence debate from two perspectives (formal and an empirical).[23] By

---

[18] Feyerabend first met Schrödinger at the 1949 *Europäisches Forum Alpbach* (*abbr.* EFA), cf. *FFY1*, Feyerabend to Popper, 1 November 1948-2: 64n3. They also met at the 1953 EFA, cf. *FFY1*, Feyerabend to Popper, 10 September 1953-4: 72, and n2. On Feyerabend and Schrödinger at the *EFA*, see Hacohen 2000: 314–315; Kuby 2016. In 1955, Feyerabend, seeking a recommendation, asked Popper to contact Schrödinger on his behalf. Popper replied: "I shall of course write to Schrödinger if you want me to but it is infinitely preferable that you should write to him yourself. I know that he remembers you, and it is a general rule that people prefer to be approached directly by the person most immediately concerned, rather than through an intermediary" (*FFY1*, Popper to Feyerabend, 1 April 1955-11: 173). On Feyerabend and Schrödinger, cf. *FFY1*: 5–8, 28n24; Ch. 3: 55; Feyerabend to Popper, 1 November 1948-2: 64 and 64n3; Feyerabend to Popper, 10 September 1953-4: 72, especially 72n2, 73: Feyerabend to Popper, 28 September 1953-5: 76–77; Feyerabend to Popper, 8 October 1953-6: 79n3, 80, 80n5; Feyerabend to Popper, 1953-7: 82; Feyerabend to Popper, 31 October 1953-9: 89, 89n4; Feyerabend to Popper, 2 October 1954-18: 132; 19. Feyerabend to Popper, 3 October 1954-19: 140; Feyerabend to Popper, 19 March 1955-8: 166; Feyerabend to Popper, 25 March 1955-10: 170; Feyerabend to Popper, 1 April 1955-11: 173; Feyerabend to Popper, May 1955-13: 175; Feyerabend to Popper, 27 May 1955-16: 178; Feyerabend to Popper, 28 October 1955-29: 201; Feyerabend to Popper, 1956-4: 216; Feyerabend to Popper, 11 February 1956-10: 223; Popper to Feyerabend, 11 November 1958-13: 300, 300n6; Feyerabend to Popper, 18 November 1958-14: 302; Ch. 7: 476; 478-47. See *below*, Ch. 2, Feyerabend to Feigl, 28 Juhn 1957-2: n14; Feyerabend to Kuhn, 1961-3.

[19] On Feyerabend and Rosenfeld, cf. *FFY1*, Ch. 2: 30; Feyerabend to Popper, 1 November 1948-2: 64n3; Feyerabend to Popper, 14 January 1957-1: 246; Feyerabend to Popper, 27 March 1957-4: 253; Feyerabend to Popper, 30 March 1957-5: 256; Feyerabend to Popper, 7 April 1957-7: 260; and see *below*, Ch. 3, 1959-3; 1961-4; Feyerabend to Peat, 7 September 1993-2; 1993-3; Feyerabend to Kuhn, 1961-3: especially n13.

[20] On Feyerabend and Pryce, cf. *FFY1*, Feyerabend to Popper, 1 November 1948-2: 64n3; Feyerabend to Popper, 14 October 1955-27: 196n2; especially Feyerabend to Popper, 28 October 1955-29: 199n1; Feyerabend to Popper, 6 February 1956-7: 219; Feyerabend to Popper, 11 February 1956-10: 223; Feyerabend to Popper, March 1956-14: 228n4; Feyerabend to Popper, 7 April 1957-7: 261. See *below*, Feyerabend to Peat, 7 September 1993-2.

[21] On Feyerabend and Frank, cf. *FFY1*, Feyerabend to Popper, 24 December 1953-15: 97n5; Feyerabend to Popper, 3 October 1954-19: 134; Feyerabend to Popper, 28 August 1955-25: 191n2.

[22] Cf. *FFY1*, Ch. 1: 2.

[23] On the protocol sentence debate, see Uebel 2007.

drawing from research in the psychology of perception and elaborating on the development of logical empiricism in northern Europe,[24] the first part of the work develops a critique of 'subjective theories of basic sentences', that is, approaches attempting to ground basic sentences in "the [immediately] given". Specifically, Feyerabend discusses 'sense data' and Schlick's '*Konstatierungen*'. In the second part of his dissertation, Feyerabend groups various accounts under the heading of 'objective theories of basic sentences', which includes Carnap's and Neurath's physicalism, and which is where he situates his own proposal of a "theory of reactors" [*Theorie der Reaktoren*]. Feyerabend argues that while subjective theories correctly focus on the production of basic sentences, they give the wrong account of their epistemological function, because they ground the meaning of basic sentences in their phenomenological occurrence (in the cause or appearance of the observation). By contrast, Feyerabend takes his theory of reactors to give an entirely empirical theory of the production of basic sentences, whose meaning is only then established in a separate, subsequent act of interpretation, so that what a sentence means will depend on which theory is used to interpret it. Anticipating his realism and pluralism to come, Feyerabend also criticizes Carnap's assertions that physicalism implies a reduction of "qualitative sciences" (*Qualitätswissenschaften*) to physics. He objects to this purported implication of physicalism by stressing the autonomous "ontological status" (*Realwert*) of each discipline's theories.

'On the Interpretation of Scientific Theories' (1960). Feyerabend's second engagement with Carnap took place in the mid-1950s. This time, Feyerabend published criticisms of Carnap's more recent proposals. Specifically, between 1954 and 1958, based on the ideas developed in his doctoral thesis, Feyerabend tried to find fault with Carnap's evolving double language model for explicating scientific knowledge. A brief and somewhat opaque critique of Carnap's (1939) model appeared in (FEYERABEND 1954e, §4) and was reworked in more detail in a German publication (FEYERABEND 1955b).[25] The paper examines the function and form of 'coordinating definitions' understood as semantic rules in Carnap's model, and once again, Feyerabend argues that this leads to absurd consequences. Instead, according to Feyerabend, the meaning of central terms used within a theoretical framework is holistically defined by the system itself, while observational terms used to test any such system can only be characterized causally, as utterances produced under certain conditions, which then can be variously interpreted depending on which theory is used for that purpose.

In the summer of 1955, Feyerabend's arguments were discussed by Arthur Pap and Michael Scriven at the MCPS. Their critical remarks, highlighting misunderstandings, confusions and obscurities in Feyerabend's paper, brought a temporary end to a controversy that in all likelihood never even reached Carnap's attention (see *below*, Ch. 3, Sections 1–3). The following year, however, Carnap's proposal of an updated version of his model, complemented by a novel criterion of empirical

[24] Cf. MANNINEN–STADLER 2010 (eds.).

[25] Cf. CARNAP 1939.

meaningfulness (CARNAP 1956) sparked a new and lively rebuke from Feyerabend. At the turn of 1957, Feyerabend tried to expose what in his view were "obvious defects" of Carnap's proposal in a short 'Critical Note' (see *below*, Ch. 3), which he sent to the prestigious journal *Mind* for publication as well as to Feigl, who circulated it among the MCPS's affiliates. This triggered an intense debate between Feyerabend, on one side, and Carnap, Hempel and William Rozeboom, on the other, with Feigl in the middle, editing and forwarding the cross-communications to the participants in the debate. Feyerabend's insistent critique of Carnap's views unfolded throughout the first half of 1957 and totaled nine items including all the replies and rejoinders. Eventually, Feyerabend acknowledged his misrepresentation of Carnap's views and withdrew his note for consideration by *Mind*, thus terminating the second controversy (see *below*, Ch. 3, Sections 4–13). Yet, shortly before admitting defeat, Feyerabend resumed the first controversy with a new 'Critical Note', which was intended, among other things, to clarify the opaqueness of FEYERABEND 1955b and to provide a reply to the criticisms that it had received at the MCPS two years earlier. By the late spring of 1958, Feyerabend developed this 'Critical Note' into his "On the Interpretation of Scientific Theories" (FEYERABEND 1960b), which was delivered to Feigl shortly after its completion, possibly as early as September 1958, and then circulated by Feigl to members of the MCPS group. This time, however, Feyerabend's attempt to criticize Carnap's approach was not honoured with an answer either from Carnap or from Hempel. Rather, the job of dealing with it was left to Rozeboom, who by early November 1958 offered his critical remarks in a memorandum responding to Feyerabend, unfailingly followed by Feyerabend's reply, which finally brought the first controversy to a close (see *below*, Ch. 3, Sections 14–16).

In general, Feyerabend's qualms revolved around the very idea of "partial interpretation", central to Carnap's reconstruction of scientific theories in the double language model. More specifically, Feyerabend emphasized, once again, that in Carnap's model, any improvement or change in the means of observation that affects the empirical testing of a theory would be mirrored in the systematic reconstruction, via correspondence rules, of the relation between the partially interpreted calculus formulated in the theoretical language, on the one hand, and experience described in the fully interpreted observation language on the other. Such improvements would thereby also result in a change of the theory. In other words, according to Feyerabend, Carnap's view allegedly implies that, for example, building better microscopes would change the meaning of the theories about the cells they are used to observe.[26] Such a paradoxical consequence was meant to undermine a "positivistic", instrumentalist-leaning and conservative understanding of scientific theories, and to promote indirectly their "realistic" and progressive understanding, according to which theoretical language is meaningful independently of the casual mode of its production. For Feyerabend meanings are not fixed by experience. They change

[26] Feyerabend makes the same point when he argues against formal and 'objective' accounts of explanation more generally in ERE. Cf. FEYERABEND 1962b: 93–95.

with experience as theories about what they describe and explain improve, which was the view that Feyerabend defended throughout his formative years.

These controversies with Carnap, who was irritated by Feyerabend's confrontational attitude were detrimental to Feyerabend's reputation among Feigl's MCPS group. As a result, before his first visit to the US in mid-August 1957, and despite Feigl's benevolent mentorship, Feyerabend's name was already notorious in logical empiricist quarters for his harshly critical boldness and seemingly opinionated arrogance. However, after Feyerabend took up a position at the University of California, Berkeley in the fall of 1958, both Carnap and Hempel had a chance to meet Feyerabend in person, at which point they re-evaluated their initial negative impressions that had resulted from their epistolary exchanges. When Carnap became personally acquainted with Feyerabend in the spring break of 1959, during a lecture tour that brought Feyerabend to southern California, Carnap immediately found Feyerabend a congenial, easy-going and stimulating discussion partner in the philosophy of physics. Carnap invited Feyerabend's critical comments on a projected work on entropy, which Carnap had drafted in 1952–54 as two separate papers (and which was only posthumously published as CARNAP 1977). Feyerabend and Carnap met again in February 1960, during a small and intense three-day restricted conference in Santa Monica sponsored by the MCPS that included Feigl, Paul Meehl and Grover Maxwell. It seems likely that on that occasion they discussed the objections that Feyerabend and Maxwell had raised against Carnap's views (and logical empiricism more generally) with Carnap himself.

"Explanation, Reduction and Empiricism" (1962). In the early 1960s, Feyerabend once again redeployed the basic Carnapian distinction underlying Feyerabend's earlier physicalist 'theory of reactors' (1951). This time Feyerabend developed an account of observational meaning that combines the critical rationalist maxim to try to increase testability with what he took to be insights he attributed to Wittgenstein's contextual theory of meaning. In *ERE*, Feyerabend again traces his 'pragmatic' or 'causal' theory of observation to the earlier protocol sentence debate, and in particular, to the papers in which Carnap himself, under pressure from Neurath, had come to accept the advantages of a physicalist language for describing sentences used to test theories. Feyerabend argues that later developments within logical empiricism, as exemplified by Carnap's own change in views, represent a step back, relative to the progress that the protocol sentence debate had achieved towards an 'objective' or 'causal', or 'empirical' theory of observation sentences.[27] In this way, Feyerabend uses early pre-war logical empiricism to criticize "modern" or "contemporary" post-war logical empiricism Unsurprisingly, therefore, *ERE* also resumes Feyerabend's attack on Carnap's "well-known double-language model".[28] According to Feyerabend, Carnap's

> tacit withdrawal from the pragmatic theory of observation is one of the most surprising features of modern empiricism. It is responsible for the fact that this philosophy, despite the

[27] See FEYERABEND 1962b: 35.

[28] FEYERABEND 1962b: 41.

> apparent progress that has been made since the thirties, is still in accordance with the assumption that observational meanings are invariant with respect to the process of explanation. (1962b: 42)

According to Feyerabend, Hempel's account of explanation, Nagel's account of reduction and Popper's empiricism share this same undesirable feature as Carnap's double-language model.[29] They all imply meaning invariance of the observation language, when in fact such meanings do and should change through scientific revolutions as better theories replace their conceptually incompatible predecessors.

Throughout the first half of the 1960s, Feyerabend continued to highlight what he took to be two erroneous underlying assumptions of contemporary empiricist accounts — the principles of deducibility and of meaning invariance — finding them historically inaccurate, as well as methodologically undesirable, because their conceptual conservativism would stultify scientific progress, not promote it. By reconstructing textbook cases of scientific progress, Feyerabend proposed his theoretical pluralism as a more descriptively apt and methodologically desirable strategy for promoting progress. In the process, he argued that the conceptual frameworks that shape experience are replaced, not corrected and explained, by a universal theory that offers a better corroborated, but conceptually incompatible interpretation of reality.[30]

While Feyerabend's idea of incommensurability can be seen to echo some aspects of aspects of Carnap's distinction between external and internal questions (Carnap 1950), such similarities should be regarded as a case of independent and superficial convergence, rather than the result of historical influence. Although some ideas in Carnap's work may seem to resonate with some of Feyerabend's views, one should keep in mind that Carnap and Feyerabend had deeply different approaches to science, language, and philosophy.[31] Carnap's "logic of science" resulted from and is justified by his analysis of scientific language, whereas Feyerabend's approach resulted from and is justified by his analysis of scientific method. In his early philosophy, Feyerabend argued for his methodological views about science ultimately on ethical grounds. For Feyerabend, science is what we make it and what we make of it. Conversely, Feyerabend's historical turn, which begins in the mid-to-late 1960s, was certainly partly inspired by, and deeply resonates with, Neurath and Frank's "behavioristics of scholars", both in its attention to pragmatic factors affecting scientific practice and in its wariness of formalization as a privileged tool of philosophical inquiry.

It is possible that Feyerabend and Carnap met only once after the 1962 conference in Santa Monica, at the 1964 European Forum Alpbach, together with Feigl, among others. In any case, they remained on friendly terms until Carnap's death, as evidenced by Carnap's letter of recommendation in December 1966, which was instrumental to Feyerabend's appointment at the Free University of Berlin in 1968

---

[29] On Feyerabend's criticism of Popper's empiricism, see Oberheim 2024, Oberheim and Hoyningen-Huene 2024.

[30] See Feyerabend 1962b: 92–95

[31] See especially Oberdan 1990.

(see *below*, Ch. 2, Section 11). Feyerabend's developing views, in contrast to Carnap's, can be exemplified by his re-appraisal of Carnap's 'Principle of Tolerance'. While denouncing it as sterile in one context, Feyerabend would also group Carnap among the "splendid exceptions" to the "much more rigid attitude" of most "meta-scientists" and would judge Carnap's 'Principle of Tolerance' as being "pretty close to my own anarchism" (FEYERABEND 1977: 362). Though a careful reader of Carnap, it seems that Feyerabend, like many of his contemporaries, had not initially properly understood Carnap's project of linguistic engineering centred around proposals of formally improved conceptual tools that could eventually prove fruitful for progress in scientific research. An alternative, more radical reading suggests that Feyerabend's critique of rational reconstruction was based on the belief that the Carnapian formal approach wouldn't prove fruitful for promoting progress because of meaning variance, and because it was too far-removed from the messiness and informal nature of scientific concept-building. In any case, Feyerabend seems initially to have missed Carnap's emphasis on the possibility of producing wholly new concepts, rather than just refining inherited old ones.

In sum, despite Feyerabend's close personal proximity to leading logical empiricists, Feyerabend was highly critical of logical empiricism's main tenets. Carnap's double language model of logical reconstruction of scientific theories was his main targets in the 1950s,[32] to which he added Hempel's theory of explanation, Nagel's theory of reduction and Popper's empiricism, which were subjected to repeated attacks in the 1960s.[33] The climax of Feyerabend's repeated assaults on logical empiricism was reached with the idea of incommensurability as set out in *ERE*, which is considered by some as "[t]he most influential critique of Logical Positivism's analysis of scientific observation".[34] Notwithstanding a consolidated tradition of studies in the origins and development of logical empiricism, the peculiar role of Feyerabend as both a sort of insider and as a critic of the movement has so far failed to be satisfactorily analysed in all of its historical and theoretical details.[35] Whether this might be partly due to the radical theses Feyerabend appeared to be championing or to the bad reputation his increasingly anti-academic personality gave him, it seems clear that the lack of better documented evidence as to Feyerabend's relations to logical empiricism has constrained the understanding of a crucial period of twentieth century philosophy of science.

[32] Cf. FEYERABEND 1954e; 1955b; 1958a, CARNAP 1936; 1937; 1938; 1939; 1956.

[33] Cf. FEYERABEND 1962b; 1963a; 1965a; 1965b; 1965c; HEMPEL-OPPENHEIM 1948; NAGEL 1949; 1961; OBERHEIM 2024.

[34] OBERDAN 1990: 25.

[35] Noteworthy, though rather dated, exceptions on the theoretical side are HARRÉ 1959; BUTTS 1966; TOWNSEND 1971; and OBERDAN 1990. STADLER-FISCHER (eds.) 2006 mark a turning point on the historical side.

1.2 Feyerabend, Bohm and the Foundations of Quantum Theory

David Bohm (1917–1992) was an American-British quantum physicist and philosopher of Jewish descent.[36] He was born in Wilkes-Barre, Pennsylvania and attended Pennsylvania State University, graduating with a B.S. in 1939. He began his graduate work in physics at the California Institute of Technology, then transferred to the University of California at Berkeley, to work with J. Robert Oppenheimer.

Although Oppenheimer, who was Bohm's mentor and doctoral supervisor at the time, wanted Bohm and his work on scattering to be part of the Manhattan project at Los Alamos, because Bohm was deemed a security risk (a close colleague was potentially a spy), he was initially denied security clearance. His research on scattering toward his thesis was promptly classified, and so he lost access to, and was forbidden to discuss, his own work. Oppenheimer arranged for conferral of his degree without an official submission and defence in 1943. Bohm did eventually contribute directly to the Manhattan project. His calculations at the Y-12 facility in Oak Ridge, Tennessee were used to enrich the uranium for the bomb dropped on Hiroshima in 1945.

In 1947, Bohm became Assistant Professor at Princeton University, where he worked closely with Albert Einstein who became his new mentor. In 1951, Bohm was suspended for refusing to collaborate with the House Un-American Activities Committee. He was arrested and charged for having invoked his fifth amendment right against self-incrimination, based on a Mcarthy-era law that was later to be declared unconstitutional by the Supreme Court. Oppenheimer advised Bohm to leave the country. With letters of recommendation from Oppenheimer and Einstein, Bohm joined the University of São Paulo, Brazil (1951–55), and then moved to the Technion in Haifa, Israel (1955–57) before becoming Research Associate at the University of Bristol (1957–61), where he and Feyerabend briefly became close colleagues in 1957, before Feyerabend left for Berkeley in 1958. Bohm then became Professor of Theoretical Physics at Birkbeck College, University of London (1961–83) and Feyerabend left for Berkeley in 1962.[37]

Bohm made a wide range of significant contributions to physics. He established the foundations of plasma theory. He made major advances in quantum theory and relativity theory, and he tried to develop a revolutionary new conceptual framework based on what they have in common. Bohm's contributions to philosophy are also of the highest order and span an equally impressive range.[38] While he is most widely known for his notorious 1952 (deterministic) hidden-variables proposal, his general

[36] For biographical information on Bohm, cf. Peat 1997.

[37] On Feyerabend, Kuhn and Berkeley, see *below*, section 1.3.

[38] The wide range of interconnected philosophical proposals that Bohm considered has often been emphasized: "His philosophical reflections ranged from the role of causality and chance in modern physics to the role of creativity and dialogue to enhance human sociability on our planet and included reflections on consciousness" (Freire Junior 2019: vii).

methodology and its development in the 1950s and early 1960s, has received considerably far less attention.[39] His philosophy of mind[40] and his general philosophy promoted by the Dalai Lamma (which includes political philosophy and practical philosophy, as for example, his views on conflict resolution and the potential to improve the human condition through what has been dubbed 'Bohm dialogues')[41] have taken on somewhat of a cult status in some circles. All his diverse ideas in and about science and philosophy were developed as part of an attempt to establish a revolutionarily new world based on a more humane worldview. In Bohm's terms, these ideas were an attempt to form an undivided wholeness characteristic of the qualitative infinity of nature itself.[42]

In the late 1940s, under Popper's influence,[43] Feyerabend began to develop his interest in the foundations of quantum theory. To this end, from 1949 to 1952, Feyerabend visited either Sweden or Denmark every year. These trips gave him the opportunity to study trends in scientific philosophy in those countries, and he "got

[39] For Bohm's methodological views in the 1950s, based on the idea of infinite levels, cf. BOHM 1957a. For Bohm's methodological views in the early 1960s, based on the qualitative infinity of nature, cf. *below*, Ch. 8; and BOHM 1961; 1962; 1963a; 1964. For Bohm's methodological views in the late 1960s, cf. BOHM 1966: Ch. 25 "The Falsification of Theories": 94–99; BOHM 1968 (on science and art); BOHM 1969 (on science as perception). On Bohm's methodological views in the early 1970s and his search for a new order in physics, cf. BOHM 1971a; 1971b; 1972; 1974. On Bohm's methodological views in the late 1970s and the 1980s, see Bohm on wholeness, fragmentation and the implicate order in BOHM 1976. Cf. BOHM 1977; 1978; 1980; 1985; 1986a; 1986b; 1986c; 1987a; Bohm 1988; BOHM-HILEY 1985; BOHM-HILEY 1993; KRISHNAMURTI–BOHM 1985.

[40] For Bohm's philosophy of mind, cf. BOHM 1976; 1986: 1987a; 1990.

[41] Cf. BOHM-FACTOR-GARRETT 1991; BOHM 1996.

[42] According to Bohm, "What is very probably needed instead is a qualitatively new theory, from which both relativity and quantum theory are to be derived as abstractions, approximations and limiting cases. The basic notions of this new theory evidently cannot be found by beginning with those features in which relativity and quantum theory stand in direct contradiction. The best place to begin is with what they have basically in common. This is undivided wholeness. Though each comes to such wholeness in a different way, it is clear that it is this to which they are both fundamentally pointing. To begin with undivided wholeness means, however, that we must drop the mechanistic order" (BOHM 1980: 223).

[43] Feyerabend met Popper in 1948 at the EFA and they quickly became close, with Popper taking on the role of mentor, and even father figure. See especially COLLODEL 2016. On Feyerabend and the EFA, see *FFY1*, Ch. 1: 25–26n22. Before 1949, the EFA was known as the *Internationale Hochschulwochen* (*abbr.* IHA) *des Österreichischen College* (*abbr.* ÖC) [Alpbach International Summer Seminar of the Austrian College Society]; cf. *FFY1*, Ch. 1: 25–26n22. The IHA was an interdisciplinary, international conference that has been held annually in the Tyrolean village of Alpbach since 1945. The event usually lasted approximately three weeks and took place in the late summer. The first International Summer Seminar was organized in Alpbach under the sponsorship of the French occupying forces by Simon Moser and Otto Molden. Its unforeseen success encouraged Otto and Fritz Molden to establish the ÖC. On Feyerabend and the IHA and the ÖC, see *FFY1*, Ch. 1: 25–26n22 and 26–27n23; Feyerabend to Popper 8 October 1953-6: 78n1 and 80n5. Feyerabend participated in the IHA/EFA as secretary and shorthand writer in 1948 and 1949, and probably also from 1950–52. He also later co-led seminars at the event in 1956 see *FFY1*: Ch.1: 26–27n23.

several valuable ideas for [his] own subject".[44] More specifically, in 1949, Feyerabend travelled to Denmark on a scholarship from the Danish Ministry of Education. A year later, he travelled to Sweden on a scholarship from the Swedish Student's Association, and then in 1951, he returned to Denmark through an invitation to the first "Nordisk Sommeruniversitet" (Scandinavian Summer University) in Askov, from the 11–26 of August on the general theme 'The causality problem'. This provided him with the opportunity to interact directly with Bohr and his collaborators.[45] In this way, as he had done earlier with Wittgenstein, Feyerabend went straight to the source of the issues in which he was interested. This led to his participation in an event that shaped his work in the foundations of quantum theory for the better part of the next two decades:[46] Bohr and the Bohrians' initial reaction to Bohm's 1952 "hidden variables" proposal.[47]

As Feyerabend recounts it, he had approached Bohr directly after Bohr had given a talk, to ask for details:

> Bohr repeated his old explanations [...], but with renewed vigour, for he had just heard of David Bohm's apostasy.[48] 'Can you understand that?' he asked with a puzzled look on his face. Unfortunately, he was soon dragged away to another meeting. (FEYERABEND 1995: 78)

[44] *FFY1*; Ch. 3, §352. Feyerabend may be referring to his ideas on meaning change, cf. FEYERABEND 1995:78. Feyerabend also actively attended the second *Nordisk Sommeruniversitet* [Scandinavian Summer University (*abbr.* NSU)] (general theme: "*Mennske og miljö*" [People and the Environment]), which took place in the Norwegian village of Ustaoset in the summer of 1952. Cf. *FFY1*, Ch. 3, §4: p. 54. The NSU is an ongoing institution that was founded in August 1950 with the aim of organizing a cultural event modelled after the EFA. See Hermansen 2000, <http://www.nsuweb.net/wb/>. On Feyerabend and the NSU, see FFY1: Ch. 2, n37; and *below*, Part 2, Ch. 6; Peat 1997: 129.

[45] *Nordisk Sommeruniversitet*.

[46] Cf. FEYERABEND 1961c; 1962a; 1968;1969.

[47] See BOHM 1952a; 1952b. Feyerabend first met Niels Bohr at the 1951 NSU, which was in Askov (Denmark), 11–26 August 1951, on the general theme 'The causality problem'. For background and details, see *FFY1*: Ch. 2: 30n37. On Feyerabend and Bohm, see *FFY1*, Ch. 2: 32n41.

[48] On Feyerabend and Bohm, see *FFY1*, Ch. 2: 32n41; Feyerabend to Popper, 16 May 1954-13: 124; 2 October 1954-18: 133n.4; 19 March 1955-8: 165; 7 April 1957-7: 260; 15 December 1957-13: 271-272; 1958-2: 275; 28 January 1958-3: 277; 12 March 1958-4: 279; 1 May 1958-5: 280; 7 June 1958-8: 286; 22 March 1959-5: 314; 26 October 1959-12: 328–239; 29 April 1961-6: 361; 17 August 1961-9: 369n18; 4 October 1967-2: 419-420n4; Ch. 7 "Philosophy and the Physicists (1955)": 479; Ch. 8 "Observationally Complete Theories: Some Observations on Quantum Theory (1958)": 490; and see *below,* Feyerabend to Feigl 1957-2; June 1957; "Report on the conference on philosophical foundations of physics", September 27, 1957; "Research Project on the Logic of Measurement, Probability and Confirmation, with Applications in Atomic and Quantum Theory"; Feyerabend to Hempel, Rec. 10-25-61; Part 2, Ch. 7 and 8; Part 3, Feyerabend to Kuhn 1961-2; Feyerabend to Kuhn 1961-3. Cf. *FFY3,* Feyerabend to Agassi 1956-5: 1957-9; 12 March 1958-5.(3): 1958-7.(5); 2 April 1958-8; 1958-9.(5); 1958-12; 22 June 1958-13; 2 November 1958-17; 1967-4; 25 November 1967-15; 30 November 1967-16; 20 July 1968-16; Feyerabend to Watkins 6 September 1960-2; 17 December 1967-9; 7 January 1968-2; 17 April 1968-8; 1971-1; Feyerabend to Lakatos 9 March 1967-4; Feyerabend to Musgrave, 25 July 1967, IL 14/10 (2 ss., H) in *FFY3*, Feyerabend to Lakatos 13 July 1967-17. Cf. FEYERABEND 1995: 78, 88, 92.

To Feyerabend, it "seemed that, for [Bohr], the sky was falling down".[49] According to Feyerabend, Bohr was "neither dismissive, nor shaken. He was amazed".[50] After Bohr left the room, the discussion continued. Some of those present argued that the objections to Bohm's theory were not at all conclusive. As Feyerabend put it:

> The Bohrians for a while tried to close the holes in the Bohrian fashion. When they did not succeed they said: 'But von Neumann has proved …' – and that ended the discussion. Nobody said another word though I had the impression that none of the Bohrians would have been able to explain what it was that von Neumann had proved and how he had achieved it.[51] It seems that Bohr himself never used von Neumann as a crutch. (Ch. 7, Feyerabend to Peat, 30 August 1993–4)

From the moment that he learned of it from Bohr, Feyerabend became keenly interested in Bohm's attempts to develop a 'hidden variables' interpretation of quantum theory in the face of von Neumann's *alleged* proof that no deterministic completion by 'hidden variables' can be consistent with the principles of quantum mechanics.[52]

In November 1951, directly after obtaining his Ph.D. in philosophy in Vienna, Feyerabend applied for and was eventually awarded a British Council scholarship to study the foundations of quantum theory in London for the academic years 1952–53, under Popper's supervision.[53] Von Neumann's proof and the corresponding measurement problem became major topics of discussion between Feyerabend and Popper throughout the 1950s and early 1960s, and a main focus of Feyerabend's publications on the foundations of quantum theory through to the late 1960s. In the early 1950s, prominent physicists began publishing responses to Bohm's 1952 proposal. Pauli went first, promptly responding to Bohm in 1952, rehearsing his earlier

---

[49] See *below*, Feyerabend to Peat, 1993, note that 'falling down' was changed to 'falling in' by Peat.

[50] Feyerabend was responding to Peat's suggestions, see Peat 1997:129. See *below*, Part 2, Ch. 7.

[51] Feyerabend is referring to the purported proof of the impossibility of hidden variable interpretations of quantum mechanics that was proposed by the Hungarian-American mathematician and polymath John von Neumann (1903–1957) at the beginning of the 1930s. See Neumann 1932: 2–3 [ix-x], pp. 109 [209–210], Ch. 4, § 1–2. Although the proof was shown to be flawed by the German mathematician and philosopher Grete Hermann (1901–1984) in 1935 (see Hermann 1935), it was largely– though not unanimously– considered sound for the following three decades, until its flaw was rediscovered by the British physicist John Stewart Bell (1928-1990) in the mid-1960s (see Bell 1966). On von Neumann's proof, see Jammer 1974: Ch. 7; Caruana 1995; Gattei-Agassi 2016. On Feyerabend and Neumann's proof, see *FFY1*, Ch. 2, 30–31n37; Feyerabend to Popper, 28 October 1955-29: 201; 1955-33: 206–207; 23 December 1955-34: 208; 18 January 1956-2: 213; 18 January 1956-3: 214; 30 January 1956-5: 217; 4 February 1956-6: 218; 8 February 1956-9: 221; 11 February 1956-10: 225–226; 5 April 1956-15: 230; 14 April 1956-16: 231; Popper to Feyerabend, 16 April 1956-17: 232; 18 April 1956-18: 234; Feyerabend to Popper, 9 May 1956-19: 235; 14 January 1957-1: 244–245; 21 January 1957-2: 248–250; 28 January 1957-3: 252; 27 March 1957-4: 253–254; 30 March 1957-5: 257; Popper to Feyerabend, 1 April 1958-6: 259; Feyerabend to Popper, 1965-4: 412; Feyerabend 1995: 92–93, 106, 108.

[52] Cf. Bohm 1952a; 1952b; Neumann 1932.

[53] See Collodel 2016.

objections against Louis de Broglie's earlier pilot wave proposal.[54] Einstein published a criticism of Bohm's theory a year later, which he regarded to be essentially the same as Pauli's earlier criticism of de Broglie,[55] and Heisenberg published papers explaining his reasons for being unconvinced by Bohm's proposal in 1955, and then again in 1958.[56]

Once in London, Feyerabend attended Popper's seminar and he concentrated on two topics: foundations of quantum theory (von Neumann's proof and its relation to Bohm's proposal), and Wittgenstein's *Investigations*.[57] According to Feyerabend's recollections, "Von Neumann was not easy; I worked my way through his book, page by page, and eventually wrote a critical review".[58] At that time, he did the same with Wittgenstein's *Investigations*.[59]

Popper, like Bohm, was also a vocal critic of Bohr's 'official view'.[60] Popper was lecturing that: "the official view – backed by a proof of von Neumann's– was that the theory was complete in the sense that any attempt to introduce so-called 'hidden variables' – for example, values of the momentum and position which are more precise than the principle of indeterminacy permits– must lead to contradictions".[61] As Popper saw it: "David Bohm has recently published some papers which, it appears, refute the official view in this point. These papers were not very well received".[62] Popper was very clear that he thought that Bohm had provided: "A calculation which in [his] opinion is a striking refutation of the complementarity view".[63] At least since the late 1940s, Popper had been defending "Indeterminacy in Quantum-Mechanics and in Classical Physics",[64] and was in the early stages of

[54] Cf. De Broglie 1927; 1928. On early criticisms of Bohm's theory, see Myrvold 2003. Bohm had been *unaware* of de Broglie's pilot-wave proposal when independently 'rediscovered' in 1952. De Broglie's pilot-wave model led to Schrödinger's formulation of wave mechanics, after which the pilot-wave model was abandoned in favour of the quantum formalism.

[55] See Einstein 1953. Basically, Einstein criticized Bohm's proposal because it was inelegant (lacking certain symmetries), to which Bohm (who accepted Einstein's criticism) responded that it was just supposed to be a pilot pilot-wave theory, in the sense that it was not proposed as a final theory, but just as a proof of concept, implying that a more satisfactorily developed pilot-wave theory would recover those symmetries. Cf. Myrvold 2003.

[56] Cf. Heisenberg 1955; 1958.

[57] Cf. Wittgenstein (1953); Feyerabend 1956: [46–48]; Feyerabend 1954a; 1954b; 1955a. Feyerabend had a copy of the manuscript that became the *Investigations* from Elizabeth Anscombe, see Collodel 2016.

[58] Feyerabend 1995: 92–93. Feyerabend is referring to his review of von Neumann's book *Mathematical Foundations of Quantum Mechanics* (Neumann 1932), see Feyerabend 1958d: [294–298].

[59] See Feyerabend 1954a; 1954b; 1955a.

[60] For Popper on Bohm, see *FFY1*, Ch. 6 "A Defence of Free Thinking in Quantum Theory (1953)": 459, 472n33, n34, n47; 474n26E; Popper to Feyerabend, 5 May 1958-6: 281.

[61] Popper in *FFY1*, Ch. 6 "A Defence of Free Thinking in Quantum Theory (1953)": 459.

[62] Popper in *FFY1*, Ch. 6 "A Defence of Free Thinking in Quantum Theory (1953)": 459.

[63] Cf. Popper in *FFY1*, Ch. 6 "A Defence of Free Thinking in Quantum Theory (1953)": 472n33.

[64] Cf. Popper 1950a; 1950b.

developing his own realistic, indeterministic propensity interpretation of quantum theory.[65] Popper was praising Bohm:

> *most* of the younger physicists have been brought up to believe that [...] all that matters us ⟨is⟩ the formalism and its application. An exception is David Bohm who wrote an elementary textbook of the theory which was unorthodox in its unparalleled attempt to think out matters in detail in physical terms. (It is hardly an accident that he left the camp of orthodoxy not very long after the publication of this enchanting book.) (Popper in *FFY1*, Ch. 6 "A Defence of Free Thinking in Quantum Theory (1953)": 468)

In this way in 1953, Popper seems to have been contributing to the myth that von Neumann's proof had purportedly demonstrated that any hidden variables account is impossible until Bohm had shown otherwise.[66] However, as Feyerabend points out in his first publication on quantum theory published a year later in 1954, von Neumann's proof is addressed to a specific class of hidden variable theories of which Bohm's was not one.[67] On this, both von Neumann and Bohm agreed.[68] De Broglie had confessed that he did not initially realize the limited domain of validity of von Neumann's theorem and had overstated its generality.[69] In any case, this seems to have contributed to the general misunderstanding, Popper's included. Feyerabend clearly seems to have thought so: "In this way von Neumann's book has created a good deal of confusion among physicists (cf. e.g., de Broglie's statement in the Introduction to *La Physique quantique restera-t-elle indéterministe?*) and, naturally, among philosophers,"[70] as it seems likely that this last remark was meant to include Popper.[71]

In this period, Feyerabend was trying to develop his own criticism of Bohm's 1952 'hidden variables' interpretation. By 1954, Feyerabend was circulating drafts of his own criticism of Bohm's proposal: "I am sending two articles I wrote. [...]. You [Popper] will find the presentation and criticism of Bohm's interpretation on the last three pages".[72] Feyerabend was encouraged by Margenau, who thought Feyerabend's criticism of Bohm's 1952 'hidden variables" proposal was similar to Pauli's objection: "He [Margenau] had got hold of my article on Quantum Mechanics

---

65 Cf. Popper's contribution to the 1957 Colston conference, Popper 1957.

66 De Broglie seems to have inadvertently started this myth with respect to von Neumann's alleged proof, which he later admitted, and tried to correct. See de Broglie 1953.

67 See Feyerabend 1954c: [25–45].

68 Cf. Myrvold 2003.

69 See Dieks 2017: 30n42.

70 Feyerabend 1958d: [298], italics inserted. Cf. de Broglie 1953.

71 Feyerabend explicitly pointed out that "Bohr himself did not commit this mistake" (Feyerabend 1958d: [298n]).

72 *FFY1*, Feyerabend to Popper, 2 October 1954-18: 133. Cf. *FFY1*, Feyerabend to Popper, 6 May 1954: 124. The first article to which Feyerabend is referring could be one of several possibilities: Either "*Der Beobachtung Prozess in der Quantenmechanik*" [The Observation Process in Quantum Mechanics], KP 423.5 (23 ss., T/H), or a draft of Feyerabend1954c: [25–45], or a draft of Feyerabend 1954d. Cf. *FFY1*, Feyerabend to Popper, 1954-17, 129n3. The second article is "Some Remarks concerning Bohm's Quantum-Mechanics in terms of 'Hidden" Variables', KP 423.4 (10 ss., T/H). Cf. "David Bohm's 'Quantum Mechanics'", HF 08-28-04 (28 ss., T/H).

and seemed impressed because, as emerged from his letter, Pauli had proffered similar objections against Bohm".[73] Unknowingly foreshadowing his own proposed solution to the measurement problem, Feyerabend wrote to Popper: "I think that Bohr's solution [to the EPR paradox] is essentially the correct solution if we do not interpret quantum-mechanics in a conservative way.[74] But Feyerabend's attempts to criticize Bohm's 1952 proposal were soon superseded by Bohm and Vigier's new 1954 'hidden variables' proposal in their "Model of the Causal Interpretation of Quantum Theory in Terms of a Fluid with Irregular Fluctuations" (BOHM-VIGIER 1954).[75] This is how Brownian motion took centre stage in discussions of the foundations of quantum theory. According to Bohm and Vigier, there are two kinds of Brownian motion. Classical Brownian motion was explained by Einstein as a stochastic process, when the predictions he deduced from the kinetic theory were confirmed by Perrin. In this way, Einstein showed that Brownian motion results from shaking atoms pushing particles that are small enough around.

According to Bohm and Vigier, the second kind of Brownian motion is the indeterminacy of the motion of electrons. Electrons behave like Brownian particles because they are bombarded stochastically from all sides by a surrounding sub-quantum 'fluid' in which they are suspended. In other words, just as atoms cause stochastic motions on the classical level, an underlying sub-quantum 'fluid' causes stochastic motion of electrons on the quantum level. In this way, Bohm and Vigier were proposing a causal theory that explains the confirmed results that had been predicted by quantum mechanics, but as the result of a lower-level process – and not as the result of some purported irreducible indeterminacy implied by quantum theory. The existing theory was supposed to be a limiting case of their proposal in the way that classical mechanics is related to relativity theory.

In October of 1955, Feyerabend began the next big step in his work on the foundations of quantum theory, when he committed to lecture on it in Bristol. He wrote to Popper, "I have announced a special lecture for next term the preliminary

[73] *FFY1*, Feyerabend to Popper, 19 March 1955: 165. Cf. "I am sending an article which I wrote when I returned from Alpbach and which I would like to have published in your journal in case you think it good enough" (*FFY1*. Feyerabend to Popper 2 October 1954: 133). Feyerabend may be referring to his "David Bohm's 'Quantum Mechanics'", HF 08-28-04 (28 ss., T/H). The journal he is referring to is *The British Journal for the Philosophy of Science*. Cf. *FFY1*, Popper to Feyerabend 2 January 1954-1: 99–100n2. However, there are some discrepancies to be noticed concerning the identification of this paper with the article Feyerabend is referring to here. In particular, no copy of the long paper mentioned above can be found in the Popper Collection. Moreover, the copy that Feyerabend sent to Herbert Feigl only a short time later (cf. *FFY1*, Feyerabend to Popper 3 October 1954-19: 135–135n4) is annotated with the hand-written remark: "The first article I wrote in English. The English is therefore not very good". However, Feyerabend had already written a much shorter paper in English on the same topic, which he judged to be too long to be published in *The British Journal for the Philosophy of Science*. Cf. *FFY1*, Feyerabend to Popper, 16 May 1954-13: 124.

[74] *FFY1*, Feyerabend to Popper, 1954: 204.

[75] Cf. BOHM-VIGIER 1954.

program of which I am enclosing".[76] Topics to be discussed specifically included the 'Quantum-theory of measurement' and de Broglie's theory and Bohm's 1952 interpretation.[77] Specifically, Feyerabend was to present Bohm's response to von Neumann's proof, as well as the objections raised to von Neumann's proof by Grete Hermann, as well as Einstein's criticism of Bohm's views and Bohm's response to Einstein.[78] Feyerabend then began his lectures on the foundations of quantum physics in Bristol in early 1956. Physicists regularly attended including Pryce (who was soon to become Bohm's boss in the next year), with whom Feyerabend openly sparred during the lectures.[79] This must have felt like a heavy lift for Feyerabend, who had switched to Popper's point of view with respect to von Neumann's proof at the last minute:

> I gave my first lecture to the physicists, I was very nervous (there were about 60 people), Prof. Pryce interrupted me several times, twice I gave in although I was right– but on the main point I convinced people. And this main point was that it does not follow from any presentation of the uncertainty-rel. that atomic particles cannot have sharp position + sharp velocity at the same time! For since I had left you your insistence on this point set my mind working with the result that I have changed most of what I am going to lecture in favour of your point of view. This I did about 3 days before the lectures started and I found thinking along your lines most exciting + fruitful. I discussed at length Neumann's proof. Do we conclude, so I asked, that a throw with a die gives a blurred result? The theory of Brownian movement leads again to classes which obey (1) and (2) [not sharp and pure] – nevertheless we say that Brownian particles have sharp p and q at any time of their movement. This led to a long argument between Pryce and myself by which I finally convinced him that there are no reasons whatever to believe in the queer nature of electrons. I.e. I did not actually convince him, but left him less dogmatic […] And this is certainly desirable. (*FFY1*, Feyerabend to Popper, 11 February 1956-10: 223)

This seems to be the first instance that Brownian motion appears in Feyerabend's writings, clearly following Bohm and Vigier's (1954) suggestion in the foundations of quantum theory. However, Brownian motion as part of a general methodological argument for pluralism only clearly enters Feyerabend's published work after the Colston conference in 1957.

Shortly after Feyerabend switched to Popper's point of view on von Neumann's proof and the idea that there is no *special* problem of quantum indeterminacy, he published "*Eine Bemerkung zum Neumannschen Beweis*" [A Remark on von Neumann's Proof] (Feyerabend 1956).[80] Popper immediately privately accused Feyerabend of plagiarism,[81] insisting Feyerabend write the editors and have them add a note giving himself and Agassi credit for the short paper. Feyerabend initially

---

[76] Cf. *FFY1*, Feyerabend to Popper, 14 October 1955-27: 196.

[77] Cf. *FFY1*, Feyerabend to Popper, 28 October 1955-29: 200–201.

[78] Cf. *FFY1*, Feyerabend to Popper, 28 October 1955-29: 200–201; 6 February 1956-7: 219.

[79] Pryce and other physicists regularly attended Feyerabend's lectures. Cf. *FFY1*, Feyerabend to Popper, 6 February 1956-7: 219; 11 February 1956-10: 223.

[80] Cf. Feyerabend 1956. English translation "A Remark on von Neumann's Proof (1956)": [46–48].

[81] See Collodel 2016.

agreed to write to the editors and add such a note, as Popper insisted, but then he thought better of it,[82] and told Popper it was too late to add such a note (which it was not).[83] By adopting Popper's strategy in the foundations of quantum theory, however briefly, Feyerabend (so it seemed to Popper), was getting out ahead of Popper's own propensity interpretation strategy. In any case, it was a short-lived reversal for Feyerabend, who quickly returned to what he had been arguing before with respect to von Neumann's proof. By January 1957, Feyerabend made his discomfort with respect to his published 'Remark' very clear in a letter to Popper: 'if I am right your and Joske's [Agassi] criticism [of von Neumann] which I have adopted myself [in Feyerabend 1956] is wrong'.[84] It now seemed to Feyerabend once again that "there is much more behind [von Neumann's] proof than you [Popper] were inclined to admit".[85]

In 1957, two crucial events took place that laid the foundations for Feyerabend's philosophical development to come: the publication of Bohm's book *Causality and Chance in Modern Physics* (1957a) and the 1957 Colston conference.[86] Together these events turned 1957 into a pivotal year for Feyerabend in two converging areas of his work: in the foundations of quantum theory and general methodology. In the

[82] A close look at their correspondences shows that Feyerabend excitedly sent Popper two consecutive letters outlining the four the main points of the short note, suggesting to Popper that he read an article by Fényes (Fényes 1952) that he included with the letter. The explicit goal of the short 'Remark' is to replace Fényes's complex considerations with more straightforward considerations. In this way, these letters may raise some questions about the legitimacy of Popper's accusation of plagiarism in that case, as Feyerabend (at a minimum) brought Fényes's derivation of the whole of quantum theory on the basis of the theory of Markoff-processes into the discussion, which is what Feyerabend used in (Feyerabend 1956) to clarify what Popper claimed to be his point in Feyerabend 1956, and given that Feyerabend explicitly cited Popper on the dice example, as used to illustrate the position ("I owe to Prof. K. R. Popper (London) the remark that NP did not reach its goal because it is valid not only for QM, but also for the game of dice") that there is nothing special about indeterminacy in quantum theory beyond the indeterminism on the classical level. Cf. Feyerabend 1956: [47]. Popper was understandably irritated given that he interpreted Feyerabend as pre-empting the originality of Popper's propensity interpretation of quantum theory, which Feyerabend first presented publicly on Popper's behalf at the 1957 Colston conference. Cf. Popper 1957; *FFY1*, Feyerabend to Popper, 30 January 1956-5: 217; 6 February 1956-8: 220.

[83] Cf. Collodel 2016 and *FFY1*.

[84] *FFY1*, Feyerabend to Popper, 21 January 1957-2: 249.

[85] *FFY1*, Feyerabend to Popper, 21 January 1957-2: 248.

[86] The 1957 Colston conference was the ninth symposium of the Colston Research Society, called "Observation and Interpretation", which took place in Bristol from 1–4 April 1957. It gathered together some of the most prominent quantum physicists and British philosophers of science of the time. Contributors and Participants included E. W. Bastin, D. Bohm, F. Bopp, M. Fierz, H. J. Groenewold, E. H. Hutten, S. Körner, D. M. Mackay, M. Polanyi, M. H. L. Pryce, L. Rosenfeld, Sutton J.-P. Vigier, and J. O. Wisdom. See Bohm 1957b. Feyerabend was directly involved both in the organization of the event and in the editorship of the symposium proceedings, on behalf of Körner and Pryce, who were the official organizers of the symposium. On Feyerabend and the ninth symposium of the Colston Research Society, cf. *FFY1*, Feyerabend to Popper, 1957-1: 244–247; 1957-2: 250; 1957-4: 253; 1957-5: 256–257; 1957-6: 259; 1957-7: 260–261; 1957-8: 263; 1957-9: 264–265; 1957-11: 267–268; Feyerabend 1995: 108. Cf. Kožnjak 2018.

foundations of quantum theory at the conference, Feyerabend initially delineates his own strategy for handling the measurement problem, and quickly learned that Bohr did not belong to the Copenhagen school that Feyerabend was trying to criticize. In general methodology, Feyerabend witnessed how Bohm and Vigier used the Brownian motion example as part of a methodological defence of their 1954 hidden variables proposal.

From Bohm's book, Feyerabend had learnt a lot about Bohm's views on methodology, including Bohm's dialectical views on the limited validity of universal theories and Bohm's dialectical view of the relationship between determinism and indeterminism.[87] Feyerabend also learnt more about Bohm's views in the foundations of quantum theory, most especially, how, according to Bohm, the motion of an electron is a second kind of Brownian motion. Moreover, the 1957 Colston conference then afforded Feyerabend the opportunity to clarify his understanding of Bohm's views, while also serving as a testing ground for Feyerabend's peculiar views on meaning variance and methodology.[88] As we shall see, at the conference, Feyerabend vigorously supported Bohm and Vigier against what he took to be an unjustifiable dogmatic philosophical conservatism of the Copenhagen interpretation at the conference. Moreover, Bohm clearly supported and encouraged Feyerabend's suggestions about meaning change and methodology. In this way, Bohm and Vigier's use of the Brownian motion example to defend their proposal on methodological grounds eventually led, in *ERE*, to the development of Feyerabend's methodological argument for theoretical pluralism and his test-model for comparing incommensurable theories based on the example of Brownian motion.[89]

Feyerabend had been directly involved in the organization of the 1957 Colston conference and he was instrumental in preparing the manuscript for the publication of the conference proceedings.[90] He was hoping to convince Bohm, Vigier and Pauli to attend.[91] As Feyerabend explained in a letter to Popper:

---

[87] See VAN STRIEN 2019; 2024.

[88] The "Colston Research Society" was founded in 1899 as the "University College Colston Society" with the aim of fostering and assisting the University College, Bristol and it was instrumental in founding the University of Bristol in 1909. From 1948, the Society has been committed to sponsoring and supporting an annual academic symposium at the University of Bristol. See BARNETT 1957 in KÖRNER-PRYCE (eds.) 1957: v. Cf. <http://www.bristol.ac.uk/pace/committees/colston-research/>; *FFY1*, Feyerabend to Popper, March 1956-14: 228n1.

[89] See OBERHEIM 2024. The Brownian motion example becomes central to Feyerabend's methodological argument for theoretical pluralism in FEYERABEND 1962b: 65–66. For details and further references, see OBERHEIM 2006: 254n353.

[90] Specifically, Feyerabend "undertook the arduous task of supervising the recording of the discussions from tape-recorder to the printed page" (Korner 1957 in KÖRNER-PRYCE (eds.) 1957: vii). In addition to giving his own paper (FEYERABEND 1957a), Feyerabend also read Popper's paper (see Karl R. Popper, "The Propensity Interpretation of the Calculus of Probability and Quantum Theory", KP 48.25), which was delivered *in absentia*. Cf. *FFY1*, 29 May 1956-21: 238n5. Cf. POPPER 1957.

[91] "Bohm will perhaps come [to the 1957 Colston conference], so will Pauli" (*FFY1*, Feyerabend to Popper, March 1956-14: p. 228. Pauli did not attend, see KÖRNER-PRYCE (eds.) 1957.

Vigier is a pupil of de Broglie [who is] trying to work out an interpretation of quantum mechanics which is similar to Bohm's (both have also collaborated and written a joint paper [Bohm and Vigier 1954] where electrons and similar particles are regarded as a kind of fine-grained dust which is carried along by a fluid (the ψ-field!) and whose movement is also influenced by a kind of Brownian movement.[92] (*FFY1*, Feyerabend to Popper, 14 January 1957-1: 244)

In the lead up to the conference, Feyerabend finished his own paper and read Bohm's *Causality and Chance in Modern Physics* (1957a):

I have been ill most of the last six weeks and this is the reason why I did not write before. I just managed to finish my paper for the congress and I shall send it to you as soon as it is in its final form- I think I really got something and I would like to hear your opinion [...] Did you see Bohm's book? [*Causality and Chance in Modern Physics*]. Very valuable observations (a violent attack upon the Bohr-Heisenberg interpretation) a popular exposition of his, Vigier's, de Broglie's attempts at a causal model – but many obvious blunders.[93] (*FFY1*, Feyerabend to Popper, 27 March 1957-4: 254)

Before the conference, the scientific community had taken the attitude that Bohm had been "slayed philosophically as well as physically".[94] The general view was that Bohm had been doubly excommunicated: from the Church of the Copenhagen school and from the USA by McCarthyism. Feyerabend and Popper, by contrast, were both initially very excited by Bohm's work. According to Freire Junior, Rosenfeld, who saw the causal interpretation as metaphysical regression:

mobilized colleagues wherever he could to take up the fight against the causal interpretation. He appealed to his professional connections as well as companions sharing ideological ties with Marxism. He pushed Frédéric Joliot-Curie—a Nobel prize winner and member of the French Communist Party—to oppose French Marxist critics of complementarity; advised Pauline Yates— Secretary of the "Society for cultural relations between the peoples of the British Commonwealth and the USSR"—to withdraw her translation of a paper by Yakov Ilich Frenkel critical of complementarity from *Nature*; asked *Nature* not to publish a paper by Bohm entitled "A causal and continuous interpretation of the quantum theory" and advised publishers not to translate one of de Broglie's books dedicated to the causal interpretation into English.[95] (Freire Junior 2015: 37–38)

Feyerabend later may be seen to allude directly to the situation in *ERE* without mentioning Bohm or Rosenfeld specifically by name:

Today, of course, the elimination takes the more refined form of a refusal to publish (or to read) what is not in agreement with the accepted doctrine. However, this liberalism applies to physical theories only. It does not seem to apply to political theories. (1962b: 69fn85)

[92] Cf. "The second paper of the meeting will be Vigier's a pupil of de Broglie, collaborating with Bohm in his attempts at a causal model of QM. The chairman is Rosenfeld" *FFY1*, Feyerabend to Popper, 27 March 1957-4: 253.

[93] Cf. *FFY1*, Feyerabend to Popper, 30 March 1957-5: 256. A few days later, Feyerabend eagerly asked again: "Have you seen Bohm's book 'Causality and Chance in Modern Physics'" (*FFY1*, Feyerabend to Popper, 1 April 1957-6: 259).

[94] Einstein to Born, 26 November 1953, in Born 1971: 207.

[95] Cf. del Santo 2022; del Santo and Krizek 2019; van Strien 2019; 2024.

As some recall, "the conference was organized for the purpose of setting Bohm straight [Michael E. Fisher, personal communication], but it turned into more of a coming-out party for Bohm".[96] Today, the conference is often taken as the turning point in the redemption of Bohm's reputation as a physicist, and as the end of the hegemony of the Copenhagen interpretation, as it seems to have re-established the legitimacy of questions about the foundations of quantum theory.[97]

At the conference, Feyerabend witnessed Bohm and Vigier go head-to-head with Rosenfeld and Pryce, who was soon to become Bohm's boss in Bristol a few months later. Rosenfeld gave a talk attacking Bohm, but not by engaging with his theory, but by seeming to argue against such interpretations of quantum theory in general.[98] This did not sit well with the audience, even some of the members of the Copenhagen school seemed uncomfortable. Feyerabend witnessed Bohm being derided by Rosenfeld and Pryce, but it was Rosenfeld who was often openly on the defensive, claiming that Feyerabend and others were misrepresenting his views.[99] Rosenfeld's views were repeatedly criticized by Vigier, Bohm, Feyerabend, and others. Feyerabend suggested that the dispute between Bohm and his opponents boils down to Bohm's suggestion to dispose of the quantum of action, whereas his opponents take the view that "the quantum of action will have to be an essential and unmodified part of any future theory".[100] Moreover, Feyerabend emphasized that the "discussion and the remarks that Professor Bohm has made in his lecture greatly clarified the import of von Neumann's proof. He has rightly emphasized that the statement that hidden variables do not exist is a theorem of quantum-mechanics rather than a statement whose truth has been proved absolutely and once for ever".[101] As Feyerabend described it to Popper:

> this is only a short note on the course of the conference. It was a success in the sense that the discussions were very informative, well-conducted and the papers very interesting. Körner's paper was a great success. The most obvious thing about the congress was the split between Bohrians (Pryce, Rosenfeld) and the Bohmians (Vigier, Bohm) which led sometimes to riotous discussions in which the Bohmians, fighting *like crusaders* and bringing up the controversy whenever possible, sometimes at quite irrelevant places, were violent sometimes nearly to the extent of being abusive. (*FFY1*, Feyerabend to Popper, 7 April 1957-7: 260)

---

[96] Howard 2004: 679.

[97] Cf. Kožnjak 2018.

[98] Cf. Rosenfeld 1957b. Rosenfeld is taken to epitomize the positivistic 'shut up and calculate' attitude of the Copenhagen school. At the conference, Rosenfeld objected to ascriptions of positivism to him and of the view "that things only exist in so far as they could have been observed" (Rosenfeld, in Körner-Pryce (eds.) 1957, "Discussion": 186). Cf. Rosenfeld, in Körner-Pryce (eds.) 1957, "Discussion": 148; Feyerabend in Körner-Pryce (eds.) 1957, "Discussion": 48.

[99] Rosenfeld often tried to correct what he took to be distortions of his views, cf. Rosenfeld in Körner-Pryce (eds.) 1957, "Discussion": 52, 53, 59, 179, 184.

[100] Feyerabend in Körner-Pryce (eds.) 1957, "Discussion": 48. Cf. Rosenfeld in Körner-Pryce (eds.) 1957 "Discussion": 52

[101] Feyerabend in Körner-Pryce (eds.) 1957, "Discussion": 48.

However, the real scandal, for Feyerabend, was that Bohm's interpretation of quantum theory was dismissed *despite* the acceptance of the fact that it was empirically indistinguishable with the existing theory. From Feyerabend's perspective:

> It is the great merit of Bohm's first model to have shown that a formulation of quantum theory is possible which is empirically equivalent with extended quantum mechanics as well as consistent (and which therefore satisfies all those criteria which according to Niels Bohr are sufficient for accepting it as an adequate theory), *and which is yet not observationally complete*. He has thereby shown that the 'epistemological turn' of modern physics is not 'imposed upon us' by experience, as some physicists seem to believe, but is the result of a choice (which in the case of the adherents of the Copenhagen interpretation consists in the acceptance of the Heisenberg program). (*FFY1*, Paul K. Feyerabend, "Observationally Complete Theories: Some Observations on Quantum Theory (1958)": 490, italics inserted)

Here we can see the close connection between Feyerabend and Bohm's strategies[102] and how Feyerabend was trying to criticize the existing quantum theory as 'observationally complete' in support of Bohm and Vigier's 1954 hidden variables proposal. According to Feyerabend, observationally complete theories imply that what cannot be observable cannot be real. Feyerabend argues that the Copenhagen interpretation of quantum mechanics shares this undesirable feature with Berkeley's idealism.[103]

The deep impact these events had on shaping Feyerabend's theoretical pluralism to come could not be clearer. Feyerabend concluded:

> The exclusion of alternatives is then simply a measure of expediency: their invention not only does not help, it even hinders progress by absorbing time and manpower that could be devoted to better things. The [logical empiricists'] consistency condition [new theories must be consistent with established views] eliminates such fruitless discussion, and it forces the scientist to concentrate on the facts which, after all, are the only acceptable judges of a theory. *This is how the practicing scientist will defend his concentration on a single theory to the exclusion of empirically possible alternatives.*[104] (FEYERABEND 1963a: [90–91], italics inserted)

Bohm characterized "the usual quantum mechanics," as "closer to the epicycle point of view",[105] and Feyerabend seems to have agreed:

> Now although there have been lots of discoveries after 1930 I do not perceive any jump as regards the fundamental philosophical attitude and I just do not see how parity, anti-matter, second quantization modify the probabilistic interpretation, or complementarity and all the rest. Some of the developments even seem to me like ever so much epicycles. (*FFY1*, Feyerabend to Popper, 17 August 1961-9: 367)

---

[102] According to Feyerabend "even of the great David Bohm himself [...] says that he is closer to Bohr than anyone else." See *below*, Feyerabend to Hempel, 20 April 1968. Feyerabend and Bohm can be seen to be working together in this way.

[103] Cf. Feyerabend's "Observationally Complete Theories: Some Observations on Quantum Theory (1958)" in *FFY1*, Ch. 8: 487–491.

[104] Cf. DEL SANTO AND KRIZEK 2019; DEL SANTO 2022; KUBY 2021; VAN STRIEN 2019; 2024.

[105] Bohm in KÖRNER-PRYCE (eds.) 1957, "Discussion": 56. Mackay objects that *Bohm's strategy* seems like just adding more "epicycles", Mackay in KÖRNER-PRYCE (eds.) 1957, "Discussion": 50.

At the conference, Bohm and Vigier defended their proposal from the accusation that they were introducing pernicious metaphysics, by drawing attention to the methodological implications of the Brownian motion case (Perrin's corroboration of Einstein's predictions of Brownian motion). Bohm and Vigier were proposing 'hidden variables' explicitly in the hope that their new theoretical framework would eventually lead to new insights and to the development of *new kinds of experiments* that could be used to confirm it. Moreover, they argued that new proposals like their hidden variables may lead to new kinds of experiments *in a way that may otherwise not be directly discoverable*, just as in the case of Brownian motion, where kinetic theory led to new kinds of experiments that were otherwise unavailable, eventually leading to a scientific revolution that limited the validity of the second law of thermodynamics. More generally, paving the way for Feyerabend's argument for pluralism, as Bohm later put it, his "new interpretation constituted a broadening of the conceptual framework, permitting the introduction of a wide range of possible modifications, that could not even be expressed in the Copenhagen interpretation".[106] According to Bohm, this can be helpful because "very often a different interpretation lends itself to changes and generalizations which *are not possible* in the first interpretation".[107]

The first time Feyerabend alluded to the methodological implications of Brownian motion in print was at the 1957 Colston conference,[108] where Feyerabend indirectly brings up the example of Brownian motion as a test of the atomic theory and classical phenomenological thermodynamics, introducing the episode into the discussions. According to Feyerabend:

> The first thing that comes to one's mind when one hears the objections of Bohm's opponents,[109] are objections which were raised, in the nineteenth century, against the atomic theory. Those objections were much more formidable, as they seemed to reveal a contradiction in the idea of a reduction of thermodynamics to the mechanics of a collection of atoms. [...] nevertheless the atomic theory was finally successful *and could even be used for showing that the phenomenological account could not be universally correct.* (Feyerabend in Körner-Pryce (eds.) 1957, "Discussion": pp. 48–49, italics inserted)

This episode subsequently became a reoccurring topic mentioned by many participants at the conference. The crucial methodological point is that the apparently anomalous phenomenon only became decisive evidence after the atomic theory (kinetic theory) was used to make successful predictions about its motion. This example becomes Feyerabend's main illustration of a crucial experiment between rival incommensurable, universal theories as an indirect refutation from *ERE*

---

[106] Bohm 1962: 270.

[107] Bohm in Körner-Pryce (eds.) 1957, "Discussion": 60, italics inserted.

[108] The discussion took place after the second session of the conference, in which Bohm and then Rosenfeld gave their papers. Cf. Körner-Pryce (eds.) 1957: 33–64. Feyerabend would have seen Bohm use the episode in this way in Bohm 1957a, Ch. 5. Cf. Bohm 1957b; Rosenfeld 1957b.

[109] Feyerabend seems to mean Rosenfeld.

(1962b) through *Against Method* (1993), which he repeatedly used to illustrate his methodological argument for pluralism.[110]

Rosenfeld clearly understood Feyerabend's suggestion and Bohm and Vigier's defence of their 'hidden variables' proposal in this way:

> As regards the atoms and the atomic theory which has been invoked by Bohm and also by other people[111] as an example of a theory with hidden parameters which has been successful, I have to say this: of course there are atoms and therefore people who spoke of atoms before it was possible for us to observe them were very lucky. (Rosenfeld in KÖRNER-PRYCE (eds.) 1957, "Discussion": 54).

Bohm responded to Rosenfeld, making a point that had a lasting impact on Feyerabend's formative development:

> I want to bring up this question of the discovery of atoms. Again I believe this was not a question of luck [as Rosenfeld had just suggested]. For the foundation of the atomic theory consisted in the fact that *it was necessary to explain certain difficulties of the large-scale theory*". (Bohm in KÖRNER-PRYCE (eds.) 1957, "Discussion": 57, italics inserted).

Here Bohm explicitly points out that an alternative was *necessary* to demonstrate that a mere apparent anomaly actually did refute the existing theory. This is the main methodological point of the Brownian motion example, which initially becomes central to Feyerabend's methodological argument for theoretical pluralism in *ERE* where Feyerabend develops Bohm's point that classical phenomenological thermodynamics could not have discovered that Brownian motion refutes it *by itself*, that is without the kinetic theory, which was necessary to make the predictions that, when confirmed, limited the validity of the second law of thermodynamics. Feyerabend uses this example to illustrate his test-model for comparing rival incommensurable theories in contrast to Popper's falsificationism,[112] and to show how incommensurable rivals can increase each other's empirical content.[113]

In this way, the attempt by Rosenfeld to exclude Bohm and Vigier's views directly motivated Feyerabend's work in methodology in the early 1960s, such as Feyerabend's 'plea for tolerance in matters epistemological' on Bohm's behalf.[114] Feyerabend tried to explain 'how to be a good empiricist' (FEYERABEND 1963a) to explain why physicists should take proposals like Bohm's more seriously, arguing that new incommensurbale theories can increase the empirical content of established theories and thereby promote progress. In other words, part of the motivation for Feyerabend's early 1960s papers was to help his friend, David Bohm, by

[110] In *Against Method*, Feyerabend initially cites (BOHM-BUB 1966) for the idea that alternatives are sometimes necessary to stage crucial tests and the example of the Brownian particle to illustrate the point. Cf. FEYERABEND 1975: 43fn9.

[111] Rosenfeld clearly means Vigier, who also develops the example in this way in the discussions, and Feyerabend who initially alluded to it to criticize what he took to be Rosenfeld's dogmatic rejection of Bohm's proposal.

[112] Cf. FEYERABEND 1962b; 2024; OBERHEIM-HOYNINGEN-HUENE 2024.

[113] Cf. FEYERABEND 1962b: 65–67 [72–73].

[114] Title, FEYERABEND 1963a: [78–103].

showing why new proposals such as Bohm's should always be welcome as they can lead to unforeseen progress that would otherwise be impossible. In this way, the example of Brownian motion, as part of an argument for methodological argument for theoretical pluralism, was developed directly out of Bohm's book and Bohm and Vigier's suggestions in the discussions at the conference, and then went on to take centre stage in Feyerabend's publications on methodology from *ERE* (1962b) through *Against Method* (1993).

Feyerabend also used the conference discussions as a testing ground for his peculiar views on meaning variance and methodology as later set out in *ERE* (1962b), when Feyerabend presented his basic idea that once we have a new theory "we have to use it throughout, for the description of experience as well as of anything else".[115] Feyerabend's idea is that after a successful crucial experiment, we have to admit that the earlier view was mistaken, and replace its concepts with new concepts better suited to explaining the laws used to test it.[116] Feyerabend casts the Copenhagen interpretation as committed to "describing experience not in terms of the best theory available [relativity theory] but in terms of some previous theory [classical theory] which has become so intimately connected with our ways of experiencing things that we can hardly imagine things being different".[117] Feyerabend uses the example of the concept of length and the special theory of relativity to explain what he means:

> Take the special theory of relativity. According to this theory the concept of length is a relative concept in the sense that you cannot attribute length to a given object without at the same time referring to some co-ordinate system. Now before the special theory was introduced, length was regarded as a property of an object.[118]

According to Feyerabend, in this way, the concept of length changed from a property of an object that is independent of a coordinate system to a relation that requires the specification of an object within a coordinate system, and the new concept should be used to describe the results of any measurements of any 'length'.

Bohm explicitly agrees with Feyerabend on this point: "I agree with what Dr. Feyerabend has said about using a new theory to modify the definition of older concepts".[119] In a striking anticipation of Feyerabend's 'ever-expanding ocean of alternatives' understanding of progress (Feyerabend 1965b: 224–225) and Feyerabend's contrastive account of empirical content,[120] Bohm illustrates his own views on meaning variance substituting 'position' for 'length':

> I want to raise the question of how we are to define a thing or the conception of a thing, such as position and so on. This was raised by Dr. Feyerabend. I believe that we must define a property such as position in terms of all the concepts that are available. During the Middle

[115] Feyerabend in KÖRNER-PRYCE (eds.) 1957, "Discussion": 52.

[116] Feyerabend in KÖRNER-PRYCE (eds.) 1957, "Discussion": 52.

[117] Feyerabend in KÖRNER-PRYCE (eds.) 1957, "Discussion": 53.

[118] Feyerabend in KÖRNER-PRYCE (eds.) 1957, "Discussion": 53.

[119] Bohm in KÖRNER-PRYCE (eds.) 1957, "Discussion": 55.

[120] Cf. FEYERABEND 1962b: 67 [73].

> Ages or before that time people were not very successful in defining positions accurately; then, when geometry developed, they were better able to define positions. Then with the development of mechanics and wave optics, it was possible to do this still better, for one could use mechanical and optical parameters and so on. And now if we should come to a sub-quantum level then we could define position in terms of the parameters of the sub-quantum level. In other words, the conception of position like all conceptions, may have infinite possibilities as we go deeper and deeper into nature and find out more and more of what is there. (Bohm in KÖRNER-PRYCE (eds.) 1957, "Discussion": 57)

Bohm thinks that "this is really a fundamental way by which progress in science is made: difficulties of conceptions at a certain level are resolved by suggesting a new conception".[121] This sounds very much like Feyerabend's understanding of revolutionary scientific advance as change of worldview.

However, Bohm adds a dialectical twist that Feyerabend is not willing to accept. For Bohm, the new conception and concepts are dialectically related to the opposite views that preceded them: "This again is characteristic of our concept. Each particular concept finds its reflection in its opposite and can be approximated to an arbitrary degree in its opposite".[122] Bohm repeatedly emphasized the dialectical nature of his methodology at the conference:

> I think that it is part of the cause of the difficulty inherent in the conception of probability that it does not apply with complete correctness any more than the conception of perfect determinism applies in this way. Rather, both of these conceptions must be regarded as opposite approximations to the same problem. (Bohm in KÖRNER-PRYCE (eds.) 1957, "Discussion": 57)

Moreover, Bohm also ties his characterization of the positive results of his new approach as presented in his paper at the conference very closely to his dialectical methodology:

> from one conception you can deduce the opposite conception as an approximation, so that in a certain sense one conception implies the other. This is what happens here: from the continuity of the phases of the particles that are coming out of the background of the sub-quantum-mechanical level I deduce precisely the Einstein-relation and the de Broglie-relation for energy and momentum [...] I do not wish to say by this that only continuity is a correct concept and that discontinuity is wrong; I want to say that both conceptions are correct but that each one reflects only a certain side of a phenomenon. Each one is reflected into the other, so that you can deduct the one from the other as an approximation. (Bohm in KÖRNER-PRYCE (eds.) 1957, "Discussion": 55)

Bohm wanted to explain why there are different kinds of particles of different masses, why they interact in the way they do, and why they have the properties they have. Feyerabend specifically asked what Bohm meant by derivation of these equations, and Bohm clarified by qualifying that he had not meant to suggest that they achieved or even sought a formal deduction of the equations of the existing quantum theory from their new proposal, but rather that they could deduce, and thereby recover, the successful predictions derived from the equation as an approximation

121 Bohm in KÖRNER-PRYCE (eds.) 1957, "Discussion": 57.

122 Bohm in KÖRNER-PRYCE (eds.) 1957, "Discussion": 57.

or a limiting case from a theory with a different ontology that no longer takes the quantum of action as fundamental.[123] This anticipates Feyerabend's views on the lack of deductive relations between incommensurable theories.[124]

At the conference Feyerabend also tried to develop his own outline of a strategy to handle the measurement problem, which he used to try to criticize the 'Copenhagen interpretation'.[125] For Feyerabend, the measurement problem and the irreducible indeterminacy implied by the existing quantum theory gradually turned into a methodological criticism, not of Bohm's strategy to re-explain the empirically testable results of quantum mechanics, but of the existing quantum theory itself as an observationally complete theory. Both Bohm and Feyerabend wanted to see the existing quantum theory replaced by a causal theory that allowed for a realistic interpretation that was not observationally complete.[126] According to Feyerabend:

> *For a realist the solution of the problem of duality is not to be found in alternative interpretations of a theory which in all probability is nothing but a predictive scheme anyway, but in the attempt to devise a completely new universal theory.*[127] (FEYERABEND 1958c: p. [60])

And Feyerabend clearly meant to support Bohm's proposal when he emphasized that:

> Only the invention of a new set of ideas which boldly oppose appearances and common belief, and which attempt to explain both in a deeper way, can then lead to further progress and to the continuation of rational argument. (FEYERABEND 1958c: p. [72])

After his talk at the conference, Feyerabend found out he had mistakenly thought Bohr belonged to the Copenhagen school, while in fact the outline of the strategy Feyerabend had proposed was the same type of strategy Bohr himself had proposed: "I discovered that I had been following a line which Bohr had taken against von Neumann long ago, but without showing in detail how a better theory of measurement would have to look like".[128] After the conference, Feyerabend wrote to Popper:

> Do you know that Bohr was *extremely critical* of both Heisenberg's and von Neumann's book because of the account of the 'reduction of the wave-packet' given in those two books? What I tried in my paper on measurement was to make use of this criticism as well as of Bohr's suggestion to regard measurement as an *irreversible process* which is repre-

---

[123] Cf. "With regard to the *derivation* of the Dirac equation, this was meant in the sense of an approximation or a limiting case" (Bohm in KÖRNER-PRYCE (eds.) 1957, "Discussion": 55).

[124] Cf. OBERHEIM 2024.

[125] Feyerabend presented an outline of his strategy in the foundations of quantum theory in FEYERABEND 1957a; 1957b. He used it to criticize the 'Copenhagen interpretation', which after 1957 he was careful to qualify so as not to include Bohr: "Bohr himself did not commit this mistake" (FEYERABEND 1958d: [298n1]).

[126] Cf. Feyerabend's "Observationally Complete Theories: Some Observations on Quantum Theory (1958)" in *FFY1*, Ch. 8: 487–491. Cf. Bohm in "Discussion" in KÖRNER-PRYCE (eds.) 1957, "Discussion".

[127] Cf. "In writing this paper I have learned much from discussions with Profs. D. Bohm, S. Körner, and K. R. Popper" (FEYERABEND 1958c: [50n76]).

[128] *FFY1*, Feyerabend to Popper, 7 April 1957: 261.

> sented, in a *schematic* way, by the 'reduction of the wave packet'.[129] (*FFY1*, Feyerabend to Popper, 21 July 1957-11: 267–268)

Feyerabend's strategy, following Bohr and much like Bohm, was to advocate deterministic laws, realistically interpreted. He identified the source of the indeterminism in the measurement procedure itself, concluding that macroscopic measurements of microscopic processes are limited to superpositions of quantum properties. As Feyerabend explained in his correspondence with Hempel, in contrast to Popper he "want[ed] our laws to remain precise, reversible, deterministic but want to put the imprecision and sloppiness into the method rather than into the laws."[130]

At the 1957 Colston conference, Feyerabend defend von Neumann's proof in his paper "On the Quantum-Theory of Measurement" (FEYERABEND 1957a), where he discusses Heisenberg and von Neumann's assumption of 'quantum jumps' and the difficulties to which it leads when incorporated into a theory of measurement. Feyerabend tried to show that these difficulties arise because of the incompleteness of their accounts, and he tried to provide a sketch or outline of a more complete account of the theory of measurement that does not contain mysterious 'quantum jumps'. Versions of the paper were promptly published both in German and English in 1957.[131] After having learned that Bohr did not belong to the Copenhagen school, Feyerabend was often careful to point this out in his subsequent criticisms of it.[132]

At the time, Feyerabend was also very critical of Popper's propensity interpretation strategy in the foundations of quantum theory, both before and after it was presented to the public for the first time by Feyerabend on Popper's behalf at the Colston conference.[133] According to Feyerabend, as he wrote to Popper, Popper's propensity interpretation:

> does not contain any contribution towards the main problem of quantum-mechanics [...] Or if it does, this contribution clearly contradicts the Neumann proof which is not changed by introducing the propensity-interpretation instead of the usual interpretation in terms of relative frequencies (which von Neumann uses). (*FFY1*, Feyerabend to Popper, 27 March 1957-4: 254)

---

[129] Feyerabend is referring to FEYERABEND 1957a.

[130] See *below*, Feyerabend to Hempel, 20 April 1968, Part 1.

[131] Cf. FEYERABEND 1957a; FEYERABEND 1957b. On the history of the composition of FEYERABEND 1957a, cf. *FFY1,* Feyerabend to Popper, 27 March 1957-4: 254.

[132] Cf. Feyerabend's 1958 review of VON NEUMANN 1955 (FEYERABEND 1958d: [298n1]). Cf. Feyerabend's paper on complementarity, FEYERABEND 1958c: [49–73].

[133] According to Feyerabend: "The fact that the probability-statements of quantum-mechanics, whether they are now interpreted in terms of relative frequencies or in terms of propensities imply, in the case of the single electron, that it cannot have definite momentum and position at the same time. That means: Whether a physicist adopts now the frequency interpretation, or your interpretation, he would IN BOTH CASES add some further arguments in order to show that the single system cannot be in a well defined state. More especially: Von Neumann's proof is valid in the frequency-interpretation as well as in the propensity-interpretation. But if this is true, then your paper is not conclusive" (*FFY1*, Feyerabend to Popper, 30 March 1957-5: 256).

Feyerabend was quite clear that, once again, he thought that von Neumann's proof was correct in this respect.[134]

Just a few months after they initially met at the infamous 1957 Colston conference, Bohm and Feyerabend then became close colleagues, when Bohm took up a research fellowship at the University of Bristol in August.[135] At the time, Feyerabend had been a lecturer in philosophy of science at the University of Bristol since 1955. Feyerabend and Bohm appear to have immediately bonded over their shared dislike for Pryce, who was head of the department of physics.[136] From 1957 on, Feyerabend championed Bohm, as an ostracized outsider, over the establishment.[137] For his part, Bohm appears to have found a kindred spirit in Feyerabend, "Discussing philosophy with [Feyerabend] made his otherwise intolerable situation in Bristol more tolerable".[138]

Bohm's unorthodox views on the foundations of quantum theory became a major topic of discussion for Feyerabend and Popper, and Bohm's acceptance of a position at the University of Bristol from 1957–61 added a new substantial member to Feyerabend's otherwise rather limited discussion group in Bristol.[139] In this way, Feyerabend brought Bohm in contact with Popper's work and acted as a go between:

> *Would it be possible for you to send proofs of your book + especially also of the chapter containing your criticism of determinism. Bohm is discussing his contribution with me and I really think that he should read what you have to say on the matter.* Can you suggest any way in which this can be done? I gave Bohm all things I have from you, *Open Society, Three Views*, the Determinism-paper, your *Scientific Autobiography*, he read it, but he did not get

[134] According to Feyerabend, "if within von Neumann's proof (which I think is correct and which I must discuss with you as soon as you return!) your probability is used instead of another interpretation, the proof still excludes hidden variables" (*FFY1,* Feyerabend to Popper, 27 March 1957-4: 254).

[135] Bohm held a research fellowship at the University of Bristol until 1961, when he became Professor of Theoretical Physics at Birkbeck College in London. See Peat 1997.

[136] Feyerabend wanted to perform some of Felix Ehrenhaft's experiments, but he was too afraid to ask Pryce to let him into the lab: "Yesterday I had a private discussion with Pryce, the physicist here. He listens to my lectures. I did not yet dare to ask him for allowance to perform some experiment in his laboratory – what I would very much like to do" (*FFY1*, Feyerabend to Popper, 6 February 1956-7: 219). On Feyerabend and Ehrenhaft, see *FFY1*, Ch. 2: 29, especially 30n35; Feyerabend to Popper, 1 November 1948-2: 63, 64n3; Feyerabend to Popper, 14 October 1955-27: p. 196; ("There is a very definite hope to repeat Ehrenhaft's experiments here") Feyerabend to Popper, 23 December 1955-34: 211; Feyerabend to Popper, 6 February 1956-7: 219n4. See Feyerabend 1977: [170–172]; Feyerabend 1978: 110–111 (= Feyerabend 1975²: 276–277); Feyerabend 1979a²: 219–220; Feyerabend [1995a]: pp. 66–67; Feyerabend [1995b]: 117). Cf. Oberheim (2006); Collodel (2022); Feyerabend (2022). Feyerabend's trepidation helps explain the motivation of his 'plea for tolerance in matters epistemological' (title, Feyerabend 1963a).

[137] Feyerabend 1954c argues against determinism in the context of Bohm's 1952 hidden variables proposal. However, even then, Bohm was not a determinist, which was clarified in Bohm-Vigier 1954. Cf. Del Santo–Krizek 2019; van Strien 2019; 2024.

[138] Bohm in Wilkins 1987.

[139] On Bohm's interpretation of quantum theory, see Jammer 1974, Ch. 7.

> on with the *Logik der Forschung* and gave it up as it is in German.[140] (*FFY1*, Feyerabend to Popper, 15 December 1957-13: 271, italics inserted)

This is how Feyerabend came to act as an intermediary between Bohm and Popper after the conference: "I am enclosing a MS by Bohm which contains his more recent ideas on hidden variables and which he asked me to send to you and to Joske".[141] Feyerabend also tried to bring them together face to face: "Next week I want to bring Bohm along – it is not yet certain – in order that he may perhaps meet Karl".[142] He also wanted to bring Bohm into direct contact with other Popperians: "Dear Joske, [...] Please write in advance when you will come as I would like you to meet Bohm (perhaps he will be in your lecture, but if not I would like you to meet him privately), and also a few other people here".[143]

In the late summer of 1957, Feyerabend also helped to organize and participated in the "Conference on Philosophical Foundations of Physics" at the MCPS from 12 August to 14 September, where he began his next paper on foundations of quantum theory, which was eventually published in 1962 as "Problems of Microphysics" (1962a).

Toward the end of 1958, Feyerabend had still not written his review of Bohm's book *Causality and Chance in Modern Physics*: "I have still got Bohm to review (I am just working on that review and I hope it will be finished, and good, by the end of this month [November]".[144] After Feyerabend finally finish his review, he gave a draft to Bohm.[145] In the review, Feyerabend presents Bohm's dialectical methodology as implying that: "Nature itself forces man to speak the truth, and it also forces him to speak in such a way that his theories have objective reference".[146] He takes Bohm to be suggesting that "Knowledge is a natural process which leads to a mirroring, in the head of man, of the properties of the universe. The mirror-image may be distorted at the edges [but] this distortion does not reach into the center of the mirror which perfectly represents the situation at a certain level".[147] Feyerabend adamantly rejects what he takes to be "the epistemology behind Bohm's belief that

---

[140] This explains Bohm limited understanding of falsificationism as direct refutation in Bohm's 1959 response, see *below*, Ch. 8, to Feyerabend's 1960 critical review of Bohm's 1957 book. Cf. FEYERABEND 1960b; BOHM 1957a; 1961; 1962; 1964, in the lead up to Bohm's chapter "Falsificationism" in Bohm's book *The Special Theory of Relativity* (BOHM 1966).

[141] *FFY1*, Feyerabend to Popper, 1 May 1958-5: 280.

[142] *FFY3*, Feyerabend to Agassi, 12 March 1958-5(3).

[143] *FFY3*, Feyerabend to Agassi, 2 November 1957-9.

[144] BOHM 1957a, see FEYERABEND 1960b: [219–235]. Cf. "Bohm has been appointed Research Fellow in Bristol starting August 1957. I think that you will very much like meeting him and I am looking forward to common discussions as I am at the present moment a bit at a loss in understanding some of his ideas" (*FFY3*, Feyerabend to Agassi 1956-5). Cf. *FFY3*, Feyerabend to Agassi 2 November 1958-17.

[145] Bohm's response to Feyerabend critical review is dated 1959. See *below*, BOHM [1959], Part 2, Ch. 8.

[146] FEYERABEND 1960b: 335 [232].

[147] FEYERABEND 1960b: 335 [233] italics inserted.

every theory, however absurd it may seem at first sight, has some kind of truth in it and correctly mirrors what exists in the universe".[148] Feyerabend explained why he rejected Bohm's dialectical methodology: "I do not believe that this account of knowledge is a correct one […] clearly Hegel's notorious *'Alles Vernünftige ist wirklich'* is at work.[149] Feyerabend was also very clear about which side he was on, declaring: "I cannot accept Bohm's methodology of caution [...] I prefer [...] the methodology of falsification as it has been developed by Popper".[150]

Bohm countered with a hefty 38-page response to Feyerabend's critical review with the lengthy title "On Certain Epistemological and Cosmological Problems Raised in Dr. Feyerabend's Review of Causality and Chance in Modern Physics" (1959).[151] In responding to Feyerabend's critical review, Bohm discusses his 'notion of the qualitative infinity of nature',[152] especially in the treatment of the relationship of human thinking to the cosmos, and of the question of what is the meaning of truth in the context of theories that do not have a completely universal domain of validity.

Bohm explicitly develops his own methodology in contrast to falsificationism throughout the essay.[153] He begins by succinctly presenting his dialectical view of progress, as illustrated by the relation between relativity theory and Newtonian mechanics.[154] He rejects Feyerabend's characterization of him as a dialectical materialist who believes that the course of history is somehow pre-determined by historical forces,[155] and he emphasizes that he never meant to suggest there is a necessity in how history will actually develop.[156] Bohm's response anticipates several central aspects of Feyerabend's criticism of falsificationism that he wrote and rewrote over

---

[148] Feyerabend 1960b: 337 [233].

[149] Feyerabend 1960b: 337 [233], italics inserted. German re-translation of the first part of Hegel's principle "What is rational is real; And what is real is rational [Was vernünftig ist, das ist wirklich; und was wirklich ist, das ist vernünftig]" (HW 7: 24); see also Heine's famous reformulation of the relevant part of Hegel's principle: "All real is rational [Alles was ist, ist vernünftig]" (Heinrich Heine, Briefe über Deutschland, Frgament 19844/45). Feyerabend rejected what he took to be Bohm's dialectical materials on a similar basis. See *below*, Section 1.3; Part 3, Ch. 8.

[150] Feyerabend 1960b: 339 [235]. Cf. *ERE* 1962b; Oberheim 2024.

[151] See *below*, Bohm [1959], Part 2, Ch. 8. Cf. Bohm 1957a and Feyerabend 1960b.

[152] For Bohm's views on the qualitative infinity of nature, see Bohm 1957a, Ch. 5.

[153] See *below*, Bohm [1959], Part 2, Ch. 8.

[154] See *below*, Bohm [1959], Part 2, Ch. 8.

[155] Cf. Bohm-Peat (1987).

[156] On Bohm and Hegel and binary oppositions, see for example Bohm-Peat 1987: viii, 299f. Compare Peat's description of Bohr's belief that, "while the opposite of truth is an error, the opposite of a great truth is another great truth" (Peat 1997: 184). According to Peat, "this maxim even influenced the way Bohr wrote his scientific papers. Bohr would agonize about the best way to express a particular idea in a given sentence; then, when he had written it down, he would attempt to express its exact opposite in the sentence that followed" (Peat 1997: 184–185).

the course of the next year (1960),[157] and which eventually became *ERE*. For example, Bohm argues that falsificationism cannot account for relating and testing framework theories against each other. This clearly exposes the Bohmian nature of Feyerabend's criticism of falsificationism in *ERE*, where Feyerabend argues that falsificationism cannot take comparing universal theories into account.[158] In anticipation of Feyerabend's *ERE*, Bohm declared that as a methodological rule, falsificationism has only a limited validity:

> the methodology of falsification (like all other theories) has its *limits of applicability*, and that it is perhaps a special case of a proper method. (Bohm [1959], Part 2, Ch. 8, italics inserted)

This is a harbinger of Feyerabend's later suggestion that new proposals should be either inconsistent with or incommensurable with the existing theory.[159] In the former case, when theories can and should be tested by falsification. In the latter case, when theories are incommensurable, they should be compared according to Feyerabend's test-model of a crucial experiment based on Bohm's Brownian motion example. In this case, the new theory showed that the existing theory could, by itself, *never* have discovered that the phenomenon refutes it just because it is based on an erroneous ontological principle (according to energeticism there are no atoms).[160]

There is also a deep affinity between Feyerabend and Bohm's ideas about subjective contributions to what counts as the facts. Bohm makes it very clear that he thinks facts are *manufactured* or constructed with the help of theories, and that without theories for that purpose, there can be no such facts. Bohm suggests that "there may be a large range of facts that we simply do not look for, because there is

[157] Feyerabend had finished a substantially complete draft of *ERE* 1962b by January 1961, cf. Paul K. Feyerabend, "Explanation, Reduction, and Empiricism", KP 423.4 (66 ss., T). This is the final draft version of *ERE* 1962b, which was later retyped at the MCPS. See *below*, Ch. 9, Feyerabend to Kuhn, 1961-2; and Paul K. Feyerabend, "Explanation, Reduction, and Empiricism", HF 08-22-01 (139 ss., T). An earlier draft of this paper, which elaborated upon an unfinished and much shorter piece sent by Feyerabend to the MCPS in early 1960, was circulated among Feyerabend's colleagues and friends in mid-summer of 1960. In fact, most of the papers planned to be published in Vol. III were submitted to Feigl and Maxwell, the editors of the volume, by October 1959, see *FFY1*, Feyerabend to Popper, 18 February 1959-2: 310n12. However, the publication process was severely delayed partly due to Maxwell's health conditions. See *FFY1*, Feyerabend to Popper, 7 April 1961-4: 357n5. The complete manuscript version of the volume was delivered to the University of Minnesota Press before September 1961, but the volume was issued only in May 1962. See *FFY1*, Feyerabend to Popper, 1 February 1961-1: 351n2.

[158] Cf. "It seems to me that this theory [Popper/Watkin's view, i.e. falsificationism] *can well deal with* all the problems that arise when T and T′ are *commensurable*, but inconsistent inside D′. It does *not* seem to me that it [falsificationism] can deal with the case where T′ and T are incommensurable" (*ERE* 1962b: 93 [92], italics inserted). This was no minor contretemps, but the main conclusion of Feyerabend 1962b, and it is what led to Feyerabend's break with the Popperian school, in 1967 see Oberheim 2024.

[159] Cf. Feyerabend 1962b: 91–92 [91].

[160] Cf. Feyerabend 1962b: 65–66 [71–73].

no room in our general framework of thinking to express these facts".[161] Bohm even provocatively claims that "Facts of this kind [retrodictions] *must* escape our conceptual net as long as we think only in terms of predictions of future events on the basis of initial conditions".[162] Bohm emphasizes that what counts as the facts will depend on the theories used to interpret experience, and that, "guided by different conceptions, one is led to seek different kinds of facts, some of which may be possible in a given field, *and some not*".[163] Feyerabend develops these ideas as what he takes to be the most significant insight of his early philosophy: his rejection of the 'autonomy principle', according to which "facts exist and are available independently of whether or not one considers alternatives to the theory to be tested".[164]

Bohm's 1959 response to Feyerabend's critical review should be seen as the missing link between Feyerabend's endorsement of Popper's falsificationism in that review and Feyerabend's criticism of Popper's falsificationism in *ERE* (1962b). Feyerabend's immediate reaction to Bohm's criticism of Popper's position was that it is inadequate:

> Bohm wrote 38 pages reply to it, containing a long criticism of your position but I am afraid this reply is very unclear and unsatisfactory. (*FFY1*, Feyerabend to Popper, 26 October 1959-12: 328–329)

Feyerabend's more considered reaction appeared a few years later in *ERE*, where he first explicitly criticizes Popper's empiricism (1962b).[165]

After Bohm left Bristol in 1961, Feyerabend had very little contact with him: "I saw him [Bohm] again, twice, once in Berkeley, and once in Zuerich (1986)".[166] Feyerabend began to revise and withdraw some of his criticisms of Bohm's views in his "Dialectical Materialism and the Quantum Theory" (Feyerabend 1966c). In the long run, some of Feyerabend and Bohm's most basic ideas appear to have continued to converge.[167]

Feyerabend's reversal with respect to Bohm and Popper's methodology becomes ever more explicit in the 1970s. It can be clearly seen in a footnote added at the very end to the republication of his critical review[168] of Bohm's book[169] in the first volume of Feyerabend's collected papers:

> *Added 1980.* My criticism [of Bohm's methodology] is a paradigm case of abstract reasoning, and shows all its weaknesses. The argument assumes (a) that methodological stipula-

---

[161] Bohm's 1952 proposal involves retrodiction, which Popper discussed at length in "Complementarity and the Two-Slit Experiment" (Popper [1953a] in *FFY1*, Ch. 5) and in "A Defence of Free Thinking in Quantum Theory" (Popper [1953b], in *FFY1*, Ch. 6).

[162] See *below*, Bohm [1959], Part 2, Ch. 8, italics inserted.

[163] See *below*, Bohm [1959], Part 2, Ch. 8, italics inserted.

[164] Feyerabend 1963a: [78–103].

[165] Cf. Oberheim 2024.

[166] See *below*, Part 2, Ch. 1, Feyerabend to Peat, 9 July 1993-2.

[167] Cf. *Wissenshaft als Kunst* (Feyerabend 1984) and *The Conquest of Abundance* (Feyerabend 1999b). Cf. Bohm 1968; Peat 1997.

[168] Feyerabend 1960b.

[169] Bohm 1957a.

> tions can be introduced and enforced independently of what goes on in the world and (b) that developing them is just a matter of imagination, not of real relations that might hinder or further the work of the theoretician. In a word - it is assumed that the domain of knowledge-construction and knowledge-modification is unencumbered by the laws of nature, it is a special domain in which the laws of nature are suspended, but whose products can transform nature. One only needs to formulate this assumption to see its absurdity. Yet philosophers of science who praise methodological rules because they circumvent some recondite logical problems (such as 'Hume's problem') and regard them as adequate on those grounds alone make this absurd assumption. (FEYERABEND 1981a: 235n20.

Feyerabend clearly seems to have meant Popper.

1.3 Feyerabend and Kuhn

The precise circumstances surrounding exactly when and where Feyerabend and Kuhn first met face to face are not well-documented. The documents in their respective archives do not seem to be of much help with respect to these specific questions. They do, however, offer many significant and valuable insights into the precise circumstances with respect to *how* they first came to meet, and *how* and *why* their relationship began to develop in the early 1960s as it did. Moreover, they lay the groundwork for understanding why there was little if any personal contact between Kuhn and Feyerabend from the early 1960s to the early-1980s, which is something that seems to have gone largely unnoticed by the academic community at the time. Kuhn and Feyerabend were seen as close allies during a period in which they had almost no personal contact. After a very brief biography introducing Thomas Kuhn, the following section will attempt to explain the circumstances surrounding how they came to meet and then why their relationship quickly devolved from an intellectual love at first sight into that of an estranged couple, who had once had a brief, but deep and tumultuous intellectual affair.

Thomas S. Kuhn (1922–1996) was an American historian and philosopher of science of Jewish descent. He was educated at Harvard University, where he studied physics earning his B.Sc. in 1943, his M.Sc. in 1946, and his Ph.D. in 1949. He continued his studies in the history of science as Harvard Junior Fellow (1949–52) and started his academic career at the same institution as Assistant Professor of the history of science (1952–56). Later he joined the Department of Philosophy of the University of California, Berkeley as Associate Professor of the history of science (1956), teaching courses for the Department of History, where he was eventually promoted to Full Professor of the History of Science in 1961. He moved to Princeton University as M. Taylor Pyne Professor of Philosophy and History of Science (1964–79), and then concluded his academic career at the Massachusetts Institute of Technology as the Laurance S. Rockefeller Professor of Philosophy (1979–91).

Feyerabend and Kuhn in Berkeley in the early 1960s. Feyerabend first met Kuhn after Feyerabend had been invited by the Department of Philosophy of the University of California, Berkeley to spend the 1958–59 session as a visiting Associate Professor, while Kuhn, who had been a member of the campus faculty for two years, was on leave. Feyerabend had only recently turned 34 when he moved to Berkeley in mid-September 1958. He had just begun his academic career, having been appointed junior lecturer of philosophy of science at the University of Bristol in 1955. Feyerabend had submitted three letters of recommendation to secure his first post. One was written on his behalf by his Ph.D. supervisor, Viktor Kraft. Another was provided by Karl Popper from his recently established chair at the LSE as Professor of Logic and Scientific Method, whose active support went well beyond a mere statement of appreciation.[170] A third letter of recommendation came from the prominent quantum physicist and Nobel laureate Erwin Schrödinger.[171] Berkeley's philosophy department had been searching for a philosopher of science since at least the mid-1950s and Feyerabend's name was chosen for the temporary post – possibly to take him on as a trial period – following a recommendation of Herbert Feigl that was lukewarmly supported by Alfred Tarski, one of the most respected and influential professors on campus.[172]

In its quest for a philosopher of science, the Berkeley philosophy department had already started making inquiries with respect to Kuhn in September 1955, when Kuhn was an assistant professor, teaching basic courses in general education and advanced courses in the history of science at Harvard University. As Kuhn had no concrete prospects of a tenured position, he was half-heartedly considering alternatives. Stephen C. Pepper – the Chairman of the department since 1953 – had approached him directly in February 1956, and in the early spring, and after careful consideration, Kuhn was offered an appointment as Assistant Professor of the History of Science from the academic year 1956–57, jointly by the Department of History and the Department of Philosophy. Kuhn accepted Berkeley's proposal, teaching a reduced load of two courses for each department in his first year and the standard amount of six courses – three for each department – in his second year. Following a recommendation made by both departments in the fall of 1957, Kuhn was promoted to Associate Professor with tenure in July 1958. During the winter, he worked with his fellow historian of science A. Hunter Dupree on a comprehensive history of science course program to be offered at Berkeley from 1958–59.[173]

[170] Kraft and Popper were friends and fellow anti-inductivists.

[171] Feyerabend's first paper was written in the summer of 1948 in preparation for the upcoming public lecture by Schrödinger on 20 August, is an anti-thesis to Schrödinger's views on intelligibility. Cf. FEYERABEND (2016b); KUBY 2016.

[172] Cf. FISCHER 1999: 79; FISCHER 2006: 5; *FFY1*, Feyerabend to Popper, 12 March 1958-4: 279n1.

[173] On the whole, ten courses – seven in history (105A and B: History of Scientific Thought and Technique; 178A and B: History of Science and Technology in American Society; 204: Seminar in the History of Science; 276A and B: Seminar in the History of Science in American Society) and three in philosophy (127 A and B: Problems in the Development of Physical Science; 220: Seminar in the Relations of Science and Philosophy) – "available for credit within the major programs of undergraduate and graduate students in both the Departments of History and Philosophy" with "[i]n addition, the history of science [...] available as an examining field and as an area for dissertations to graduate students in history" (DUPREE-KUHN 1958: 173), but not yet in philosophy. Dupree was a Harvard historian of science who had followed a career path analogous to that of Kuhn, moving from the Gray Herbarium at Harvard University to the Department of History at Berkeley on a visiting appointment as an Assistant Professor of History in 1956–57, being confirmed in that position the following year and finally promoted to Associate Professor in 1958.

Kuhn had eventually decided to accept the invitation to spend a year at the "Center for Advanced Studies in the Behavioural Sciences" (*abbr.* CASBS), which had been extended to him in 1954, when the institute was founded by the Ford Foundation.[174] Kuhn's absence from Berkeley brought about a temporary opening in the philosophy of science disciplinary area, which helps explain why Berkeley made the offer of a visiting appointment to Feyerabend. Kuhn's absence for that session also makes it unlikely that Kuhn and Feyerabend met face to face before Kuhn's return. Indeed, right at the time when Feyerabend was settling in Berkeley in the late summer of 1958, Kuhn and his family were either moving, or had already moved to Palo Alto, which is about 30 miles southward across the San Francisco Bay, where the CASBS is located on a hill overlooking the Stanford University campus.[175] By the time Kuhn returned to Berkeley about a year later in 1959, Feyerabend had been promoted to Associate Professor of the philosophy of science, finally bringing the Berkeley philosophy department's long search to an end,[176] and was leaving for Minneapolis, where he spent the fall semester of 1959–60 as a visiting scholar at the MCPS. Moreover, if there ever were a time before 1960, however briefly, when both Kuhn and Feyerabend did find themselves in Berkeley at the same time (which seems less unlikely in the late summer of 1959 than a year before), the fact that Kuhn's main office was located in the Department of History makes the chances that they happened to accidentally stumble across each other in the corridors of the notoriously labyrinthic Dwinelle Hall also seem rather unlikely.[177] Be that as it may, there is no evidence that Feyerabend and Kuhn either became acquainted with each other or entertained any significant conversation until several months after Feyerabend's arrival in California.

During his stay at the CASBS in 1958–59, Kuhn began work on what became *SSR* (Kuhn 1962), which had been commissioned to him in the early 1950s by Charles Morris for the second volume of the *International Encyclopedia of Unified Science* (*abbr. IEUS*).[178] The project had originally been conceived by Kuhn in

[174] Cf. Andresen 1999: 864–864; Hufbauer 2012: 439–443.

[175] Cf. Andresen 1999: 66.

[176] Berkeley's quick final move was triggered by an analogous offer made to Feyerabend by Cornell University in the late winter of 1959, cf. *FFY1*, Feyerabend to Popper, 22 January 1959-1: 306; 18 February 1959-2: 309; 2 March 1959-3: 311; 1959-4: 313; 22 March 1959-5: 314–315; *Feyerabend's Formative Years. Vol. 3. Feyerabend among Popperians* (forthcoming, abbr. *FFY3*), Feyerabend to Agassi, 1 January 1959; Feyerabend 1995: 113–115.

[177] While this is highly conjectural, the hypothesis that Kuhn's main office was in the history department and not in the philosophy department is based on the firm fact that the letterhead of a 1960 letter by Kuhn to Morris reads "Department of History" (Kuhn to Morris, 13 October 1960, RC 088-47-07), though further circumstantial evidence could be invoked in its support. The Department of Philosophy was in Dwinelle Hall in the late 1950s.

[178] Cf. Morris to Carnap, 2 March 1959, RC 088-70-19; Kuhn 1953: 458; Kuhn 1962: vii, viii; Kuhn 1977: xviii-xix; Kuhn 1997: 291–292, 295. Plans to include a monograph on the history of science within the *IEUS* date back at least to 1937 (possibly even to 1935), when Neurath commissioned Federigo Enriques to write one. Afterwards, a monograph on the sociology of science also became part of the project. It was initially assigned to Louis Wirth (in the 1940s), then later to

1947, while he was "a graduate student in theoretical physics [at Harvard University] within sight of the end of [his] dissertation".[179] Kuhn had started developing this project more concretely in the paper "The Function of Measurement in Modern Physical Science" (KUHN 1961), which was sketched in 1956, extended in 1958, and eventually published three years later.[180] Kuhn took advantage of the opportunity afforded to him by CASBS to finally give some shape to his long-planned monograph. In July 1959, at the end of his fellowship, he had succeeded in taking further steps forward in the project by reflecting thoroughly on the work he had done over the previous years. During this process, besides gaining crucial insights on the general approach as well as the specific structure of the projected monograph, he developed two significant papers: The first paper, "The Essential Tension" (KUHN 1959),[181] was his contribution to the third University of Utah Research Conference on the Identification of Creative Scientific Talent that had taken place in Salt Lake City in June 1959. The second paper was a draft called "The Function of Thought Experiments," which he only managed to finalize and publish several years later.[182] However, the concrete results that he obtained with respect to the monograph were rather scarce, limited as they were to just two drafts of its first chapter.[183] It would still be another three years before "the first full published report" of the original project that he had conceived some fifteen years earlier reached the bookshelves in 1962.[184]

When Feyerabend was back in Berkeley from Minneapolis, soon after the first half of January 1960,[185] Kuhn had just sent the materials that he had produced during 1958–59 to Morris with whom he was renegotiating the terms of his commitment with the *IEUS* editors. They eventually came to an agreement, setting a deadline for showing significant progress at the end of the summer.[186] In early

I. Bernard Cohen (in the 1950s), and still later (apparently) to Aldo Mieli. Eventually, the surviving editors of the *IEUS* – Charles Morris and Rudolf Carnap – Neurath having passed away in late 1945 – embraced Cohen's suggestion of Mieli, who was his younger colleague at the Harvard Society of Fellows, as the most promising choice. Cf. MORRIS 1960; UEBEL 1991: 639n34; REISCH 1994: 175; FERRIANI 1995: 174; KUHN 1997: 283, 291–292; PINTO DE OLIVEIRA 2007: 151–152.

179 KUHN 1962: v. Cf. KUHN 1977: xi, xii, xvi; KUHN 1997: 289, 292.

180 Cf. KUHN 1977: xvii-xviii, KUHN 1997: 295; BUCHWALD-SMITH 1997: 365; HUFBAUER 2012: 441, 451.

181 Cf. KUHN 1977: xviii.

182 Cf. KUHN 1964; KUHN 1977: xx.

183 Cf. HUFBAUER 2012: 450–451.

184 KUHN 1962: v.

185 Cf. Feigl to McDiarmid, 17 August 1959, HF 01-40-14; *FFY1*, Feyerabend to Popper, 1960-3 "Address until 20. January 1960": 339; Feigl to Carnap, 13 January 1960, HF 02-69-51.

186 In the second half of August 1959, Morris asked Kuhn how he was coming along with his work on the manuscript (Morris to Carnap, 18 August 1959, RC 088-70-16/2). However, at the end of December, he still had not yet received any reply from Kuhn (Morris to Carnap, 20 December 1959, RC 088-70-16/1). Kuhn only responded to Morris at the turn of 1960, illustrating the work that he had done thus far and presenting him with a problem concerning the submission dates for his projected monograph: Kuhn's plans were now to publish a long essay in the *IEUS* that would

September, after having worked on the draft with unprecedented concentration, Kuhn's efforts proved to be substantially successful as he finally had a complete, though still rough, first draft:

> I write to report on the results of my summer activity. As I promised, all but the first three weeks, spent on vacation, were devoted to a draft of the monograph. Though I am not yet entirely out of the woods and have some problems to raise with you, I think that you will be pleased with the outcome. [...]. At two o'clock in the morning of Sunday, September 11th I put the last word on a first draft and breathed a long sigh of relief. Parts of the draft are still very crude (for the last three-quarters of the draft I attained a writing pace that I have never been able to sustain before, and there are even a few parts that I have not yet reread myself). I hope, however, to do some preliminary polishing in the immediate future. If that goes well, I shall have a preliminary version dittoed for circulation to you and a few friends before Christmas. On the basis of reactions gained in that way, I shall be able to do final revisions during the spring or, at the latest, over next summer. I think that at last all the major hurdles have been crossed. (Kuhn to Morris, 13 October 1960, RC 088-47-07)

Yet, despite Kuhn's best intentions, he was unable to finalize *Proto-Structure*[187] before the first half of April 1961, when he finally submitted a copy of a preliminary version of *SSR* to Morris.[188] Probably concurrently, or soon afterwards, further

---

summarize the results of a book on which he was working (cf. Morris to Kuhn, 4 January 1960, RC 090-70-15). However, Morris, with Carnap's approval, proposed a different solution, which would have brought the monograph to printer in a shorter time: an introductory (instead of a summarizing) essay to the major work (under the suggested title of "Prolegomena to Interpretation of Scientific Revolutions"), not necessarily long (in compliance with the style of the other monographs in the *IEUS* series). This would incorporate the material that Kuhn had already written and ideally expand it in the direction of the main topic of the series (the unity of science), which then could be completed by the end that summer, and finalized by the end of the year (cf. Morris to Kuhn, 4 January 1960, RC 090-70-15; Carnap to Morris, 28 January 1960, RC 088-70-13; Morris to Carnap, 4 February 1960, RC 088-70-12). In early March, Kuhn replied to Morris, assuring him that he would devote the summer of 1960 to the *IEUS* monograph. After receiving Morris's approval, Kuhn also submitted his manuscripts, together with his monograph project, to Carnap, from who, he received an enthusiastic response and some encouraging support. Cf. Morris to Kuhn, 19 March 1960, RC 088-70-09; Carnap to Kuhn, 12 April 1960, RC 088-47-07; Carnap to Morris, 12 April 1960, RC 088-70-08; Kuhn 1962: viii; and Kuhn 1977: xix.

187 This is Hoyningen-Huene's name for this draft-manuscript. Cf. Hoyningen-Huene 2006: 610.

188 "Here at last is a partially revised and now dittoed version of the manuscript I pushed through last summer. It still needs annotation and some further polishing. In the process I am sure to find many passages that need pruning, extension, reorganization, or something else of the sort. But I think that, until the last two sections, you will find the manuscript fairly clean. Those two, and particularly the last, do need a lot more work. I had hoped to rewrite them entirely before sending the draft to you and to a few others for criticism. But time has now run out of <sic> me. I shan't get back to this project before mid-July at the earliest. At that point I hope to put the whole thing down for publication and would be glad for any advice I can get meanwhile" (Kuhn to Morris, 16 April 1961, RC 088-70-04). Cf. Morris to Carnap, 21 January 1961, RC 088-47-05; Kuhn 1977: xix. It should be noted that the existence of a *Proto-Proto-Structure* – i.e., the draft manuscript of *SSR* that Kuhn completed in September 1960 – partly confirms Kuhn's 1994 vague recollection to the effect that "he had finished a mimeographed draft of [*SSR*] in the fall or early winter of 1960" (Hoyningen-Huene 1995: 353). The draft to which Kuhn referred could well be *Proto-Proto-Structure*, though this was not the version of the *SSR* draft that was eventually mimeographed, namely *Proto-Structure*. (Notice however that Kuhn's words, privately delivered in a letter, were

copies (either identical or slightly amended) were distributed among a select group of Kuhn's friends. The subsequent revision process extended over the summer when Kuhn also delivered the paper, "The Function of Dogma in Scientific Research" (Kuhn 1963) to the *Symposium on the History of Science* that took place at the University of Oxford from the 9 to 15 July.[189] It continued in the fall until eventually, in late 1961, the final version of *SSR* was "substantially complete".[190] Kuhn put the date "February 1962" at the bottom of its preface of this final complete draft.[191] The *IEUS* editors accepted it promptly and freshly printed volumes became available in bookshops in the summer of 1962.[192]

The select group of people who were afforded the privilege of receiving a copy of the preliminary version of *SSR* and thus the opportunity to provide Kuhn with their critical remarks included his mentor James B. Conant, who received a draft manuscript soon after Morris in the spring of 1961,[193] as well as Feyerabend, who studied *Proto-Structure* closely, which is evidenced by the four long, typed letters whose content, according to Feyerabend's own words, partly reprised what he had already told Kuhn during their conversations about his views, but in which he was now able to give more detailed, at times page by page, comments on Kuhn's manuscript, and to reply to Kuhn's objections to his criticisms. Although all these letters are undated, they were all most likely written between April and November 1961.[194] Although Kuhn had the opportunity to take Feyerabend's copious criticisms of *Proto-Structure* into account before he put *SSR* in its final form, generally speaking, Kuhn seems to have made few significant changes to *Proto-Structure* in the course

reported without quoting by Hoyningen-Huene). Although Kuhn's recollection was later put into question as to its reliability (cf. Hoyningen-Huene 2006: 611), the new information that has emerged from scouring the archives at least partially vindicates his memory.

189 Kuhn introduced the paper with the following note: "The ideas developed in this paper have been abstracted, in a drastically condensed form, from the first third of my forthcoming monograph, *The Structure of Scientific Revolutions*, which will be published during 1962 by the University of Chicago Press. Some of them were also partially developed in an earlier essay, 'The essential tension: tradition and innovation in scientific research' [Kuhn 1959]" (Kuhn 1963: 347n1; cf. Kuhn 1977: xviiin6).

190 Kuhn 1977: xvii.

191 Cf. Kuhn 1962: x.

192 Cf. Morris to Kuhn, 26 March 1962, RC 088-47-03; Kuhn to Carnap, 12 April 1962, RC 088-47-02; Carnap to Kuhn, 28 April 1962, RC 088-47-01. On the reception of *SSR* by Carnap and more generally on the relation between Carnap and Kuhn's views, cf. Reisch 1991; Axtell 1993; Earman 1993; Irzik-Grunberg 1995, Friedman 1998, Pinto de Oliveira 2007, Friedman 2003; Reisch 2004; Richardson 2007; Irzik 2012. *SSR* was published by the University of Chicago Press both as issue 2 of volume 2 of the *IEUS* series, with its official yellow paperback cover, and as an independent volume, hardbound in blue cloth. Cf. Kuhn 1997: 300–301.

193 Kuhn sent *Proto-Structure* to Conant on 22 April 1961 and received his first comments in early June. Cf. Hershberg 1993, 860n84; Driver-Linn 2003: 272n2.

194 They were discovered by Hoyningen-Huene in two sets of two, on two different occasions. See Hoyningen-Huene 1995; 2006. They are reproduced, reorder and slightly corrected, in chapter 9. For details concerning why they were reordered, see *below*, Section 2.

of the final revision process.[195] Kuhn may simply have ignored most of Feyerabend's critical remarks as based on misunderstandings. In any case, he does imply that Feyerabend did not approve of his manuscript.[196]

Given the intensity of their debate, perhaps it should come as no surprise that Feyerabend was mentioned in the preface of Kuhn's masterpiece as the first of those four friends "whose contribution proved most far-reaching and decisive" and whose "reservations and suggestions" he had come to consider "extremely helpful".[197] On the other hand, in light of Feyerabend's passionate intellectual engagement with Kuhn, which is readily apparent in Feyerabend's comments on Kuhn's draft manuscript, it should also come as no surprise that Feyerabend stopped complaining (at least for a while) about the sense of isolation he was feeling at Berkeley.[198] In the fall of 1961, he wrote to Popper that he had finally "found a very good friend in the historian T. S. Kuhn with whom [he] sometimes discuss[ed] also philosophical matters".[199]

An early sign of the impact that Kuhn's work had on Feyerabend can be seen by comparing Feyerabend's essay "On the Interpretation of the Uncertainty-Relations", which was completed by the fall of 1960 at the latest and published at the end of that year only in Polish,[200] with a subsequent marginally amended version in English entitled "Problems of Microphysics" (Feyerabend 1962a), which was finalized before August 1961 (Feyerabend expected to receive proofs in November 1961 that were to

[195] Cf. Heilbron 1998: 509; Driver-Linn 2003: 272n2; Hufbauer 2012: 454. As we shall see, Kuhn was particularly busy with a new engaging project from the spring of 1961.

[196] Cf. Kuhn 1962: xii.

[197] Kuhn 1962: xii. The other three people to which Kuhn expressed gratitude are the philosopher Ernest Nagel, the atomic physicist and Kuhn's old friend H. Pierre Noyes and Kuhn's pupil at Berkeley John L. Heilbron. John W. N. Watkins also read *Proto-Structure* in 1961 (cf. Watkins 1970: 25), in all probability between June and early September when he stayed in Berkeley on a visiting appointment for the summer session.

[198] "Here I am fairly isolated. There are as it were two parties here, the ordinary language people on the one side and the axiomatizers on the other neither of whom is very attractive to me" (*FFY1*, Feyerabend to Popper, 1960-10: 349. Note these two groups are criticized by Feyerabend in *ERE* 1962b.

[199] *FFY1*, Feyerabend to Popper, 1961-10: p. 372. Feyerabend seems to have been unaware that Popper and Kuhn were acquainted with each other since early 1950 (when Popper gave his William James Lectures at Harvard University; cf. Borradori 1994: 159; Kuhn 1997: [286]) as in his letter he introduced the latter to the former as the author of "a very nice little book on the *Copernican Revolution*" (*FFY1*, Feyerabend to Popper, 1961-10: 372). Cf. Kuhn 1957. This suggests that it is unlikely that Kuhn had met Popper already in the summer of 1960, when Popper participated in the First International Congress in Logic Methodology and Philosophy of Science (Stanford University,24 August – 2 September 1960) accompanied by Feyerabend (cf. *FFY3*, Feyerabend to Watkins, 6 September 1960-2, JW 170; Borradori 1994: 159). This also provides a further clue that at the time, Feyerabend and Kuhn had not yet become intellectual partners.

[200] Cf. Feyerabend 1960b; *FFY1*, Paul K. Feyerabend "On the Interpretation of the Uncertainty-Relations: Course 21.84 Fall, 1960" Feyerabend to Popper, 1 February 1961-1: 351n2.

be printed in January 1962).[201] The first full-draft of the English version, completed in April 1961, includes a long endnote that is notably missing in the earlier Polish version. The endnote contains a brief presentation and criticism of Kuhn's alleged "dogmatism".[202] The endnote, which appears almost unaltered in the printed version of Feyerabend's essay,[203] is worth quoting in full as it summarizes the main conflict between Feyerabend and Kuhn from Feyerabend's perspective:

> As J. Agassi has pointed out to me that [sic] this principle [that a scientist should always keep an open mind and that he should therefore always consider possible alternatives along with the theory he is favoring at a certain moment] was consciously used by Faraday in his research work. Against the use of such a procedure it has been argued, by T. S. Kuhn (private communication) that the close fitting between the facts and the theory that is a necessary presupposition of the proper organization of the observational material can be achieved only by people who devote themselves to the investigation of one single theory to the exclusion of all alternatives. For this *psychological* reason he is prepared to defend the (dogmatic) rejection of novel ideas at a period when the theory which stands in the center of discussion is being built up. I cannot accept this argument. My first reason is that many great scientists seemed to be able to do better than just devote themselves to the development of one single theory. Einstein is the outstanding recent example. Faraday and Newton are notable examples in history. Kuhn seems to be thinking mainly of the average scientist who may well have difficulties when asked not only to work out the details of some fashionable theory but also to consider alternatives. However, even in this case I am not sure whether this inability is 'innate', as it were, and incurable, or simply due to the fact that the *education* of the 'average scientist' is in the hands of people who subscribe, implicitly, to Kuhn's doctrine of the necessity of concentration. My second reason for not being able to accept this argument is as follows: assume that it is indeed correct that human beings are not able at the same time to work out the details of one theory *and* to consider alternatives. Who says, then, that details are more important than alternatives which, after all, keep us from dogmatism and are a very concrete and lively warning of the limitations of all our knowledge? If I had to choose between a very detailed account of the fabric of the universe at the expense of not being able to see its limitations and between a less detailed account whose limitations however were very obvious, then I would at once choose the latter. The details I could gladly leave to those who are interested in practical application.[204] (FEYERABEND 1962b, p. 277n153)

Rather solid, though perhaps less straightforward, evidence suggests that Kuhn and Feyerabend had become acquainted sometime before the spring of 1961. In a substantially final version of Feyerabend's essay "Explanation, Reduction, and Empiricism" that dates back to the end of January 1961 at the latest (but whose proofs

[201] Cf. *FFY1*, Feyerabend to Popper, 17 August 1961-9: 366; FEYERABEND 1962a; FEYERABEND [1995a]: 372.

[202] Cf. *FFY1*, Feyerabend to Popper, Paul K. Feyerabend "Problems of Microphysics" KP 423.6; Feyerabend to Popper, 1961-7: 363n1. It should be noted that although none of Kuhn's papers are mentioned in Feyerabend's note, as he only makes indirect reference to his conversations with Kuhn, this should not be considered evidence that Feyerabend had not yet read *Proto-Structure*, as (FEYERABEND 1961a), which was certainly composed after Feyerabend had read the *Proto-Structure* manuscript, does not explicitly refer to any of Kuhn's works either. *SSR* is only explicitly cited for the first time in Feyerabend's published works in FEYERABEND 1963a: 21n* [91n26].

[203] Except for a mistake in the latter, which seems to be due to an error of transcription from the draft.

[204] The passage in square brackets to which the endnote refers is on 231.

were printed only in the late summer of that year and the collection in which it was included as late as the spring of 1962),[205] Feyerabend wrote that "[w]hile the present paper was in progress [he] had the opportunity to consult various as yet unpublished papers by Professor T. S. Kuhn (Berkeley) in which the noncumulative character of scientific progress is illustrated very forcefully by historical examples".[206] This January 1961 passage is the very first occurrence of Kuhn's name in Feyerabend's writings, and it appears unchanged in the published version of the paper.[207] Even though Kuhn's papers to which Feyerabend refers in this passage (without mentioning them by name) clearly cannot include *Proto-Structure* (as this was finalized only in April 1961), they may well be the two papers which Kuhn had written during his 1958–59 CASBS fellowship (published respectively as KUHN 1959 and KUHN 1964). Indeed, two more letters of Feyerabend to Kuhn (see *below*, Part 3, Ch. 9), this time in longhand but again undated, seem to confirm this further conjecture. In the first one, slightly more than two pages long and certainly written sometime in 1960, Feyerabend devoted a few brief comments to KUHN 1959,[208] whereas the second one, twice as long and whose date remains uncertain, focuses on a draft of Kuhn 1964. The 1960s epistolary remarks on Kuhn's "Essential Tension" (KUHN 1959) strongly suggests that Feyerabend was acquainted with Kuhn's defense of scientific conservatism by the time he developed the position he set out in January 1961 in a draft of *ERE* (1962b).

This reconstruction fits fairly well with Feyerabend's recollections of his interactions with Kuhn at the time he was working on *SSR*. In the late 1960s, Feyerabend wrote that "[i]n the years 1960 and 1961 when Kuhn was a member of the philosophy department at the University of California in Berkeley I had the good fortune to be able to discuss with him various aspects of science"[209] and, in a vivid description, he evoked their lively debates, "some of which were carried out in the now defunct *Café Old Europe* on Telegraph Avenue and greatly amused the other customers by their friendly vehemence".[210] However, on later occasions, Feyerabend has also explicitly stated that he had access to the manuscript of *SSR* at an earlier time than the spring of 1961.[211] Undoubtedly, if in these passages Feyerabend was referring to

205 Cf. Paul K. Feyerabend, "Explanation, Reduction, and Empiricism", KP 423.4; *FFY1*, Feyerabend to Popper, 1 February 1961-1: 351; see *below*, Feyerabend to Feigl, 22 April 1961, HF 03-133-05; Grünbaum to Feigl, 6 September 1961, HF 02-169-125; Ervin (University of Minnesota Press) to Feigl, 3 May 1962, HF 04-164-51; FEYERABEND 1962b.

206 Cf. Paul K. Feyerabend, "Explanation, Reduction, and Empiricism", KP 423.4.

207 Cf. FEYERABEND 1962b: 32.

208 In the letter Feyerabend refers to a paper read by Hanson in Chicago which is, without reasonable doubts, the paper he delivered to the Section L (History and Philosophy of Science) of the Annual Meeting of the *American Association for the Advancement of Science*, Hotel LaSalle, Chicago, 27–30 December 1959. Cf. HANSON 1961; FEYERABEND 1961c.

209 FEYERABEND 1970a: 197.

210 FEYERABEND 1970a: 198n2.

211 Cf. "around 1960" (FEYERABEND1995: 141), "in 1960" (FEYERABEND 1978: 117n49; where he also adds that he "discussed it extensively with Kuhn"), and even in 1959 (FEYERABEND 1976:

Proto-Structure, the eventualities there purportedly described are simply impossible chronologically speaking. On the other hand, most (though not all) of Feyerabend's claims could not simply be considered as slips of memory if they referred, in a concise (and rather sloppy) way, to the draft manuscript of *SSR* that Kuhn had completed in September 1960 (known as '*Proto-Proto-Structure*') or to the papers through which Kuhn came to develop it, such as Kuhn's "The Essential Tension" and "The Function of Thought Experiments". It should be emphasized, however, that no independent evidence in favor of the hypothesis that Feyerabend had access to *Proto-Proto-Structure* has emerged as of yet, while Kuhn's memories, however meager, tend to undermine rather than support the hypothesis, as they make reference only to *Proto-Structure*: "[Feyerabend] was at Berkeley, I gave him this draft manuscript of the book that I'd sent to Chicago".[212] More generally, it seems that any claim to the effect that Feyerabend and Kuhn met each other *personally* before January 1960 cannot be established on any firm ground based on archival sources, as there is nothing in the archive that suggests that they interacted between January 1960 and the summer of 1960. Therefore, although Kuhn and Feyerabend were formally colleagues at Berkeley since the fall of 1958, the supposition that they entertained any significant intellectual relationship before the fall of 1960 should be considered highly speculative unless more definitive documents are unearthed, or more reliable evidence comes to light.

In the spring of 1961, a noticeable shift in the evolution of Feyerabend and Kuhn's relations becomes readily apparent. Indeed, the lively debate between the two that left a lasting trace in Feyerabend's letters on *Proto-Structure* soon motivated Feyerabend to write a "concise summary of results," which he had already "explained in a more detailed fashion in 'Explanation, Reduction and Empiricism', and 'Problems of Microphysics', [and] other papers."[213] As he clarified in a note to "How to Be a Good Empiricist" (which was written between April and October 1961, i.e., exactly when the Kuhn-Feyerabend debate on *Proto-Structure* was taking place, but which was published only two years later in 1963), the "general outlook" that he was developing in the early 1960s "derive[d] from the work of K. R. Popper (London) and David Bohm (London) and from [his] discussions with both. It was severely tested in discussion with [his] colleague, T. S. Kuhn (Berkeley). It was the latter's skillful or <sic!> defence of a scientific conservatism which triggered two papers".[214] The second paper to which Feyerabend refers may be his *Knowledge*

391; this passage has no matching the English original edition, cf. Hoyningen-Huene 2005: 165n33, 174.

[212] Kuhn 1997: 310.

[213] The other papers Feyerabend mentions are, 'Problems of Empiricism' and 'Linguistic Philosophy and the Mind-Body Problem'. In the 1981 reprint, Feyerabend clarifies that the last paper was never published (Feyerabend 1981a: 83n5).

[214] Feyerabend 1963a: 8 [81–82]. The paper was prepared for the 128th Annual Meeting of the American Association for the Advancement of Science (Denver, 26–31 December 1961) and was presented on 26 December to Session I "Empiricism and the Status of Theories" of Section L (History and Philosophy of Science), chaired by Hanson and with George T. McClure and Wilfrid

without Foundations (FEYERABEND 1961a). This is the first publication in which Feyerabend refers directly to what was to become *SSR*, although even here Feyerabend does not refer to *SSR* explicitly by its title, but as "a book to be published in the near future [by] my colleague, Professor T. S. Kuhn of the University of California".[215] These lectures were delivered by Feyerabend speaking freely without a prepared manuscript and the published text is based on a slightly revised transcription of recordings by the author. They develop Popperian themes that reverberate with some of the same theses and arguments in Feyerabend's epistolary criticism of *Proto-Structure*.

The spring of 1961 was also especially momentous for Kuhn, but in different respects. At the end of March, when his painstaking monograph-project was finally on its way to fruition, Kuhn embarked on a new major intellectual enterprise, accepting the directorship of the Sources for History of Quantum Physics (*abbr.* SHQP) project.[216] In early May, only a few weeks after he had circulated *Proto-Structure,* he was notified of his promotion to Full Professor. This was not an unexpected news,[217] and Kuhn received it with mixed feelings, as much to his dismay, he also learned that the move would entail a transfer of his position entirely to the Department of History at the request of the senior members of Department of Philosophy and its Chairman (Karl Aschenbrenner). This was also accompanied by their refusal to accept history of science as an optional Ph.D. qualifying field for philosophy students.[218] Although these developments enraged Kuhn and forever spoiled his positive relation with Berkeley, the SHQP project was appealing enough to check his anger and to keep his mind busy for the following three years.[219] It also initially appeared to be a good

Sellars as discussants. However, the manuscript was circulated among Feyerabend's colleagues in Berkeley already in October and the paper had already been delivered on 30 November within the second (1961–62) Annual Lecture Series of the Center for Philosophy of Science, University of Pittsburgh. It was given again about a year later, on 23 October 1962, under the title "The Pernicious Influence of Empiricism", within the 1962–63 University of Delaware Seminar in the Philosophy of Science, in whose proceedings it was eventually published. Cf. *FFY1*, Feyerabend to Popper, 7 April 1961-4: 357; see *below*, Hempel to Feyerabend, 1961-2; 1961-3; *FFY3*, Feyerabend to Watkins, 6 October 1961; 23 October 1961, JW 710; [December 1961], JW 710; "University of Pittsburgh – The Program in the Philosophy of Science – Schedule of Lectures 1961–62", HF 09-16-01; Paul K. Feyerabend, "Research Project on the Logic of Measurement, Probability and Confirmation, with Applications in Atomic and Quantum Theory", RC 090-63-06. *FFY3*, Feyerabend to Watkins, 6 October 1961; 23 October 1961; December 1961, JW 710; April 1963/I; April 1963 /II, JW 170.

215 FEYERABEND 1961a: [61].

216 Cf. HEILBRON 1998: 510–511; HUFBAUER 2012: 453; KUHN-HEILBRON-FORMAN-ALLEN 1967; HEILBRON 1998.

217 In the early fall of 1960, Kuhn received an attractive offer of a full professorship in the history of science from The Johns Hopkins University, but he dismissed it once Berkeley made it clear that they were ready to accept his conditions for staying, among which was the acceptance of the history of science as a legitimate area of study for dissertations of graduate students in philosophy. Cf. HUFBAUER 2012: 451–453.

218 Cf. HUFBAUER 2012: 453.

219 Cf. ANDRESEN 1999: S66; HUFBAUER 2012: 453.

opportunity to collaborate with Feyerabend, who had been working on the philosophy of quantum theory, and who had signaled that he was eager to participate in the venture from the first moment he heard about it in the spring or summer of 1961 when the project was being planned. However, the prospect of Feyerabend's participation in the project was not received favorably by everyone involved, and so nothing came of Kuhn's proposal to include him.[220] As a consequence, from the fall of 1961, Kuhn and Feyerabend's relations dramatically subsided or at least did not leave any appreciable sign in the historical record until the early 1980s. Feyerabend spent the first part of his 1962 sabbatical 'in residence' giving a joint seminar with Popper, who stayed in the Berkeley campus from January to May 1962 on a visiting appointment for the spring.[221] During this period, Kuhn arranged a meeting with Popper and interviewed him together with Heilbron, although quite independently of the SHQP project.[222] This project took Kuhn to Denmark for a whole year, starting the following summer. While at the Niels Bohr Institute in Copenhagen, Kuhn received an offer from Princeton University and accepted it without hesitation, seizing on his first good opportunity to leave Berkeley and settle down in a department where the history and the philosophy of science were understood as integrated and strategically combined.[223] He devoted 1963–64, his last academic year as a professor at the University of California, to bringing the SHQP project to completion.

For Feyerabend, on the other hand, Popper's temporary presence in Berkeley also contributed to his decision to leave California and move back to England, since, as he confessed to Watkins in the early spring of 1962, "there is really nothing to keep me here."[224] This attitude quickly changed rather dramatically, as within a few weeks, Feyerabend was made Full Professor of Philosophy at Berkeley.[225] Feyerabend's promotion had been under review since 1960. It was supported by the philosophers but was opposed by some of Berkeley's senior physicists. Emilio Segré, in particular, found fault with Feyerabend's work on the philosophy of quantum theory and

---

[220] In the summer of 1961, Feyerabend wrote to Popper: "I was supposed to participate in a very interesting project to reconstruct the history of the quantum theory from interviews (much has been done in discussions and only a very small part of these discussions has reached the printed page; this is very different from the case of relativity). However I was dropped from the project as it was discovered that mentioning my name leads to violent reaction in Copenhagen" (*FFY1*, Feyerabend to Popper, 17 August 1961-9: 366–369). It is not unlikely that Feyerabend received the bad news together with an invitation to a party given by Kuhn, which Feyerabend attended the day after he wrote this letter. Cf. Feyerabend's Notebook 1960–1968, PF 3-5-1.

[221] Cf. *FFY1*, Feyerabend to Popper, 11 March 1961-2: 353; 13 April 1961-5: 359; 29 April 1961-6: 360–361; 14 May 1961-8: 364–365; 17 August 1961-9: 366–369; 1961-10: 370–373; 1961-11: 374; 1961-12: 375–376; 25 October 1961-13: 377; *FFY3*, Feyerabend to Watkins, 10 January 1962, JW 170; *FFY1*, Feyerabend to Popper, 16 January 1962-1: 378–379. Feyerabend spent the second part of his sabbatical at the Minnesota Center for Philosophy of science, which had invited him, as well as Popper, for the fall semester 1962.

[222] Cf. Heilbron 1998: 510 and 510n20.

[223] Cf. Heilbron 1998: 510–511; Andresen 1999: S66; Hufbauer 2012: 453.

[224] Cf. *FFY3*, Feyerabend to Watkins, 21 March 1962, JW 170; April 1962, JW 170.

[225] See Popper to Feigl, 24 May 1962, KP 294.6.

questioned its relevance.[226] However, Kuhn's departure from the Department of Philosophy in 1961, seems to have left a vacancy in Feyerabend's intellectual life that needed to be filled. This event set off a period of seclusion that was only partly mitigated when they split about twelve months later.[227] Feyerabend reconsidered the idea of leaving Berkeley for England, but such a major turn in Feyerabend's (intellectual and social) life only came in the summer of 1964, when he returned to Europe for the first time after six uninterrupted years in the US that had seen him rise from a lecturer to the highest rank of the academic echelon. In the first of the papers that Feyerabend came to consider as "belated after-effects"[228] of his early exposure to Kuhn's ideas, he declared that the one who "ha[d] made it clear to [him] that history cannot be dismissed out of hand as being irrelevant to the methodologist" had been Kuhn.[229]

The rift in Feyerabend and Kuhn's correspondence (1962–1983). Kuhn and Feyerabend are often very closely associated, especially because they have both been recognized as having introduced the idea of incommensurability into discussions of the nature of scientific progress in 1962. For similar reasons, they are often lumped together, as together having initiated the movement away from a logic-oriented approach to philosophy of science toward an historically oriented approach to philosophy of science. In this way, in the 1960s, they seem like two compatriots, each excited to find a brother in arms, especially with respect to meaning change, incommensurability and the relevance of history to the philosophy of science.[230]

---

226 The proposal of the Department of Philosophy was prompted by a competing offer of a Full Professorship that had been extended to Feyerabend by the Minnesota Center for Philosophy of Science at the University of Minnesota. Cf. *FFY1*, Feyerabend to Popper, 3 April 1960-4: 340; 11 April 1960-5: 341–342; 16 April 1960-6: 343; 17 August 1961-9: 366–369; 1961-10: 370–373; *FFY3*, Feyerabend to Watkins, 23 October 1961, JW 710.

227 Cf. *FFY3*, Watkins to Feyerabend, 27 August 1962, JW 170; 6 October 1961, JW 710; Agassi to Popper, 12 December 1962, KP266.26. Popper's pupil, William W. Bartley, III, spent 1963–64 as a visiting professor at Berkeley. During his two-semester stay in Berkeley, Bartley had contacts with Kuhn, whom he had known since the mid-1950s, when he had been a graduate student at Harvard. Feyerabend's correspondences with Agassi, Lakatos, Popper and Watkins testify of Feyerabend's feeling of isolation at the time.

228 FEYERABEND 1970a: 197n2.

229 FEYERABEND 1970c: 277.

230 Cf. FEYERABEND 1970a; Part 2, Ch. 8, *below*. This is Feyerabend's contribution to *Criticism and the Growth of Knowledge,* LAKATOS-MUSGRAVE (eds.) 1970, which is, for the most part, based on the proceedings of the International Colloquium in the Philosophy of Science (Bedford College, Regent's Park, London, 11–17 July 1965); see *FFY1*, 13 October 1965-3: 402n1. According to the pre-liminary conference program as of January 1965, Feyerabend was supposed to read a paper in the session "Criticism and the Growth of Knowledge" on 13 July 1965, which was designed to compare and confront Popper and Kuhn's approaches to the philosophy of science, and their views on science and its method. The session was supposed to feature Kuhn, Lakatos, and Feyerabend as the main invited speakers. See LAKATOS-MUSGRAVE (eds.) 1970, Preface; Lakatos to Popper, 21 January 1965, KP 318.7. Feyerabend, however, had expressed strong doubts about his presence at the conference to Lakatos, the main organizer of the event, as early as late May 1965. See *FFY3*, Feyerabend to Lakatos, 20 May 1965, IL 13/272a-37; Lakatos to Feyerabend, 24 May 1965, IL 13/272a-38; Lakatos to Popper, 31 May 1965, KP 318.7. At the beginning of June, Feyerabend announced his final decision not to attend the conference in London and proposed that his pro-

Feyerabend's harsh criticisms of Kuhn's philosophy did not become visible to the public until 1970, with the publication of "Consolations for the Specialist" (FEYERABEND 1970a), which draws heavily, directly from these letters. In the early 1960s, however, Feyerabend was a strident critical rationalist,[231] who adamantly defended the need for normative methodology, which he tried to justify on ethical grounds, in contrast to Kuhn's historical approach. Feyerabend's reaction to Kuhn's *Proto-Structure* was to write a series of lengthy letters acerbically attacking Kuhn's 'normal science' as a pernicious form of dogmatism, and to vehemently challenge Kuhn's use of history given the *implicit* normative implications of his account of progress. For Feyerabend: "What you [Kuhn] are writing is not just history. It is ideology covered up as history".[232] Feyerabend's persistent, caustic criticisms of Kuhn's work in these letters, especially the recurrent accusation that he was promoting a dogmatic, conservative ideology, contributed to a rift in their personal relations. As Kuhn had considered himself to be a left-leaning, progressive since school, and he seems to have felt deeply misunderstood by Feyerabend's reaction to his ideas. As we have seen, Feyerabend characterized the tone of their boisterous discussions in *Café Old Europe* on Telegraph Avenue in Berkeley as 'friendly vehemence'. Kuhn seems to have seen Feyerabend's vehemence in a less friendly light. By 1962, he seems to have had more than enough of Feyerabend's relentless attacks. Perhaps he was somewhat relieved when he left for Denmark to lead SHQP,[233] as this seems to have provided him with some respite from Feyerabend's continuous barrage of criticism of his allegedly harmful and dogmatic ideology. The fact that Feyerabend had not been welcome to participate in the project would have provided Kuhn with a convenient excuse for not writing to Feyerabend while he was in Copenhagen, given that Feyerabend had not been welcome to participate in the project. After Kuhn left Berkeley, and there is no evidence of any further correspondence between the two until the middle of August 1983, when they briefly met at Robert S. Cohen's house in Boston, and subsequently beginning in April, they exchanged a few reconciliatory letters.[234] These appear to be the first direct

jected paper could be read in his absence. He also admitted that he might not be able to submit the paper in time, due to the surgical removal of a kidney stone in mid-June, which might keep him in a Midwest hospital for two or three weeks. Feyerabend's proposal was not well-received by the organizing committee and Lakatos suggested replacing Feyerabend with Watkins in the conference session. See *FFY3*, Feyerabend to Lakatos, 9 June 1965, IL 14/3; Lakatos to Feyerabend, 25 June 1965, 1965-17. Eventually, Lakatos himself was replaced by Popper. See LAKATOS-MUSGRAVE (eds.) 1970: Preface. Feyerabend recovered from the operation only in late July. See *FFY3*, Feyerabend to Lakatos, 4 July 1965; IL 14/3; Feyerabend to Agassi, 7 July 1965; Feyerabend to Lakatos, 9 July 1965, IL 14/3; 28 July 1965, IL 13/272a-41/42; Lakatos to Feyerabend, 28 July 1965, PF 2-7-24. Feyerabend did not participate in the conference but was asked afterwards to contribute a paper.

[231] Cf. COLLODEL 2016.

[232] See *below*, Part 3, Ch. 8, Feyerabend to Kuhn 1961–2.

[233] Feyerabend warned Kuhn of the dogmatic attitude he will meet in Copenhagen, see *below*, Part 3, Ch. 8, Feyerabend to Kuhn 1961–3.

[234] Cf. below, Part 3, Ch. 8, Feyerabend to Kuhn 1983–1.

interactions between Feyerabend and Kuhn since the early 1960s, after which they were sporadically in contact in the mid-1980s. In September 1983, they had another friendly encounter, when Kuhn was in San Francisco.[235] A month earlier, in August 1983, Feyerabend sent a brief postcard to Kuhn on behalf of Paul Hoyningen-Huene[236] and a year later.[237] Hoyningen-Huene arrived in Boston on the 15th of August 1984 for his year-long visit.[238] They were then briefly reunited in Meilen on Lake Zürich.[239] At the time, mirroring the philosophical community more generally,

[235] See Augustin (ed.) 2010, Feyerabend to Hoyningen-Huene 1983–11: p. 57.

[236] Paul HOYNINGEN-HUENE was born in Pfronten/Allgäu, West Germany in 1946. He studied Physics and Philosophy in Munich (1966–1971), Mathematical Physics in London (1971–72) and Theoretical Physics in Zurich (1975). He taught in Switzerland and Germany (1976-present) and was a visiting scholar at the Massachusetts Institute of Technology with Thomas S. Kuhn (1984–85). Afterward, he published HOYNINGEN-HUENE 1989, which advances an original interpretation of Kuhn's ideas that was developed in regular, extensive discussions with Kuhn about his views. The book was enthusiastically endorsed by Kuhn in the 'Foreword' dated August 1988. Kuhn recounts that when he met him, "Hoyningen knew my work better than I and understood it very nearly as well," declaring: "No one, myself included, speaks with as much authority about the nature and development of my ideas" (KUHN 1993: xi). In 1997, HOYNINGEN-HUENE founded the Center for Philosophy and Ethics of Science at the Leibniz Universität Hannover (Germany). His views on the nature of scientific knowledge were published in HOYNINGEN-HUENE 2013.

[237] Hoyningen-Huene (see *below*) was trying to use Feyerabend's help to arrange for Kuhn to host him as a visiting scholar at MIT. He asked Feyerabend to write a letter to Kuhn on his behalf for this purpose on 14 August 1983. See AUGUSTIN (ed.) 2010, Hoyningen-Heune to Feyerabend 1983–8: p. 55; Feyerabend to Hoyningen-Huene 1983–9: p. 55; Hoyningen-Heune to Feyerabend 1983–10: 56; Feyerabend to Hoyningen-Huene 1983–11: pp. 56–58; Hoyningen-Heune to Feyerabend 1983–12: pp. 58–59; Feyerabend to Hoyningen-Huene 1983–13: pp. 59–60; Feyerabend to Hoyningen-Huene 1984–16: 63. The postcard was very brief and to the point, describing him as a 'perfectionist with humour' [Hoyningen-Huene, personal communication]. As Kuhn recounts in the 'Foreword' of Hoyningen-Huene 1993: "I first met Paul Hoyningen in mid-August 1984, when he arrived in Boston to spend a year at the Massachusetts Institute of Technology. [...] His initiative was accompanied by persuasive recommendations" (Kuhn 1993: xi). At least one of these recommendations had come from Feyerabend in the form of the postcard to Kuhn about Hoyningen-Huene mentioned above.

[238] Hoyningen-Huene had met Feyerabend in 1980 and they became friend in 1982/1983. In 1984, Hoyningen-Huene was instrumental in promoting the personal relationship between Feyeabend and Kuhn. Eventually, on 15 August 1984, Hoyningen-Huene arrived in Boston with a generous all (travel) expenses paid invitation from Feyerabend for Kuhn to visit him in Zurich to give a talk. This was initially received by Kuhn unenthusiastically, but it did temporarily reunite them, however briefly, in 1985. Cf. Hoyningen-Huene, personal communication (DATE); and Hoyningen-Heune to Feyerabend 1984–22: p. 71; Hoyningen-Heune to Feyerabend 1984–22: pp. 72–73; Hoyningen-Heune to Feyerabend 1984–23: p. 72–73; Feyerabend to Hoyningen-Huene 1984–24: p. 73; Feyerabend to Hoyningen-Huene 1984–25: p. 75; Hoyningen-Heune to Feyerabend 1984–26: p. 76; Feyerabend to Hoyningen-Huene 1984–27: p. 78; Feyerabend to Hoyningen-Huene 1984–28: p. 78; Feyerabend to Hoyningen-Huene 1985–30: p. 82. Hoyningen-Heune to Feyerabend 1985–31: pp. 83–85; Feyerabend to Hoyningen-Huene 1985-33: p. 82; Feyerabend to Hoyningen-Huene 1985–34: pp. 90–91; Feyerabend to Hoyningen-Huene 1985–36: pp. 96, 99; Feyerabend to Hoyningen-Huene 1985–36: p. 99; and below Feyerabend to Kuhn 1988–1. Cf. Hoyningen-Huene 1993.

[239] See below, Feyerabend to Kuhn 1988–1; Feyerabend 1995: pp. 152–153, 162. Kuhn and Feyerabend had also met briefly in Berkeley (see below, Feyerabend to Kuhn 1988–1).

Hoyningen-Huene was completely unaware that Kuhn and Feyerabend had only met a few times over the twenty years since they had been colleagues at Berkeley, and that a rift had opened in their personal relations since the early 1960s.[240] He had erroneously assumed that they had remained in close contact.[241]

The letters on *Proto-Structure* document the state of their relationship at a crucial juncture in its subsequent development, and they clarify Feyerabend's initial deep reservations with respect to Kuhn's work. The also help to explain the reason for the rift that opened in their personal relations, while adding substantial content to Feyerabend's subsequent published criticisms of Kuhn.[242] Feyerabend repeatedly made it very clear that it was Kuhn's mode of presentation that he found most objectionable:

> It is this bewitching way of representation to which I object most, the fact that you take your readers in rather than trying to persuade them. This manner of presentation you share with Hegel and with Wittgenstein; and with all those who say, when engaging in a political enterprise, that 'history will be our judge'. Do you really believe this?[243] (Feyerabend to Kuhn 1961–2)

From Feyerabend's point of view, the problem with this kind of approach is that it promotes a pernicious conservativism by smuggling undesirable norms that pretend to be neutral (and indisputable) facts into the historical narrative. In his letters on *Proto-Structure*, Feyerabend repeatedly recast variations of this point turning into a caustic attack.[244]

For Feyerabend, his disagreement with Kuhn was not primarily about how to describe science most accurately. It was not even primarily a factual question that can be answered by historical investigation. It is a question about choosing what we want to have, and for Feyerabend, the choice was clear:

> Also I think I have shown in my own essay that set of mutually inconsistent but factually adequate theories increases the empirical content of any element of the set and this for the simple reasons that many tests presuppose the existence of an alternative! (they are crucial tests). If this is the case then we must make the decision what do we prefer, increased empirical content of the theories we possess, or that unanimity of research and the close

[240] The ground had already been somewhat softened as Kuhn and Feyerabend had had a friendly encounter already in September of 1983, when Kuhn was in San Fransisco. See Augustin (ed.) 2010, Feyerabend to Hoyningen-Huene 1983-11: 57.

[241] Hoyningen-Huene had not known about the rift in their relations, and only learned from Kuhn about what had happened and why shortly after Feyerabend had died, while he was preparing a talk on Feyerabend and Kuhn for a conference on Feyerabend in Vienna in 1995: "*Paul Feyerabend und Thomas Kuhn". Eingeladener Vortrag am Symposium „Paul Feyerabend und die gegenwärtige Philosophie*", Vienna, Austria, 17 March 1995). Cf. Hoyningen-Huene 2000; 2002.

[242] Cf. Feyerabend 1970a and *below*, Part 3, Ch. 8.

[243] See *below*, Part 3, Ch. 8.

[244] See *below*, Part 3, Ch. 8.

> fitting produced by it in the periods which you call the normal periods.[245] (Feyerabend to Kuhn 1961–2)

For Feyerabend it is a choice between revolutionary progress, which protects us from the conformist dangers of dogmatic ideology, and Kuhn's idea of normal scientific progress, which Feyerabend thinks would just deteriorate into dogmatic stagnation – just as he fears Kuhn's *SSR* will promote the dogmatic conservativism that was already stultifying progress in quantum mechanics.

For Feyerabend, the problem with Kuhn's view is that normal science will not lead to progress in our understanding of nature by eventually leading to an anomaly that will promote a crisis and then a revolution. Instead, from Feyerabend's perspective, normal science will lead to stagnation and so is both undesirable and self-defeating: "Revolution in Permanence should be the battle cry of every empiricist".[246] Kuhn, by contrast, emphasizes that normal science, based on the shared paradigms, is required to motivate and catalyze the depth of research necessary for a crisis phase as a prerequisite to revolutionary advance.[247] It provides a period of relative conceptual stability, which is necessary for Feyerabend's cherished revolutions. Feyerabend objected to any kind of alleged necessity in the course of scientific advance, and in this significant respect, Feyerabend's initial criticism of Bohm's views (see *above*, Section 1.2) was quite similar to Feyerabend's early criticism of Kuhn:

> every assertion may be read in two ways, as the report of a historical fact, and as a methodological rule. You thereby take your readers in. You present your material in such a manner that (at least for the periods following the introduction of paradigms) history seems to satisfy the principle that ALLES WIRKLICHE IST VERNUENFTIG.[248] (Feyerabend to Kuhn 1961–2)

As we have seen (*above*, section 1.2), at about the time he wrote these letters, the methodological argument for theoretical pluralism that Feyerabend was developing was based on Bohm's (classical) Brownian motion example in support of Bohm and Vigier's quantum Brownian motion proposal. It is therefore unsurprising that these letters also contain a succinct rendition of Feyerabend's methodological argument for theoretical pluralism directed against Kuhn's account of normal science as illustrate with the Brownian motion example:

---

[245] The essay to which Feyerabend refers is probably *ERE* 1962b. Cf. Feyerabend's "Research Project on the Logic of Measurement, Probability and Confirmation, with Applications in Atomic and Quantum Theory", *below*, Part 1, Ch. 4, 1961-4.

[246] See *below*, Part 3, Ch. 8.

[247] Bohm can be seen to be emphasizing the same point in his criticism of Popper's views, see section 1.2.

[248] See *below*, Part 3, Ch. 8. German re-translation of the second part of Hegel's principle "What is rational is real; And what is real is rational [*Was vernünftig ist, das ist wirklich; und was wirklich ist, das ist vernünftig*]" (HW 7: 24); see also Heine's famous reformulation of the relevant part of Hegel's principle: "All real is rational [*Alles was ist, ist vernünftig*]" (Heinrich Heine, *Briefe über Deutschland*, Fragment 19844/45). As we have seen, Feyerabend rejected what he took to be Bohm's dialectical materials on a similar basis. See *above*, Section 1.2.

> Elaboration of the paradigm of phenomenological thermodynamics therefore [would] never be able to discover the anomaly presented by the existence of the Brownian particle. But this anomaly was discovered. It was discovered by the elaboration of an alternative account, viz. of the kinetic theory which then produced predictions that could be tested by experiment. [...] Your insistence upon faithfulness to one and only one paradigm is bound to result in the elimination of otherwise very important tests and it is bound in this way to reduce the empirical content of the paradigm you want to be accepted. It may well be – and Bohm and Vigier are definitely of this opinion – that the situation is the same in the present quantum theory. [...] Bohm points out that the limitations of the present point of view will become evident only if one has first introduced an alternative and shown that it is preferable.[249] (Feyerabend to Kuhn 1961–2)

1.4 The Collection

Feyerabend's Formatives Years is planned as a three-volume set.[250]. This volume is thematically structured into three chronologically ordered parts. The first part collects material directly relevant to Feyerabend's relation to logical empiricism into five chapters. After the introduction, Chapter 2, "The Dogmas of Logical Empiricism", is a significant paper that Feyerabend delivered on the 14th of December 1951 at the *Institut für Europäische Gegenwartskunde* in Vienna, together with a short discussion moderated by Viktor Kraft (Feyerabend's Ph.D. supervisor). This is just one week before Feyerabend's official *Promotion* (Ph.D.) in December 1951 at the University of Vienna and about half a year after he had submitted his thesis, which handles some of the same topics. The paper was posthumously published in its German original as FEYERABEND (2010). The English translation for this volume was conducted on the original manuscript by Kati Hennig, Matteo Collodel and Eric Oberheim. Feyerabend's paper has two main aims: to defend logical empiricism from unwarranted attacks, and to criticize logical empiricism based on its own assumptions. The paper is a rich source for Feyerabend scholars and historians of philosophy. The title follows Quine's "Two Dogmas of Empiricism" (QUINE 1951a), sometimes regarded as the most influential paper of twentieth-century philosophy, which had been published earlier that year. Various early influences and the germs of ideas that later became influential are readily apparent. As in subsequent chapters, extensive notes by the editors provide background information, context, and references.

---

[249] See *below*, Part 3, Ch. 8. Cf. *ERE* 1962b.

[250] The final volume, *Feyerabend and the Popperian School* will collect Feyerabend's correspondence with three prominent members of Popper's circle: Joseph Agassi (1954–1985), John W. N. Watkins (1960–1987) and Imre Lakatos (1964–1967); together with a 1960 draft of what became Feyerabend's seminal paper "Explanation, Reduction and Empiricism". The materials collected provide significant insights into how his personal and intellectual interactions shaped Feyerabend's formative years as he developed the position that came to fruition in the early 1960s; and how he subsequently developed those ideas into "Anything goes" and "Against Method".

Chapter 3 includes papers and epistolary exchanges exhibiting Feyerabend's sustained attack against logical empiricism in the second half of the 1950s. Specifically, Feyerabend targeted the developing views of Carnap, the intellectual leader of logical empiricism, which he criticized in three different stages during this time period. The first section of Chapter 3 presents the first part of the first Feyerabend-Carnap controversy, centering on Feyerabend's criticism of Carnap's views on the model of 'logical reconstruction' or 'explication' of scientific theories. It features three papers: Feyerabend's own 1955 English abbreviated version of a paper that he published in German that year; Pap's summary of Feyerabend's German paper, including Pap's own critical comments; and Michael Scriven's critical comments on the content of Feyerabend's paper. Feyerabend had been Pap's research assistant in 1953–54, when Pap was a Fulbright visiting professor at the University of Vienna and he introduced Feyerabend to Herbert Feigl, the director of the MCPS. Feyerabend's English paper was addressed to Feigl, whereas both Pap and Scriven were visiting members of the MCPS when they composed their papers in the summer of 1955. The second section of Chapter 3 presents the second Feyerabend-Carnap controversy, centered on Feyerabend's criticism of the revised criterion of empirical meaningfulness advanced by Carnap in 1956. It consists of an intense and indirect epistolary debate that took place in the first half of 1957 between Feyerabend on the one hand and Carnap and Hempel on the other, made possible thanks to Feigl's mediation. In addition to the first initial note by Feyerabend excerpted by Feigl from Feyerabend's correspondence, this section includes two replies by Hempel, two replies by Carnap, three rejoinders by Feyerabend, a note by Rozeboom on the whole controversy, as well as Carnap's final comments on Rozeboom's note. Finally, the third section presents the second part of the first Feyerabend-Carnap controversy, featuring a note by Feyerabend dating June 1957. Here we find Feyerabend reprising his earlier criticism of Carnap's views on the logical reconstruction or explication of scientific theories, now including a new application to quantum theory; Rozeboom's 1958 critical comments of Feyerabend's 1957 note, and Feyerabend's reply to Rozeboom's comments.

In addition to spelling out the early stages in the process that led to Feyerabend's more radical attacks against logical empiricism in the early 1960s, of which there is no trace in the published records, this chapter illuminates the new format for an intellectual circle that Feigl set up in the 1950s, in his attempt to build a semi-formal setting for philosophical discussion that could revive the productive intellectual climate of the Vienna Circle, after its dissolution following the Nazi's rise to power.

Chapter 4 presents (chronologically) different types of papers that document Feyerabend's collaboration with Feigl's MCPS between the late 1950s and the early 1960s. It includes two letters from Feyerabend to Feigl, in which Feyerabend substantially contributed to arranging the section on quantum theory of the six-week "Conference on Philosophical Foundations of Physics", held at the MCPS in the summer of 1957; Feyerabend's 1959 research project on the logic of measurement with special reference to quantum theory, which earned Feyerabend a NSF-funded visiting appointment at the MCPS in the 1959/60 fall semester; Feyerabend's 1961 follow-up research project on the same topic; as well as Feyerabend's unpublished

"Notes on Scientific Method", composed during his second appointment at the MCPS in the 1962/63 fall semester while Popper was also a visiting member of that institution.

Chapter 5 collects the eighteen surviving letters of the Feyerabend-Hempel correspondence, dating between 1961 and 1977. They testify to the intermittent dialogue between Feyerabend and Hempel and include significant comments by Hempel on Feyerabend's papers and insightful attempts by Feyerabend to clarify some passages of his papers that Hempel found opaque or obscure.

Chapter 6 "Excerpts from the Feyerabend-Smart Correspondence (1963)" contains a long letter that clearly shows the beginning of the transition in Feyerabend's thinking toward his later ever more anthropological approach, yet still inexorably mixed with his early methodological approach. It was recently published in English for the first time as an appendix in Feyerabend's *Philosophy of Nature*, (see FEYERABEND 2016), but it finds a more natural home in this collection, together with the few other surviving letters between the two.

Part II, "Feyerabend and Bohm" has two chapters. Ch. 7 "Autobiographical writings" (1993) begins with a letter from Bohm's biographer Francis David Peat, asking some specific questions about Bohm, and inviting Feyerabend to further elaborate in preparation for Peat's biography of Bohm (see PEAT 1997). This is followed by Feyerabend's retrospective exchange with Peat on Bohm. Ch. 8 is Bohm's 1959 paper "On Certain Epistemological and Cosmological Problems Raised in Dr. Feyerabend's Review of *Causality and Chance in Modern Physics*", previously unpublished, which responds to Feyerabend's 1960 published review (FEYERABEND 1960b) of Bohm's book (BOHM 1957a). In this captivating paper, in the process of answering Feyerabend's criticisms, Bohm explicitly tries to develop a number of thought-provoking consequences of his notion of 'the qualitative infinity of nature' in contrast to Popper's falsificationism.

Part III concerns Feyerabend and Thomas Kuhn. Chapter 9 contains a few previously unpublished letters between Feyerabend and Kuhn from 1983 to 1988 that shed light on how their relationship rekindled, in addition to the six previously published letters from Feyerabend to Kuhn, composed between 1960 and 1962, that include detailed comments on a draft of *The Structure of Scientific Revolutions* (KUHN 1962).[251] Minor corrections have been made to the previously transcriptions. As Hoyningen-Huene emphasizes, little can be determined with certainty about the four letters, especially with respect to their exact sequence. It is almost indisputable, however, that the (more than 150 page) long manuscript of Kuhn on which Feyerabend offers critical remarks is Kuhn's draft of *The Structure of Scientific Revolutions* (1962). Furthermore, it seems clear that FEYERABEND [1995a] followed the letters published in FEYERABEND [1995b] since in the former the detailed (as opposed to general) comments stop at page 82 of *Proto-Structure*,[252] while in the

[251] The letters were found by Paul Hoyningen-Huene at different times and published under his editorship in two sets. Cf. HOYNINGEN-HUENE 1995; 200.

[252] Cf. FEYERABEND [1995a]: 371.

latter they start with page 90.[253] However, Hoyningen-Huene's conjecture concerning the order of all four letters – i.e., that the sequence FEYERABEND [1995a] - FEYERABEND [1995b] - FEYERABEND [2006a] - FEYERABEND [2006b] is "very likely" – seems somewhat more problematic.[254]

---

253 Cf. FEYERABEND [1995a]: 375.

254 HOYNINGEN-HUENE 2006: 612. Feyerabend starts FEYERABEND [2006b] with the following note: "my affairs are not yet settled, and will not be settled for another six weeks at least (unless a miracle happens). This means that I am still surrounded by an ocean of noise [...] and not very capable of thinking. However I cannot much longer keep back the comments I want to make on your essay. The longer I delay their deliverance, the more rebellious they become, and the more violent" (FEYERABEND [2006b]: 619). This suggests that Feyerabend 2006b is the first of his (written) comments to Kuhn's "essay" (though it is not completely clear if this phrase actually refers to *Proto-Structure*). Moreover, Feyerabend closes both FEYERABEND [2006b] and FEYERABEND [1995a] by inviting a reply from Kuhn – "I am curious as to your reactions" (FEYERABEND [2006b]: 628); "I am looking forward to your reaction" (FEYERABEND [1995a]: 372) – and FEYERABEND [2006a] is clearly Feyerabend's rejoinder to a (missing) letter of Kuhn responding to Feyerabend's critical remarks, since in the first paragraph Feyerabend clarifies: "I only want to make some comments on your reply" (FEYERABEND [2006a]: 613). In the final paragraph of (FEYERABEND [2006a]), Feyerabend mentions "the letters" that Kuhn has already received (presumably concerning his essay) and announces a further one that "will deal with [Kuhn's] remaining items" (FEYERABEND [2006a]: 618). Although it seems that Feyerabend's phrase refers to points raised in Kuhn's reply, it could also well refer to the part of *Proto-Structure* not covered by FEYERABEND [1995a]. In any case, it cannot be excluded that those points of Kuhn's reply left over by FEYERABEND [2006a] were discussed in person between Feyerabend and Kuhn during the Feyerabend-Kuhn debates in 1960–1961. Indeed, not only does FEYERABEND [1995b] opens with the phrase "*eccola la lettera promessa*" (cf. FEYERABEND [1995b]: 372; Italian for: "here is the promised letter"), but also mentions "two points [...] which [Kuhn] conceded" (FEYERABEND [1995b]: p. 372), possibly in their conversations (though it is equally possible that Kuhn's concessions were made in his reply). Finally, the conclusion of FEYERABEND [1995b] does not convey a strong commitment to continue their discussion on the content of *Proto-Structure* in a written form, since Feyerabend writes: "I think I stop here. I shall perhaps send you another letter" (FEYERABEND [1995b]: 387). On the whole, the opening remarks of FEYERABEND [2006b] together with the subtle internal cross-references between specific pairs of letters seems to make the sequence (FEYERABEND [2006b] - FEYERABEND [1995a] - FEYERABEND [2006a] - FEYERABEND [1995b]) more natural. If this is the case, then it becomes possible to restrict the time interval within which the four letters were written to April-October 1961, on the basis of some speculative, though still reasonable, assumptions. Moreover, in his autobiography Feyerabend says that from his arrival at Berkeley up until the late 1960s, he lived in four different places: First "a smelly room in a flophouse on Telegraph Avenue"; then "a small apartment atop a set-back garage on Hillegass" that eventually "the land lady took over"; finally, "[a]fter a brief stay on Benvenue, [he] rented a huge wooden box at6041A Harwood Avenue in Oakland" (FEYERABEND 1995: 113). Feyerabend's correspondences and diaries allow us to identify the time of his relocations rather precisely. He moved out of the hotel room by early February 1959 (cf. *FFY3*, Feyerabend to Agassi, 17 February 1959; *FFY1*, Feyerabend to Popper, 18 February 1959-2: 308–310), to Benvenue Avenue in the summer of 1960 (cf. *FFY3*, Feyerabend to Watkins, 6 September 1960, JW 170; Feyerabend to Lakatos, 11 October 1960, IL 13/272a-1) and to Harwood Avenue on 15 June 1961. The reason for the latter quick relocation was the noise he experienced on Benvenue – so unbearable for him (and ironically so, given the name of the street) that he had already decided to leave in October 1960 (cf. Feyerabend's Notebook 1960–1968, PF 3-5-1). Now, in the light of these elements, it seems fair to assume that the "affairs" to which Feyerabend refers in FEYERABEND [2006b] are related to his transfer from Benvenue Avenue to Harwood Avenue and this would imply that the first of

According to the present reconstruction, FEYERABEND [2006b] was Feyerabend's excited reaction to *Proto-Structure*, written on the wake of having read it soon after its delivery in April 1960, but in less than ideal conditions given the intolerable noise at his apartment. FEYERABEND [1995a] contains his more considered reflections as well as point-by-point comments on the first part of Kuhn's mimeographed manuscript. FEYERABEND [2006a] was his rejoinder to the reply he had urged from Kuhn in the first two letters. Finally, by November 1961, FEYERABEND [1995b] summarized some of the conclusions reached during their (written and/or face-to-face) discussions and provided Feyerabend's detailed comments on the second part of *Proto-Structure.* These letters have already proven invaluable to numerous scholars, and these versions have been included in Vol. 2 in order to facilitate access to them all in one location, and together with the other primary material from which *Feyerabend's Formative Years* has been composed.

Taken together, these materials provide a fresh look at the genesis and development of Feyerabend's provocative ideas from a hitherto hardly accessible vantage point, offering behind the scenes insights into major developments in twentieth-century philosophy. Scientists, laymen and a wide range of scholars – especially philosophers, historians of science and philosophy and, more generally speaking, intellectual historians – will find this captivating material an invaluable resource that offers an inside view of Feyerabend's influential ideas and their notoriously tumultuous development. Finally, this volume will provide a significant resource for research on Feyerabend and his ideas to a growing community of scholars interested in history and philosophy of science, especially in light of the ongoing Feyerabend centennial celebrations.

1.5 Editorial Policy and Conventions

The main purpose of this collection is to make archival materials that are scattered throughout the world at various institutions easily accessible to a larger audience of both academic specialists and interested readers. Accordingly, the main criteria on the basis of which the collection has been organized follow from the aspiration to reliability characteristic of critical editions and from the ideal of faithfulness to the original documents typical of diplomatic editions.

With respect to the various correspondences collected, no attempt has been made to operate a selection of "relevant" letters or parts thereof, thereby limiting the richness of the sources.[255] There may be some letters mentioned in the correspondence

Feyerabend's letters was certainly written before mid-June 1961, but not too many weeks earlier. Indeed, if Feyerabend's expectations expressed there were approximately correct, FEYERABEND [2006b] dates to late April or early May, namely, at least a couple of weeks after *Proto-Structure* was mimeographed. On the other hand, FEYERABEND [1995b] was undoubtedly written before November 1961. Cf. FEYERABEND [1995b]: 372.

[255] With the exception of Ch. 6, see *below*.

but not included in the collection that have survived outside the archives, but this seems unlikely. By the same token, the transcriptions strive to include as much information as possible with respect to the original documents, short of facsimile but with some grains of good sense. In fact, editorial doctoring has the unpleasant drawback of blurring the size of each omission, the import of the missing information, and the specific reasons that motivated the editorial choice in every particular instance. More generally speaking, assessing the relevance of specific pieces of information presupposes the editors' full awareness of potential readers' interests. However, as the audience is meant to be kept as wide as possible, the adopted editorial policy gives to every reader the opportunity to evaluate the relevance of almost every aspect of the surviving correspondence according to their own specific and varying interests. Any other choice could be found to some extent wanting for unduly simplifying or arbitrarily circumscribing, therefore compromising, the quality and amount of information incorporated in the sources. Conversely, such policy renders the full complexity of the original documents, even though it has the apparently cumbersome consequence of including multifarious and heterogeneous topics of conversation, ranging from highly specialized scientific discussions, through academic gossip, to private affairs.

In the collection, the following editorial conventions have been adopted. For every document, an introductory unnumbered endnote indicates synthetically: the archival reference(s) of the source(s); the number of sheets constituting the original document (n ss.); whether such document is (mostly) handwritten (H) or (mostly) typed (T); if it is a (carbon) copy (c) or a draft (d); if the transcribed text is an excerpt from the original document; and, in case of letters, if the envelope is also available (e). Any further general detail concerning the original document, the reconstruction of its text and its transcription is specified in this note. As for transcriptions themselves, in general their layouts do not mirror exactly those of the original documents. This holds in particular for letters, whose layout has been normalized according to a pattern that takes into account the standard conventions for letter composition. However, as such conventions were broadly followed by the authors of the letters, the transcriptions depart only marginally from the originals. On the other hand, the distribution of a single text in different pages has been indicated as specified below.

As a rule, editorial interventions and interpolations within the transcribed text are highlighted by the use of angle brackets "⟨ … ⟩". Interpolations are limited to editorial guesses for hardly readable words, letters or punctuation signs. Exceptions to the rule are missing periods, which have been inserted without notice to avoid burdening the text with excessive annotations, thus compromising its readability. Editorial interventions include indications of page changes. These have not been given according to the original page numeration (if present), but in the form "[n|m]", where n and m stand for the sheet numbers of the original document. Original misspellings have not been corrected and the standard "⟨sic⟩" has been inserted to indicate any spelling or grammar anomaly present in the original text. The frequent German interferences, especially evident in the use of commas, are an exception for reasons analogous to those mentioned above with reference to missing periods.

Internal notes added by authors themselves to their own texts have been located at the end of the main texts to which they refer, as part of their transcriptions. Where needed, each internal note is preceded by a reference, within angles brackets, to the page in which the number or mark of the internal note appears. Editorial remarks and comments, including internal cross-references and suggestions for correct readings of authors' faulty expressions which may not immediately be evident or intuitive, have been provided in numbered editors' notes added at the end of each transcription indicated with "E".

In the adopted standard layout pattern for transcribed letters, the letterhead has been italicized and centered. It is followed by the date of the letter, aligned to the right-hand side, and by the salutation, aligned to the left-hand side. Subsequent paragraphs within the body of the letters, with the exception of the first one, have their first line indented. The complimentary close and the signature have been centered. Letterheads have been included as they usually provide indications as to the place in which letters were written.

Letters are presented in chronological order, following either the date of composition, if annotated by the author, or of sending, if the envelope is available and the postmark is readable or in the case of postcards and telegrams. If the date annotated by the author differs from the date of sending as indicated by the postmark, both dates are reported. Dates of sending thus detected are indicated within angled brackets, followed by a superscript "PM" (for "postmark"). Feyerabend's letters are often undated or only partially dated. The symbol "⟨UNDATED⟩" signifies that the date of composition or sending cannot be specified with due precision. In such cases, the position of the letter within the chronological order is derived either from clues present in the correspondence or based on other archival sources, such as other contemporaneous correspondences. Attributed dates or parts thereof are included within square brackets in the headings of the letters.

In the body of the letters, self-corrections by authors have not been indicated, with the exception of particularly significant deletions. Emphases through single or multiple underlining have the same form in the transcriptions, with multiple underlining rendered through double underlining. On the other hand, the use of italics has been reserved to handwritten additions to typed letters, such as signatures. The symbol "⟨UNSIGNED⟩" is given if the signature is missing, as is often the case with drafts or copies of letters.

Acknowledgements The project leading to this volume could not have been carried through without the generous funding from the German Research Council and the precious help from a number of collaborators including, Prof. Joseph Agassi, Prof. Friedrich Stadler, Prof. Olaf Müller, Prof. Paul Hoyningen-Huene, Prof. Jagdish Hattiangadi, Prof. Ian Jarvie, Prof. David Miller, Prof. Zev Bechler, Prof. Marij van Strien, Prof. Jordi Cat, Dr. Manfred Lube, Dr. Nimrod Bar-Am, Dr. David Peat, Dr. Daniel Kuby, Dr. Christian Damböck, Dr. Christoph Limbeck-Lilienau, Dr. Flavio Del Santo, Dr. Bryan Miller, Michael Schorner, Hans-Joachim Dahms, Flavio Del Santo, Kati Hennig, Bryan Miller, Doron Broitman, Carla Nicole Wood, Nicola Teubner Oberheim, Daniel Sirtes, and Jessica Gardenal. Special thanks go to Dr. Borrini-Feyerabend and to Prof. Agassi and to Prof. Grünbaum, for giving us access to their private archives, and to the directors and staff of the *Philosophisches Archiv der Universität Konstanz*; the Hoover Institution Library & Archives, Stanford University; the *Universitätsbibliothek, Alpen-Adria-Universität Klagenfurt*; the *Bibliothek Universität Wien*; the LSE Library Archives and Special Collections; the Archives and Special Collections, Elmer L. Andersen Library, University of Minnesota; the Archive of Scientific Philosophy, Hillman Library, University of Pittsburgh, and the MIT Libraries, Distinctive Collections for their invaluable assistance.

Part I
Feyerabend and Logical Empiricism

Chapter 2
The Dogmas of Logical Empiricism (1951)

Paul K. Feyerabend

Ladies and Gentlemen,

At the Seventh International Congress of Philosophy, in October 1930, Moritz Schlick, professor of philosophy at the University of Vienna, gave a lecture where, among other things, he stated the following:[±)]

> it appears that by establishing the natural boundaries of philosophy we unexpectedly acquire a profound insight into its problems; we see them under a new aspect which provides us with the means of settling all so-called philosophical disputes in an absolutely final and ultimate matter. This seems to be a bold statement, and I realize how difficult it is to prove its truth and, moreover, to make any one believe that the discovery of the true nature of philosophy, which is to bear such wonderful fruit, has already been achieved. Yet it is my firm conviction that this is really the case and that we are witnessing the beginning of a new era of philosophy, that its future will be very different from its past, which has been so full of pitiful failures, vain struggles, and futile disputes.

Schlick was not the only one who voiced this opinion about philosophy in the 1930s. His contempt for the philosophical tradition and his optimism with regard to the present and future of philosophy were shared by many quarters. The age of philosophy seemed just about to start.

If we turn our attention to the past, we actually come across similar phenomena. In 1620, in his introduction to the *Novum Organum*, lord chancellor Francis Bacon wrote:[±±)]

> For if you look closely at the wide range of books which are the boast of the arts and sciences, you will frequently find innumerable repetitions of the same thing, different in manner of treatment but anticipated in content. , so that things which at first glance seem to be numerous are found on examination to be few. One must also speak plainly about usefulness, and say that the wisdom which we have drawn in particular from the Greeks seems to be a kind of childish stage of science and to have the child's characteristic of being all too ready to talk but too weak and immature to produce anything. For it is fertile in controversies, and feeble in results. The story of Scylla seems to fit the current state of letter exactly: she showed the face and visage of a virgin but barking monsters clothed and clung to her

M. Collodel, E. Oberheim (eds.), *Feyerabend's Formative Years. Volume 2. Feyerabend on Logical Empiricism, Bohm & Kuhn*, Vienna Circle Institute Library 10,
https://doi.org/10.1007/978-3-031-57519-8_2

loins. Similarly, the sciences to which we are accustomed have certain bland and specious generalities, but when we get to particulars (which are like the general parts), so that they may bring forth fruit and works from themselves, disputes and scrappy controversies start up and that is where it ends and that is all the fruit they have to show." [1|2]

Descartes, the founding-father of modern philosophy according to popular histories of philosophy, was sharply attacked by the renovators of philosophical analysis of his time as a mythologist and a muddlehead,±) just in the same way as he used to attack his predecessors:±±)

> Regarding philosophy, I shall say only this: seeing that it has been cultivated for many centuries by the most excellent minds and yet there is still no point in it which is not disputed and hence doubtful. I was not so presumptuous as to hope to achieve any more in it than others had done. And considering how many diverse opinions learned men may maintain on a single question – even though it is impossible for more than one to be true – I held as well-nigh false everything that was merely probable.

The same spectacle can be found yet again in antiquity. Aristotle's philosophy is widely known today only for its negative aspects and is considered as a stage of philosophy that has to be overcome. And pointing out the errors in his conclusions seems to be a sophomoric pleasure that is appreciated by some neo-positivists.±±±) Yet, Aristotle himself was rather critical towards his own past in the same way. Concerning some of his predecessors who had discussed philosophical questions, he wrote that "they did so vaguely, inadequately, and rather like untrained troops in action who run about all over the field and may often get in some good blows, though not as the result of any skill." ±±±±)

It is apparent then that the history of philosophy was repeatedly pervaded by the spirit of reformation. Philosophy seems to be in a state of permanent crisis. For this reason, the reformatory spirit of logical empiricism here embodied by Schlick loses most of that uniqueness that it presents to those who are only aware of one particular trend within the history of philosophy.

Such a retrospective view, however, can also teach a second lesson: Every complex criticism of the past is accompanied by an attempted re-construction and the latter presupposes that there are problems that were unknown in the past. Ionian natural philosophers, to whom we owe the first attempts to describe the world as a whole, approach their task quite naively. It is as if they stick to things without acknowledging enough that between objects and their description stands the faculty of cognition, which makes the relation between the description and the described objects anything but unproblematic. This problem is then emphasized by the Sophists who put it at the center of their argumentation. At first, nature dominates; just as man dominates now; and at last the individual. Aristotle tried to develop criteria for a kind of knowledge that could claim certainty, while taking subjectivity into account. Bacon admonished that this kind of Aristotelian knowledge does not have any predictive power; it is only relevant after the fact, and it classifies being according to what happened.±±±±±) New instances are necessary in order to establish a foundation for the actions of technicians. Descartes pointed out that every kind of knowledge, [2|3] if it is to be of value for argumentation, must have a certain form in addition to being grounded either in experience or in reason, and in his pre-metaphysical period[±)] he reflected on this form extensively.

What kind of new discoveries – we may ask – have logical empiricists to show for themselves that they behave with such confidence? What kind of new problems have they seen?

On this, a few preliminary remarks: The philosophical trend that we quoted Schlick for at the beginning is known in the literature, sometimes infamously so, as "logical empiricism", "logical positivism" and "logical neopositivism". Its infamous character has two reasons. First, as we have seen, the advocates of logical empiricism condemned a great number of traditional and sacred (as it were) problems in such a self-confident manner that it must have triggered anguish to their guardians. Furthermore, their procedure applied an extremely sophisticated logic that was only known to its full extent to very few of their opponents. This logic provided an instrument for the destruction of the old as well as for the construction of the new. In a paper entitled "Widerlegung der Metaphysik durch logische Analyse der Sprache" ⟨sic⟩,[±±)] Carnap tries to prove, with special reference to Heidegger as an example (even though this is a rather bad example), that many philosophers come into conflict with their own language when they ask their questions. With the same analysis, in a positive step, he attempts to clarify the nature of mathematical propositions and of scientific propositions in general. Russell's *Principia Mathematica* and even more clearly his *The Principles of Mathematics*, as already earlier in Frege's, exemplifies the art of analysis in the formal disciplines. Carnap's *Der Logische Aufbau der Welt* was supposed to accomplish the same with respect to the empirical sciences.

Therefore, we have to distinguish between the following three aspects:

1) a specific method that should enable us to find out about the sense [*Sinn*] of complicated sentences, that is at the same time a research tool and a norm *and plays an important role in logic.*[1)] ⟨sic⟩

2) A negative application of this method to traditional philosophy, a diagnosis that many investigations are meaningless [*bedeutungslos*] and that philosophy in its original form is not a scientific, but rather an artistic, activity.[±±±)]

3) A positive application to the sciences. General and more specific statements about the nature of scientific sentences are thereby obtained. In the literature, the more general statements have been labeled as "Theses of Logical Empiricism".[±±±±)] These theses, or that method, or both together, constitute the bulk of the new philosophy, while the specific meaning criteria [*Sinnkriterien*] constitute its most important findings.

We shall set aside, for the moment, the consideration that the restriction of the scope of this method to the sciences is already highly problematic. In fact, not all advocates agree on this point. Reichenbach[±±±±±)] explicitly states: Scientific philosophy "constructs the theory of knowledge by the analysis of the results of science," but [3|4] he was scolded by his reviewers.[±)] Generally, there seems to be no consent at all concerning the domain of application of this method and there are at least three different trends.[±±)] We shall set aside the discussion of this question here and focus solely on the results that the application of this method within the sciences has brought about. Our purpose then is to show:

1) how this analysis works
2) what specific results it has brought about in mathematics, physics, etc.
3) to what extent these specific results are problematic
4) that the theses at hand could only be obtained on the basis of a particular prejudice with respect to their specific interpretation and that, hence, *within this interpretation*, they should be rather referred to as dogmas.[+++)]

Now, the faultiness of the results might be due to faults in the method applied. Therefore, we would have to discard the method as well and nothing else would remain of logical empiricism. This is the conclusion that speculative philosophers are far too hastily inclined to draw. But: The fallaciousness of the results may have its cause in the fallaciousness of the viewpoint of the method but it may also have it in the fallacious application of the method. And it will be shown that too narrow an application, i.e. the prejudiced use of this important tool, leads to the dogmas. The method itself remains unaffected and once more proves its full power against all ingenious, visionary but unreasonable philosophizing. However, because of the generality that results from the exclusion of all unjustified restriction, it converges towards the philosophical tradition in a meaningful way, though without merging into it. And this last and valuable difference that remains seems to be the truly new element of logical empiricism.

I shall now illustrate what I have just described.

Bacon, Descartes, and Schlick find themselves in agreement with respect to one point. They regard the philosophical past as something that has to be overcome. But the reasons why they consider this necessary are different. Bacon blamed philosophy for its uselessness, which makes it completely fit for arguing, less so for acting. He refers to the mechanical disciplines that always move step by step towards newer perfection and whose description exhibits that kind of concreteness that is necessary if an agreement is to be established.

Descartes likewise had the clash of opinions before his eyes. Yet, facing this situation, he was interested in something else: This clash should not be left behind as useless, but should be put on a secure foundation. Only one opinion can be the true one. Which criteria are there to find out this opinion? We are looking for a standard that would allow us to decide what is right and what is wrong with regard to a possible solution. In any case, however, we expect that there exists a correct answer. And this presupposes that the question asked, the disputed problem, can actually be answered. Is this condition fulfilled at all times? [4|5]

It has been presumed ever since that it is not always possible to arrive at an answer by means of human reason. There are questions that cannot be answered simply because the power of reason is insufficient for that purpose. Such questions that cannot be answered through the power of the human mind alone can perhaps be answered with the help of irrational instances. The answer is then presented to reason by a force that is more powerful than the capacities of reason and which goes beyond them. Thomas Aquinas was convinced that the existence of God, of its properties and of the creation *ex nihilo* are provable, but that the temporality of the creation of the world is not. This was first declared to be true by means of the Holy Scriptures. F. Bacon held that the existence of God and the specific nature of evil can be proven, but again with regard to the positive nature of our responsibilities and to the properties of God, he held that the guidance of the Holy Scriptures is inevitably necessary. Locke knew that there are truths that contradict reason, that correspond to reason and also truths that are simply above reason.

However, the specifically theological interpretations of this tense relation [*Spannungsverhältnis*] between question – activity of reason – answer by reason or by an irrational instance can be set aside. Then, the relation of arbitrary answers to a question and to the domain of reason becomes apparent. But regardless of how we interpret, one point is here confirmed: The double and manifold relation of the answers to a question. The problem creates at the same time a tension that, according to its nature, mobilizes reason, the Holy Scriptures or other instances (e.g. coffee grounds).

And here now our investigation starts: I draw a random figure on the blackboard

and add that I will now try to find an answer to it. Then you will rightly object that there is no problem yet and you will add that the drawing is actually meaningless [*bedeutungslos*]. But with this objection the tense relation just mentioned is dissolved. We cannot give an answer, not however because nobody knows it (no God and no scholar), but because nobody asked a question in the first place.

The drawing on the board is a very trivial case. The decision will be more difficult as soon as I write words on the board, for example: "Is inflation yellow?". Almost everybody will object that this is nonsense indeed, that inflation has no color. But, since this obviously is a grammatically correct sentence [*Satz*], there might be someone who still tries to find an answer. We can predict his fate, however: He will strive in vain and should he really succeed in giving an answer, we can reject it *a priori*. We will accuse its author of not understanding his own language.

This case is also still relatively simple. But it carries a certain historical meaning: Locke[±)] tried to show that the problem of free will is due to an error of this kind.

Now the simplicity, in the latter case, is based on the fact that we recognize the senselessness [*Sinnlosigkeit*] of the sentence just by looking at it. But there are strings of words and even entire chains of arguments that seem to be completely flawless at first glance and that reveal their nonsensical character [*Unsinnigkeit*] only through the difficulties [5|6] to which they ultimately lead. In this case the nonsense [*Widersinn*] can only be recognized through its consequences. And afterwards, finding its actual source is a very difficult task. The so-called logical paradoxes are an example. I shall take a simple case.[±)]

All words will be separated into two classes. All those words that describe their own verbal shape [*Wortgestalt*] may be called autological words. Example: "pentasyllabic" (because the sentence "'pentasyllabic' is pentasyllabic" is a true sentence). In general, all words "w" are autological then, if

"'w" is w' (Sentence S)

is a true sentence.
All other words are heterological. (Definition D).
Here a clear division of words into two groups is established and it seems that there will be no further problems.

If I now ask to which of these two groups the word "heterological" belongs, the result will be blatant nonsense [*Widersinn*]. To begin with, we assume that "heterological" is autological. It follows that this assumption does not have the form of sentence S and, from definition D, that "heterological" must be heterological. This consequence, however, has again exactly the form of sentence S, so that "heterological" is autological, etc.

So far we have the following results:

(het) aut implies (het) het (1)

(het) het implies (het) aut (2)

Definition D means that every word is either autological or heterological. Thus, (het) aut .v. (het) het. (1) reads now

$$\overline{\text{(het) aut}} \cdot\text{v.}\ \text{(het) het.}$$

and from these two formulas (het) het follows. Hence, according to (2), (het) aut follows. But since "(het) het" and "(het) aut" contradict each other, then a sentence of the form "it rains and it does not rain" follows from this apparently simple and transparent argument. Contrary to the way in which it is usually presented,[±±)] the contradiction here at hand is not the same as it arises from a *reductio ad absurdum*. In the latter case, a contradiction results from a certain axiomatic system and an additional assumption A. This contradiction indicates the inconsistency of A with the fundamental axioms and can be dissolved either by dropping A or by changing the axioms. In our case, however, no assumption was made. The contradiction does not result from the inconsistency of the assumptions made; it rather follows from the inferential chain itself. This is now an entirely new affair: It is not any assumption whatsoever, but the means of description that we employ that show inconsistencies. Now, it is very easy to see that within a world that contains a contradiction everything becomes possible. If the sentence "It rains and it does not rain" can be legitimately stated, then any kind of sentence can be stated. But then if that is the case, the formal apparatus employed in the argumentation fails to fulfill the most important function of a language: To determine certain differences in the world.

It is neutral towards what is and is not in the world, and, therefore, it also ceases to be a language. We should then describe what was disclosed by the paradoxes not as a discovery of contradictions within a language but as the abandonment of language. And hence we arrive at a significant point in the discussion of the current question: ⟨sic⟩ [6|7]

The following argument is frequently made: The language we have begun to speak turns out to be riddled with fractures in certain areas. If catastrophes are to be avoided, then we have to circumscribe these areas in some way (e.g. by specifying rules that safely guide us through the critical cases). How we circumscribe them is in some respects arbitrary. The language domain [*Sprechgebiet*] we obtain after the circumscription is different from the language domain we started from. This difference, however, is not a "natural" one, but the result of a new construction. We have constructed a new language where we are no longer faced with those paradoxes that made the old language such a poor tool.

From here we move forward to the claim that the language used is completely arbitrary and that it is therefore a system of conventions that can be approved, but that can also be rejected. The conventional interpretation of logic and mathematics receives significant support from the procedure for avoiding paradoxes.±) We have to investigate if this appeal is justified.

First of all, this appeal presupposes the comparison between old languages and new languages in a way that is, as such, wrong. This is because it is actually not about re-constructing, but about having to remain within the limits of language at all. The series of conclusions that leads to the paradoxes is not, as we have already mentioned, part of an "old language" that has to be amended, rather it does not have the character of a language at all: A sign system that is indifferent towards true and false can by no means be called a language. The paradox has not displayed the imperfection of a certain language; it has led us from language into nonsense. How we should proceed is determined by the linguistic parts of the initial calculus. We do not suddenly start to speak Chinese because of the paradoxes. It is completely arbitrary that we do not do so and did not do so right from the start. The problem under investigation, however, is not the problem of what particular language to choose, but how to use a particular language correctly once it is chosen. The correction aims at leading back to language. Therefore, they do not create anything new with regard to objects, but only with regard to the reflecting consciousness. (The same happens, by the way, if someone walks around a house having seen only its front side and discovers that a blue flower is painted on the white walls of its back side. Something remains the same for him nonetheless: It is still the same house he is looking at.)±±) In any case, if reconstruction were the only important aspect, it would not at all be clear how the problem at hand should be solved. The connection with the given problem gets lost.

Thus, we will have to describe the procedure in a different way: In the series of conclusions that we presented earlier, we missed the language we started from. We discover the imperfection not of language but of our knowledge of language (just like the wanderer we mentioned before did not discover the imperfection of the house but the imperfection of his knowledge of the house). Now we try to grasp more clearly the nature of the spoken language. Yet, this implies that we presuppose that a language somehow already exists, even if not as an empirical matter of fact. The mysteriousness of such a mode of existence should not result in the idea that it does not actually exist (like many empiricists suggest). The entire procedure **[7|8]** is illustrated in the right way by the examination of the boy in Plato's *Meno*.±)

A second issue leads us back to the starting point. The paradoxes showed us that the nonsensical character of a question can come to light only after thorough argumentation. And it showed us that even the apparently clearest formulation does not provide us with absolute certainty with regard to the sense of the problem. "Omne est verum, quod clare et distincte percipio"[1E] is a bad rule in philosophical matters. It pretends to provide certainty when, in fact, there is only the illusion of it. Thus, every question can be doubted in principle not so much with regard to the correctness of an offered answer, but mainly with regard to its legitimacy as a question. The elimination of such doubt is, however, the presupposition for the acknowledgement of any kind of instance. If it turns out that it is simply impossible to establish this kind of certainty – should it be perhaps even proven that the problem itself is not meaningful [*sinnvoll*], then any answer would lose its value, even if it is offered by the highest authority. Yet, being entitled to examine any problem, human reason is entitled, again, to consider any problem as within its scope. If once it appeared that reason had humbly to wait for the solution of certain problems, it now turns out that such a hesitant attitude is neither justified nor allowed: Either everything is subject to the judgment of reason or nothing is.

Now, this has significant consequences: All fields of knowledge are now subjected to a single discipline, no matter with what kind of requirements they approach the reflecting mind: A discipline that critically determines to what extent the laws of language have been violated, on the basis of the nature of the linguistic means applied, and that logical empiricists have identified with philosophy without further ado. "All philosophy is a critique of language," says Wittgenstein.[++)] G. E. Moore, his teacher, said the following: "It appears to me that in Ethics, as in all other philosophical studies, the difficulties and disagreements of which its history is full are mainly due to a very simple cause: namely, to the attempt to answer questions without first discovering precisely *what* question it is which you desire to answer."[+++)] Looking back to the history of philosophy, Schlick came to the following conclusion:[++++)] "The hint that man has, after all, solved his most stubborn problems … is small solace to the philosophic connoisseur; for what he cannot help fearing is that philosophy will never get so far as to pose a genuine problem." And again Wittgenstein:[+++++)] "Most propositions and questions, that have been written about philosophical matters are not false, but senseless [*unsinnig*]. We cannot, therefore, answer questions of this kind at all, but only state their senselessness [*Unisnnigkeit*]. Most questions and propositions of the philosophers result from the fact that we do not understand the logic of our language. (They are of the same kind as the question whether the Good is more or less identical than the Beautiful.) And so it is not to be wondered at that deepest problems are really no problems."

It is certain that the evaluation of the philosophical past with the help of the newly developed instruments of language analysis was somewhat hasty and unjust: First, because only a few of its advocates knew the history of philosophy well enough to be able to issue a justified judgment about it; but mainly because it was not distinguished between a meaningful [*sinnvolle*] problem and a meaningful formulation of a problem. As even a growing number of logical empiricists have come to acknowledge,[++++++)] in between the logically correct formulation of a question and complete nonsense, lie all those accounts that intend to make a **[8|9]** correct statement but fail to do so because they lack proper means of expression. Here, a historian of philosophy does not have to conduct much logical analysis. He rather has to find out what better formulation is to replace an incorrect, and hence *a fortiori*, senseless [*sinnlos*], one. He has to focus more on the mimicry of language than its logical structure. Even if nothing is conveyed in a strict sense, a lot can still be said.

However, from the perspective of a strict theory of science, the area of language mimicry [*Sprachmimik*] (as we might conveniently call it) will be considered only as a passing stage toward a final formulation. But even if logical empiricism underestimated the significance [*Bedeutung*] of the different phases of discovery, it nevertheless succeeded in providing proper conditions for scientific accounts.

So far we have tried to clarify the relation between the analysis of meaning [*Sinnanalyse*] and philosophical as well as special scientific issues with the help of a few examples. Two questions remain to be answered⟨:⟩

How do we find out about the meaning of judgments and once we have found a meaning criterion [*Sinnkriterium*] – how do we find out about the correctness of an account within a certain domain?

Which judgments are meaningful [*sinnvoll*] and what meaning [*Bedeutung*] do they have?

The answer to the second question is partly contained within the main theses of logical empiricism. It is the main issue of our investigation. The manner of the investigation is itself an example that can clarify how to answer the first question. This question is much more difficult to answer than the second question; in fact, it is so difficult that Wittgenstein has characterized it as senseless.[±)]

The Sense of Scientific Judgements

In empirical philosophy, two classes of judgments are often distinguished: analytic and synthetic judgments. The former, such as the principle of contradiction in logic, do not say anything about facts. The latter need to refer to some kind of observational results for their justification. In 1748 Hume wrote about this:[++)]

> All objects of human reason or enquiry may naturally be divided into two kinds, to wit, relations of ideas and matters of fact. Of the first kind are the sciences of Geometry, Algebra and Arithmetic, and in short, every affirmation, which is either intuitively or demonstratively certain. … Propositions of this kind are discoverable by the mere operation of thought, without dependence on what is any where existent in the universe. … Matters of fact, which are the second object of human reason, are not ascertained in the same manner, nor is our evidence of their truth, however great, of a like nature with the foregoing. The contrary of every matter of fact is still possible.

Hume regarded these two kinds of judgment as the only meaningful [*sinnvoll*] ones and the disjunction analytic-synthetic as exhaustive. About this, he wrote:[+++)]

> When we run over libraries, persuaded of these principles, what havoc must we make? If we take in our hand any volume; of divinity or school metaphysics, for instance; let us ask, does it contain any abstract reasoning concerning quantity or number? No. Does it contain any experimental reasoning concerning matter of fact and existence? No. Commit it then to the flames: For it can contain nothing but sophistry and [9|10] illusion.

This disjunction was maintained by logical empiricism with one difference: They tried to determine more precisely the nature of analytic sentences, of the sentences of mathematics and of logic, and to further specify what was meant by a fact. In the latter case, it was emphasized that every synthetic sentence must somehow lead to an observation. Moreover, it was assumed that only simple qualities, impressions, colors, etc. were actually decisive in an observation, even if the field itself seems to be much more complex. "Now, if an expected sense-datum constitutes a verification, what was asserted must have been about sense-data."[+)] This relation, however, involves various complications: A feeling can be very easily and immediately detected, but in order to confirm the existence of Neptune, complicated calculations, a theory of the observational apparatus, etc. were necessary. According to a widespread theory, however, even in such a case at the end of the investigations stands a sense-datum. Thus, the difference between Neptune and a feeling consists, first, in a difference between the phenomena at issue themselves and second, also in a difference in the length of the inferential chains leading from a statement about the object in question to the observational instances. In the case of feelings, the construction of the inferential chains is reduced to a minimum, whereas in the case of Neptune it exhibits a remarkable complexity.

And yet, the number of sense-data is limited because the threshold of variance has a finite value in every domain of observation.[++)] We can also reduce the link between an object and the domain of sense-data to a finite number of inferential methods [*Schlussweisen*]. (It is the task of logic to find out these inferential methods.)

Thus for every object it is possible in principle to describe its nature accurately by specifying the sense-data and the type of construction. The sentences of science become typographical abbreviations for complicated constructions, which however are simple insofar as every construction always contains elements from the same domain in ever different modifications. A sentence, however, is meaningful if and only if

a) it can be constructed in the specified way (in which the specification of all the inferential methods available as well as of all the accepted sense-data is presupposed)

b) if its truth is already clear with reference only to the inferential method applied; hence, literally speaking, if the construction is completed without relying on any sense-datum.

a) and b) display the form into which logical empiricism transformed the Humean disjunction. The following two sentences

1) Every sentence is either analytic or can be determined by experience;
2) Every sentence that can be determined by experience can be pictured as a construction based on sense-data in the specified way;

these two sentences roughly encompass the content of the three main theses of logical empiricism.[+++)]

Now, it is further assumed that numerous classes of analytic sentences can be traced back to logical sentences. If we take this assumption in its most radical form it follows from the main theses that there are only factual states of affairs [*Sachverhalte*] and sense-data. They are the material from which we build, in a more or less artistic way, the edifice of science. Factual states of affairs that cannot be constructed in such a way cannot be presented in meaningful sentences. They have rather only a pseudo-existence in quixotic and imaginative brains. [10|11]

Today the falsity of the radical thesis has become quite evident, even if not uniformly throughout the entire literature. In order to investigate even the more moderate claim, however, the concepts that occur in the main theses need more substantial content. First we have to find out how we have to define "analytic" exactly before we may criticize the main theses. Then we will see that what can be criticized are not so much the main theses themselves but a too narrow interpretation of their central terms. The main theses only provide a framework that interprets the nature of our world if properly filled with conceptual material. Therefore, we shall start with the discussion of the concept "analytic".

The Logical Sentences

Consider a simple example: The sentence "Today it rains or it does not rain" is true in any case. Why is that? Obviously it does not matter if we talk about rain or about other things. The perpetual truth of this sentence is not connected with its talking about the rain. Therefore, we could have also written: "It is hot or it is not hot here" and this sentence is again true in any case. Since specific content does not matter, we can also write "p or not p", where "p" is any arbitrary sentence. The only words that remain stable are now "not" and "or". They have to be somehow responsible for the permanent truth of the sentence. But how? In general every word designates [*bezeichnen*] a certain object. "Table" designates the object directly in front of me and other objects with the same form and with the same function. "Flower" designates this colorful thing. What does "or" talk about? You might feel that this question is not meaningful [*sinnvoll*]. Or is not an object. But what other function has it if not to designate an object?

To answer this question, I will present another example: "Today I go into town AND buy a new hat." The sentence before and the sentence after AND designate clear circumstances [*Verhältnisse*]. The big sentence that is generated by the connection of the two small ones through "and" reports that both activities happen about the same time. Conversely: If one of these two activities does not occur (if one of these two sentence is false), then I have not performed the complex activity described with the use of "and" either (the big sentence is false as well). And this is the case independently of the content of the two sub-sentences – if one only talks about their truth-value and the relation between their truth-value and the truth-value of the entire sentence. Therefore, we can say: "Whenever two sentences A and B are connected through "and" a new sentence, AB is generated and, for any arbitrary A and B, this sentence is true if and only if the two single sentences are true and it is false in all the other cases". (The sentence just formulated will be referred to as sentence U).

Sentence U describes how and-sentences relate to their sub-sentences, but conversely it also informs us about the function of "and" within complex sentences.

Hence, we see that "and" cannot designate a specific object. It is a connector that allows us to build bigger sentences out of smaller sentences so that the truth-value of the big sentence can be calculated from the truth-values of the sub-sentences.

Instead of the sentence U, a table of the following kind can also be used:

a	b	a.b
T	T	T
T	F	F
F	T	F
F	F	F

It shows which truth-value is ascribed to the sentence a.b if each of the sub-sentences has been already ascribed a particular truth-value. "Or" is of a similar character. Its truth-table is a bit different, though: T, T, T, F (these are the vertical values of the column a∨b, if the columns for a and b have the same values as in the And-table). With regard to sentence U the corresponding sentence reads: Whenever two [11|12] sentences A and B are connected via "or" a new sentence A∨B is generated and, for any arbitrary A and B, this sentence is false if and only if both single sentences are false and it is true in all other cases (sentence O).

We call those words that affect the transfer of truth-values in the specified way logical particles or logical constants. "Not" is a logical constant as well. But it does not connect multiple sentences; instead, it reverses the truth-value of a certain sentence.

Consider now the proposition "I will read or go for a walk". I know whether this sentence is true only after I have experienced whether the activities reported in it actually happened or not. However, in the case of an Or-sentence it is already sufficient to know that one of the sub-sentences is true in order to be able to determine the truth of the entire sentence, as it can be easily seen from sentence O. But in any case empirical information is necessary.

In the case of the sentence "p or not p", however, empirical information is no longer necessary. p can be true or false and the entire sentence is true in any case. Now, a sentence that is already true because of its specific structure (i.e. because of the particular way in which logical constants are used in it) is called a logical sentence. To be more precise: A logical sentence is a true sentence containing only logical constants, apart from variables. Or: A logical sentence is a sentence that is true, and remains true, whenever any of its components, except for its logical particles, are replaced.[±)]

Analytic Sentences

Thus, the concept "analytic" can be generally defined in the following way:
A sentence is an analytic sentence if
it is a logical sentence or

it is constructed in such a way that it can be transformed into a logical sentence by replacing some of the expressions occurring in it with their synonyms.[++)]

It is important to note that the elucidation just given is not the only possible one. In recent literature, there are various other accounts of the concept "analytic" as well.[+++)] In the event that the basic theses concerning the use of "analytic" in the meaning just specified turn out to be flawed, this can also be due to an incorrect interpretation of "analytic" itself. Thus, the falsity of the basic theses cannot be shown so easily.

Here we investigate the usefulness of the characterization just mentioned with regard to the basic theses. We show that "non A1" (where A1 refers to the definition of "analytic" presented above) and "empirical" do not coincide; that therefore the basic theses are wrong if we define "analytic" as A1; in such a way that we present a sentence that is neither logical, nor A1, nor empirical – but still not a senseless sentence. We show that this sentence is not senseless by proving that the significance [*Sinnhaftigkeit*] of the sentence that is singled out is a necessary condition for the possibility of determining the analytic character of a sentence according to the definition just given.

According to the conditions given in the preceding paragraph, sentence U now fulfills all the requirements to make it a meaningful sentence. As an important element of metalogic, it provides information about the use of a constant. We put it at the basis of our analysis. **[12|13]**

Sentence U is considered by some logicians to be an empirical sentence.[+)] This is also the point of view of linguists: Sentence U states how "and" is used and this is a specific statement about the properties of the English language. It could be discovered tomorrow that there is an exception in the conjunction "Franz eats apples and Fritz sits under a tree" and this would be no surprise considering the numerous exceptions that permeate grammar.

This interpretation is justified. Now we simply claim that we do not understand the sentence in the specified way. With sentence U we do not intend to express an empirical fact. Rather, we speak the language in which the "and" itself occurs and we want to learn about its rules, not so much because of its exceptions, but in order to follow them properly. Thus, we interpret sentence U as a stipulation that we must obey if we want to speak a certain language properly. And this second kind of interpretation is not far-fetched at all. A German teacher tells a foreigner, for example, "*mit* takes the accusative". In this case he does not want to convey any empirical information; rather, he stipulates a rule that one must obey. In the latter case, a deviation will not be considered as a correction of an empirical sentence but as a mistake. This presupposes that the given instruction is held as a standard for the evaluation of speaking a certain language, regardless of any empirical fact. Such a standard will be presented by describing ideal states of affairs (as it is also the case with arithmetic as a discipline where the properties of numbers are determined). Thus, in the context here considered U is not empirical.

U is not a logical sentence either. This can be easily seen by replacing all non-logical constants with variables and showing that the sentence so generated is not logical.

And finally, U is not a sentence that can be transformed into a logical sentence by replacing certain expressions with their synonyms. In this case, we only have to focus on the non-logical constants. Consider, for example, "sentence". "Sentence" could be defined in such a way that U turns into ⟨A1⟩. Within a certain language, the nature of sentences is hence such that whenever a sentence is connected to another sentence through "and", a new sentence again is created. Therefore, if we introduce "and" into a language that contains "not", "or", "it follows from", but no "and", the meaning of the word "sentence" changes. This is a quite strange consequence. Because we do not count among the specific characteristics of sentences that they could be connected in a certain way through "and". We can therefore not assume that the meaning of "sentence" changes if such a characteristic varies (just as the meaning of "house" does not change just because the architect has created a new design). If we elaborate further on this issue, we consequently arrive at a claim that has been attacked on the basis of other assumptions.[±±)] If we let the meaning of a certain term be dependent on all of its relations, then we can no longer cleave to a fixed meaning, as there is no single mode of application that would resemble another in every respect.[±±±)] We then no longer have fixed concepts, but [13|14] only constantly shifting spheres and therefore, for this reason, also no possibility of a correction.

In such a case "we can no longer correctly speak unrestrictedly of analytic statements."[±)]

The same applies to the concept "true" for which Tarski[±±)] had already tried to find a definition that would make it dependent on all possible constellations within a certain language.

But in this case there is still another objection that is to be made: If, for example, contrary to the intentions we generally associate with this word, we define "sentence" in such a way that U turns into A1, then such a definition is reformulated into another sentence W. In general, W will be considered as a definition, i.e. as a sentence that ascribes a certain meaning to a senseless term. Therefore, in W we do not yet understand 'sentence'.[±±±)] On the other hand, we understand this term in U. We face therefore the strange situation in which an expression within a certain language has sense at one time and has no sense at another. If we consider the definition as a sentence, then we arrive at contradictions: Because a sentence cannot have meaningless [*sinnlosen*] components. If we do not consider it as a sentence, it becomes evident that within a language there are non-sentence-like structures that are needed to determine the sense of other structures, but which can occur as elements within inferential procedures (in fact, definitions are used in inferences).

This in itself would already be an objection to the main thesis that allows for only analytic and synthetic sentences. But we can do without this objection since it is flawed: Elements containing meaningless [*sinnlose*] parts cannot occur in an inferential chain. If definitions occur in conclusions, then they must have the character of a sentence; and therefore they may not contain any meaningless parts.

At this point, we face an apparent dilemma: If we consider definitions as sentences, then we fail to acknowledge the fact that they actually assign a new meaning. If we do not consider definitions as sentences, then they cannot be used in chains of arguments [*Argumentationskette*] at all. We can dissolve this dilemma by distinguishing between a) the act of defining (that is one meaning of the word "definition") and b) the sentence that is created following such an act (that is a second meaning of the word "definition"). In the act of defining, a sentence is created within which a sign occurs that did not yet have a meaning. It was chosen arbitrarily to bear a certain meaning. The sentence resulting from such an act of defining, however, is not the act itself and therefore the opinion that logic and mathematics are systems of conventions or stipulations[±±±±)] is wrong: As soon as an expression is arbitrarily used within a language it gets a sense. We find this sense by an analysis of the structure of a language [*Sprachbau*] itself and not by a description of the action that led to its construction. In this respect, definitions differ from other kinds of sentences only with regard to their history, not with regard to their "inner nature".

These conclusions are of importance. Reference to definitions plays a significant role for analytic sentences. We see what remains: A series of sentences from which the synonymity of expressions can be extracted. Yet, these sentences are not logical sentences. They also cannot be only A1, if we want to extract the information we need from them. [14|15]

In order to identify the synonymity of two expressions, we have to be able to rely on a sentence whose non-empirical character can be grasped almost immediately (not just by the detour of replacing certain expressions with their synonyms).[1)] The sentence can also be A1. But if it is supposed to provide a basis for identifying synonyms, there must be another possibility to determine its non-empirical character. Therefore:

Either sentence U is empirical or it is logical or there is at least one more sentence that is not empirical, not A1, and whose non-empirical character must be detectable, though not in the way presupposed in A1.

As we found out earlier that U is not logical and it is not empirical, we proved the falsity of the basic theses, if for "analytic" A1 is used.

We arrive at a completely similar conclusion in the following way:

Logical sentences are characterized by reference to logical constants. But how can we find the logical constants? In general, they are referred to as signs that appear essentially within tautologies.[±)] But then we need a non-A1 possibility to determine tautologies. This determination is no longer purely formal. And with that we arrive at another important point. Logical inferential methods are usually praised for their formality, their independence from the content of the domain [*Gebiet*] under investigation. Now, however, the concept of form is not completely clear. The sentences "The sky is blue" and "Mr. X's nose is blue" have the same form with regard to one specific aspect. They end with "is blue". Therefore, when referring to the formality of a domain, the nature of the intended form has to be specified more accurately. In the case of logic, this again cannot happen in any other way than by reference to the logical constants. Should they be found, however, then we have to identify those sentence in which they occur and which moreover are emphasized as being true solely on the basis of their form, on the basis of a content-related, though not empirical, specificity.[±±)] Thus, the concept of a material inference regains a specific meaning.

Another example: Parts of mathematics could be translated into logical notation and were further deduced ⟨with⟩[E] a quite restricted set of axioms of logic (though only with the help of basic principles [*Grundsätze*] whose purely logical character is quite uncertain).[+++)]

(log).equiv.(math) asserts that where very long logical notations stand on the left hand side, there are simple mathematical expressions on the right hand side. These equivalencies are not empirical. But they are A1 only if we presuppose that mathematical sentences have no sense of their own or if we assume that the problem of reducing mathematics to logic is already solved. At the beginning of the investigation, logic and mathematics face each other as completely autonomous disciplines, each having their own independent sense. Logical reconstruction consists in the attempt to rebuild the edifice of one particular discipline by the means of the other. This presupposes that each of the edifices has its own structure. The sense of mathematical sentences, however, is entirely independent of whether a reconstruction has been carried out or not. Hence, the equivalencies of the type in question, though *a priori*, are not analytic (A1). [15|16]

So far we have shown the following: The basic thesis is flawed if we ascribe to the term "analytic" the meaning ascribed at the beginning of the current analysis: that is, if we consider analyticity as a purely (absolutely) formal property.

The disjunction: True on the basis of form.[+)]

True with reference to content-related circumstances is not exhaustive. Non-empirical knowledge is not yet purely formal knowledge.

But this does not tell us anything about the truth-value of the basic thesis itself. We could try to broaden the term "analytic" so that it covers the entire field of non-empirical knowledge. With regard to the older version of "analytic", now there are synthetic sentences *a priori*.[++)] But this discovery is not very exciting, considering how little we have specified our concepts in this domain. There are logicians, however, who consider the basic theses fallacious even after the expansion. It seems that the basic theses provide less secure knowledge, and they are rather the expression of a definite view about the nature of our knowledge that has not yet been completely and properly formulated.

One can draw on completely new distinctions as well. We mention E. Tranekjaer-Rasmussen as an example.[+++)] He introduced two kinds of relationships between phenomena: validity through experience [*Erfahrungsgültigkeit*] and validity through stipulation [B*estimmungsgültigkeit*]. These relations hold independently of the specific kind of phenomena between which these relations are recognized as holding. Validity through stipulation is a relation approximately resembling what one is inclined to mean by analytic. An example: We believe that there is a particular object in front of us. This object is then given to one's consciousness in a characteristic way, either through thought or observation. After that a second *object* appears and that can now have the consequence that the first object is referred to as an illusion. But this is only because the existence of the first object seems somehow connected to other objects, so that the object disappears whenever the others are missing. On the contrary, one can try to draw an inference about general situations from a few observations (for example, from the fact that the doorbell rings, it is inferred that someone is standing in front of the door). At the same time, the ringing sound "indicates" some person that has rung the bell. But in this case, evidence that nobody is standing in front of the door does not make the sound disappear. Because it is not part of the stipulations concerning the sound of a bell that it would be generated by a person pushing a doorbell button.

Now, it is not absolutely necessary that empirically observable circumstances provide the elements of both the relations just mentioned: So, the relation between a mathematical sentence and its logical equivalent is experientially valid [*erfahrungsgültig*], while the relation between the states of affairs

X is human and humans are mortal and

X is mortal

is valid by stipulation [*bestimmungsgültig*]. It would be interesting to find out how our entire knowledge could be ordered on the basis of this distinction. Because, so far, "empirical" means something like "standing in relation to phenomena of a certain kind" (to sense-data, as the positivist theory specifies it). And "analytic" is likewise determined with reference to content. The distinction just mentioned is neutral towards contents and thus stretches across the current corpus of the sciences. [16|17]

The Empirical Sentences

If analytic sentences are characterized quite generally as those sentences that maintain their truth value come what may, then this presupposes that one is properly informed about what "come what may" is. It must be possible to characterize empirical sentences positively by pointing to the specific instances that they should refer to.

This positive characterization has been one of the main problems of empiricism since time immemorial. One initially points out quite naively that empirical sentences have to refer to facts and sciences that have yet to clarify fully their object of investigation are required to have a clear relation to such facts. This presupposes, however, that one already knows what facts are. Exactly the same also applies to practical procedures and physicists can proceed with their practices without having received any lessons about the nature of facts. Difficulties turn up, however, when the nature of facts is not only used, but should also be described. Such a description, however, is the task of philosophy.

What could be more obvious than listing the facts that provide the foundation for a theory? Locke's theory of ideas contains such a list. But this does not tell us anything about the meaning of the concept "empirical". It would be strange if we wanted to call everything that occurs in Locke's theory of ideas empirical. In fact, Locke did not intend to provide an exhaustive description. His list does not specify just what qualifies as an idea, but what ideas he discovered. Therefore, his principle of discovery was different from his principle of enumeration [*Aufzahlung*].

On the other hand, one attempted to find a foundation for empirical sentences in sense-data. Thereby specific events were no longer emphasized. In fact, it is quite generally specified how an event must relate to an observer if it can qualify as a verification of an empirical sentence. We will call this characterization phenomenal since it is established with regard to the particular mode of appearance of the relevant instances. Primarily, those events are chosen whose appearance is quite immediately detectable, such as a sound, a color or a touch. For this reason, however, a second point of view gets mixed in and the confusion between both these demanded requirements of an ideal fact makes it terribly difficult to follow the argumentation of empirical philosophers with complete understanding. Berkeley had already emphasized that a statement about a sound has a more immediate and better foundation than a statement of the kind "I hear a carriage passing by".[±)] He claimed that in this case only the sound perceived could be determined with certainty, while in contrast the judgment "I hear a carriage passing by" is based on conjectures that are immediately attached to the impression of the sound, and hence are more prone to error. Now, this conclusion is no longer phenomenal. Because we do not obtain statements about simple sounds, colors, etc. essentially more quickly and more immediately than statements about complicated spatial circumstances. We can also immediately conclude whether somebody has a cheerful or a sad face as well as many other things. The uncertainty of how these statements are obtained in observation does not necessarily increase with the articulation of what is immediately given. Thus, what is the difference between more or less articulated events? [17|18]

The second point of view, that is decisive here, is the following one: if we obtain a sentence in some way, it is immediately subject to a criticism that is largely independent of the way in which it was obtained. E.g.: An observer B utters the sentence: "A red rose in front of me". If it happens that another observer makes a similar statement ⟨sic⟩, and that he himself cannot grasp the rose he has described, then he will be forced to retract his statement. The statement that was obtained directly and immediately through observation was phenomenally quite certain; it was not certain, however, in the sense that in making such a statement its truth was at once guaranteed forever. It was not functionally certain, as we would say.[±)] Why is that? We get an answer to this question by investigating the further fate of this statement. B initially uttered it. Then C realizes that he cannot see a rose and B realizes that he cannot touch his rose and, on the basis on these observations, the initially uttered statement is ultimately declared false. This shows that *the sentence* did not report on what was immediately determined in observation, but must have reported on a series of further instances. The sentence is therefore obtained – and it is not possible otherwise – in a singular situation (which is one of the instances about which it makes a statement), but it still refers to other circumstances beyond this situation and therefore can be doubted with regard to these additional circumstances. If we establish a series of all circumstances relevant to a test, then we can divide the statements into classes, each according to the instances of this series that are decisive for their test. Let the universal series be a1, a2, a3 etc. Then, for example, the sign 1, 8, 10, 17, 23 indicates that a certain statement has to be tested with regard to the elements 1, 8, etc. of the fundamental series. We call a sign that is designed in such way a real sign [*Realzeichen*] and the property of the statement it denotes its real value [*Realwert*].

If B discovers, in the way specified, that actually there is no rose there, then he can repeat his statement, but in the form "I perceived a red rose". To begin with, this sentence contains all those elements that occurred in the initial statement, therefore all those elements that were immediately detectable by observation; in addition, however, it contains a part that does not correspond to anything in observation – the word "perceived". The entire sentence can be obtained through observation in the same way as the original statement; now, however, it can no longer be refuted by any further observations. Its real value has apparently changed. "Perceived" therefore does not describe anything that can be somehow seen, heard, etc., but it determines how the sentence is to be tested in the further course of the argument.

Thus, so far we have established the following:

1) There are sentences that can be immediately obtained by observation. They are certain with regard to their genesis, i.e. they are constructed in such a way that if a sentence of the kind just specified is presented to an observer with a proper question, he can take a stand clearly and without hesitation.

2) These sentences are not certain with regard to their truth, or at least they do not have to be. In general, they describe [18|19] a plurality of instances independent of the one that led to the utterance of the sentence in the observation situation. Its real sign contains a plurality of elements.

Henceforth, we will call those sentences that have property (1) basic sentences [*Basissätze*] if the circumstances of their utterance have certain properties (they are stated in a purely phenomenal way). Then any sentence is empirical if there is at least one basic sentence whose reversal changes its truth value, either from true to false or the other way around. The problem of the relation between basic sentences to the other empirical sentences is therefore a purely logical problem and can here be left out of the picture. The main problem for the empirical sciences lies in the basic sentences themselves.

And so we return to the assumption that there are states of affairs (and corresponding sentences that describe them) whose appraisal is functionally certain in a specific way and which alone are hence suitable as a basis for the empirical sciences. The relevance of the assumption presupposes that the certainty of the empirical basis is a necessary condition of all empirical disciplines, and that in our argumentation we must, sooner or later, arrive at a point where doubt is no longer possible. But how should this demand for certainty be justified? It seems that older philosophical theories still play a certain role here. But, independently thereof, it can easily be shown that certainty would, at a certain point, annihilate science and this for several reasons. We can mention one of them immediately: The predictive power [*Prognosenfähigkeit*] of a theoretical system presupposes that several circumstances follow from one sentence. Uncertainty is a correlate of predictability [*Prognostizierbarkeit*]. As far as basic sentences are concerned, the following can be considered: A sound [*Lautgestalt*] that cannot be relinquished under any circumstance, and that, according to its own nature, cannot have any other truth value than the one that was assigned to it, has ceased to be part of a language. We find ourselves within the realm of being, which is admittedly neutral vis-à-vis true and false, and no longer within the realm of knowledge.

We may point to still another mistake within the assumption just mentioned. Unarticulated qualities, simple colors, acoustic events, etc. are being emphasized. There is the opinion that a relation holds between the phenomenal character of a statement and its functional value. Therefore, according to the nature of the language used, a sentence about color is also functionally as certain as a sentence about the countenance of another person in front of me.

That is, however, a fundamental error that leads the theory of empirical knowledge astray in many cases. It is true that in many cases there are sentences with varying degrees of certainty within a particular language. That is though a purely historical fact without significance for a theory of knowledge. However, if one aspires to sustain all philosophical wisdom on the basis of everyday language, as some philosophers are inclined to do, then conclusions of this kind are inevitable. But, where do we find information about the real value of a sentence, hence about the degree of its functional certainty, if not through an analysis of its use within a certain language?

The answer is relatively simple: Every observation sentence [*Beobachtungssatz*] stands in a twofold relationship: First, it refers to all those instances that were decisive for its genesis. If this relation is properly met, then it is called a sentence that was correctly obtained. It is connected to an instance of perception. Secondly, this sentence, however, is supposed to convey knowledge and this function can no longer be sufficiently described by phenomenal reports. Plato knew that knowledge does not arise in perception.[±)] Knowledge, however, is grounded on the hypotheses that [19|20] ⟨we⟩ have set up.

In fact, the hypotheses postulate relations between perceivable events. These relations, however, cannot be perceived. And they convey knowledge that cannot be found within perceptions alone.

As part of a hypothetical system, however, the sentence is subject to a second condition that we shall briefly illustrate: Let H be a hypothesis to be tested and let I1, I2, …, In be the instances that are relevant for the test. Furthermore, let B be a basic sentence that is used in the testing procedure. The series of its instances are K1, K2, …, Kr. The derivation then always takes place with reference to a certain parameter (mostly time, e.g. let the value of the basic sentence be determined at the same time as the hypothesis makes a certain statement). In both H(… t) and B(… t), both "t"s designates the same object. Therefore, in both cases "t" must be determined according to the same procedure. Now, it is easy to see that if a sentence R(a, b, c, …, t) is constructable, then all of the objects that the statement talks about must have the same real value. Otherwise, different testing procedures are demanded from this sentence at the same time. From this, we can conclude that H and B must have the same real value. Thus: All sentences that occur within a hypothetical system must have the same real value and this entails the second condition that must, in general, be fulfilled by a basic sentence. In the example just given, it must therefore be the case that I1=K1, I2=K2, etc., just as r = n: The real signs of the hypothesis and of the basic sentence must coincide.

On this basis, we oppose a widespread theory that wants to ascribe more certainty to basic sentences than to hypotheses.[±)] Roughly speaking, this theory claims that while an infinite number of chains of inference[*Schlussketten*] are attached to every hypothesis (which accounts for the uncertainty of hypotheses), every chain of inference has a finite length, and hence after a finite number of steps leads to a sentence that either cannot be doubted at all because the instance that led to its creation is the only one relevant for it, or that can be doubted to a much lesser extent than the hypothesis.[±±)] Now we see how we have to judge this theory: It makes it impossible to set up a relation between basic sentences and hypotheses and therefore it itself makes the test impossible. That this escaped this theory is entirely due to the fact that what is related to the hypothesis is not the basic sentence thus determined (in which even states of affairs occur as relevant), but the description of the way the basic sentence is obtained. This description, however, exhibits all those properties that a useful basic sentence must have and thereby covers up the flaws of the theory.

But, on the basis of this foundation, how can we interpret the fact that a sentence that starts with "perceived" is no longer at hand? In general, one is inclined to assume a decrease in the series of instances and thereby an increase of certainty. Yet, this interpretation is no longer possible if we consider that sentences about perceptions should be brought into relation to other scientific sentences. A **[20|21]** psychologist investigates the perceptions of organisms that can also be considered to exist in a physicalist sense. Thus, relations between physicalist statements and statements about perceptions must be possible and this presupposes once more that statements about perceptions have a real value that is the same as the real value of physicalist statements. But how then is their higher degree of certainty to be interpreted?

Wittgenstein has offered a valuable suggestion here:[±)] If we talk about perception, we do not mean an object that exhibits the same observational foundation as the perceived object, but a smaller series of instances. We rather mean a situation in which the observer finds himself if he is confronted with a certain object. His statement "I perceive X" is then no longer to be interpreted as a statement about a certain state of affairs, but as a case of the state of affairs "perception of X". The sense of statements is always initially determined independently of the way statements are produced. In observation anything can be the case. The way we use the sentence depends on the system in which it should be integrated.

As soon as the statement is no longer considered as logical, but rather as factual, as part of the flow of appearance [*Erscheinungsablauf*], the question of its truth vanishes as well. The statement about perceptions is undoubtable, but not because it is especially certain (exhibiting especially fewer instances), but because there is no longer any statement at all.

The autonomy of content and phenomenal foundation of a statement has another important consequence: Physicalism loses an important basis. The main motive for its introduction was the insight that the sentences of all sciences must be somehow related if it should be possible to establish a connection between them.[±±)] Furthermore, it was easy to see that within physics there is the highest diversity of testing procedures. Therefore, the testing procedure of physics applies to all objects.

This reasoning suffers from one mistake: It presupposes that besides physical tests there are no other specific testing procedures. If the latter is *not* the case, then according to explanations of real value, every scientific sentence must be tested with regard to every possible instance (hence, in the physically relevant, plus the remaining, instances). Universal science [*Universalwissenschaft*] exhibits a uniform real value not only with reference to physical instances. But now, what does this distinction [*Auszeichnung*] of physics consist in, which made it so valuable in the eyes of its advocates that it was chosen as the science par excellence [*kat'exochen*]?

First of all, it consist in its intersubjectivity, i.e. in the fact that physical statements can be verified by everybody, no matter what impairments their senses exhibit. If a person does not see well, he might undertake an acoustic test, etc. and thereby reestablish a connection to the statements of others. The domain of experience seems to be cut off here to a large extent. Since, at the time of the formation of physicalism, some adventurous ideas about statements that were obtained on the basis of observation still ruled, there was a tendency not, as it would be right to do, to conduct a closer examination of their actual character, but to remove them from the domain of science altogether, and this removal seemed to a large extent successful in physics. It appeared as a purely theoretical system that shows only minimal contact to the dubious given [*Gegebenen*]. In Platonic terminology: Physics combines a maximum of knowledge with a minimum of perception, and that is all that counts. **[21|22]**

Now, does this correspond to the facts? To begin with, physical statements are also tested with regard to experience, even if to quite curtailed experience. It is generally said that only coincidences are important. These are experiences of such a kind that there will be the highest agreement among all observers with regard to them. But this agreement is somewhat of an accident and it could easily happen that a similar conflict about the existence of a coincidence suddenly breaks out as it happens with regard to judgments about the genre of a work of art directly observed. Conversely, observations in the so-called humanities are directed towards very complex states of affairs. And once again, it is an accident that agreement here is not so easily possible. But there are tribes that exhibit a certainty of judgment, in this respect and in their domain, that is not at all inferior to the certainty of judgment of the occurrence of a coincidence.[±)] As soon as perception comes into play, there is the possibility of conflict, regardless of how complicated this perception appears.

On the other hand, it is always the case that the knowledge conveyed in a sentence is different from the perception that provides its foundation. In every scientific discipline, no matter what kind of phenomena it is related to, there is a theoretical part that is absolutely and completely detached from the domain of perception. Admittedly, this theoretical part has a different scope in different scientific disciplines. But once again, this lies within the nature of things and tells us about an essential feature of the world: In specific relationships, e.g., facial expressions, gestures, etc. mean anger. "X is angry" is obtained through a very simple observation, but it reports on a whole set of further events that can be revealed by suitable procedures. Thus, it already has significant theoretical content.

The disadvantages of the humanities could only be thoroughly judged if all observers were subjected to the same extensive training, since their childhood with regard to the relevant states of affairs, as physicists are with regard to their coincidences.

Even if it were impossible to attain such perfection, the thesis of physicalism would still not be justified. The impossibility of the humanities might then indeed be proclaimed, but not their identity with physics. Yet, it seems to me that we are not forced to draw such radical conclusions.

Hence, the following is essential: There is no other reason that would distinguish the physical procedure of referring to coincidences alone, except for the fortuitous fact that we have well-trained observers in physics and bad observers in the non-physical sciences. From an epistemological point of view, however, there is no difference to be noticed. And with that, the relevance also of non-physical phenomena for observation is restored: Wherever there are observers who conform to a law-like behavior with regard to certain complex phenomena (which can be seen in the law-like articulation of their statements), a phenomenal foundation for a scientific discipline is laid. And different scientific disciplines no longer differ from one another with regard to their structure, but essentially with regard to the phenomenal basis. The possibility of a particular structure upon such a basis [22|23] is the result of the choice of such a basis.

The individuality of historical events would then be a consequence of the fact that one has the tendency to make only particular observations and is not an exclusive characteristic of historical disciplines. The specific behaviors of organisms would again be a consequence of the fact that among the phenomena that are relevant for biology (about which the students of biology are trained in a very methodical way), different relations, of a phenomenal as well as of a functional kind, prevail than those that prevail among coincidences. This reference to the specific character of observation statements enables a classification of the empirical sciences that is much clearer and epistemologically more satisfying than the classification that turns out from the ultimate objects that are reached on the basis of these special observations. These objects are subject to constant change and consequently are not a suitable basis for a theory of science.

Physics, however, becomes relevant at an entirely different point: From the identity of the real value of all statements, it follows that every scientific object must also exist in the physical sense. The historian does not investigate the figures [*Gestalten*] of Napoleon's dreams as figures, rather, he investigates them as figures of a dream [*Traumgestalten*]. But, conversely, every physical object must, in principle, also be subjectable to the testing procedures of the other sciences, even if in most cases this procedure leads to rather trivial results.

Within Gestalt psychology, phenomenology and especially psychology as well as in certain domains of philosophy, the tendency to shape the domain of the given phenomenally as extensively as possible has recently become more and more clearly discernible.[±)] Only in very few cases, however, has this actually been applied to the theory of science as yet.[±±)] It seems that the future of this discipline depends essentially on developing such considerations.

After everything that we have said about empirical sentences so far, we can call a sentence empirical if and only if either it is a basic sentence or if it is connected to at least one basic sentence “b” in such a way that a reversal of “b” results in a reversal of the sentence under investigation.

Conclusion

We started with a rough characterization of the analytic procedure; we demonstrated its efficacy using the example of the discussion of the concepts “analytic” and “empirical” as well as in the critique of some of the opinions prevalent in these domains. It appeared that the main outlines of these ideas could be maintained, though in particular cases numerous details have to be corrected. And now we return to the starting point itself: The analytic procedure is not something new in philosophy. Platonic writings [23|24] provide excellent examples of philosophical analysis in their epistemological parts.

This procedure, and with it many similar ones, however, exhibits a peculiarity that has prevented the theory of science from taking the form of an exact discipline. If e.g. we talk about perception, the distinctiveness of perception statements is not explained by indicating their special features, but through the production of different sentences in which one can see (not recognize) what is characteristic about perceptions: “Perceptions have no weight”, “Perceptions cannot be seized” or through the description of circumstances in which perceptions occur and their invisibility, etc. is revealed. Thus, the theory of perception never comes to an end. And some philosophers hold the opinion that such an end is not to be expected: “It is true, we can teach a person to understand the word ‘time’ by producing examples of its use, some characteristic examples: but what we cannot do is to present a fixed formula that captures the complete and often elusive sense of the word in a magic crystal.”[±)] In addition, the quotation claims that it is impossible even just to develop a theory, so that we always have to rely on examples, although they can never exhaust the meaning. Therefore, with regard to these meanings we are in the same position as Locke was with his ideas. We can give examples, but we cannot determine the scope of the source from which all of these examples flow. From this perspective, philosophy is not a theory, but a perpetual activity.[±±)] Now we hold the opinion that within logical empiricism, a way has been found for the formation even of philosophical theories (which however, as is often their fate, will soon form a discipline of its own, the philosophy of science [*Wissenschaftstheorie*]) or, to avoid the ambiguous word “philosophical”, theories of meaning. In the literature of the Vienna Circle, the so-called “formal mode of speech” plays a particular role. With it, it is meant that a problem (for example the problem of the relation between body and mind) can be brought into a certain form, if one does not investigate the objects and their relations appearing within these problems, but rather investigates the verification procedure of the sentences that provide some information about these objects. In fact, this is the more original procedure with regard to these objects.

However, clear statements about the verification procedure can be made, which then, with regard to the meaning of the words used, establish the role of the theories or of the rules that one wants to comply to in future discussions of the problem. These theories contain a great wealth of statements in a simple way and thus actually provide the foundation for a philosophical discussion. It seems that the future of philosophy (or of the philosophy of science [*Wissenschaftstheorie*]) consists in starting to regulate the testing procedure of these kinds of theories and to systematize them. For the time being, however, there still remains the task of testing already existing theories, and theories in disguise, by investigating all possible instances with regard to their truth content.

---

[1] ±) Seventeenth ⟨sic⟩ International Congress of Philosophy, Oxford 1930, pp. 112-113.[6E]

[1] ±±) Bacon, *Novum Organum,* transl. Kirchmann, p. 37.[7E]

[2] ±) As by G. Ryle in his Concept of Mind ⟨sic⟩.[8E]

[2] ±±) Discours, transl. Reclam 3767, p. 20.[9E]

[2] ±±±) Cf. the evaluation of Aristotle in Kaila, den maenskliga kunskapen ⟨sic⟩,[10E] passim.

[2] ±±±±) Metaphysics, transl. Kirchmann[11E] I/4, p. 38.

[2] ±±±±±) Cf. K. Lewin, Aristotelische und galileische Betrachtungsweise in Biologie und Physik ⟨sic⟩, *Erkenntnis*, Bd. I.[12E]

[3] ±) In the Regulae ad dirigendum ingenii ⟨sic⟩.[13E]

[3] ±±) Erkenntnis, Bd II.[14E]

[3] 1) "Many of us … hold the view … that logic is the very foundation of philosophical and methodological investigation." "Viele von uns sind der Ansicht … dass die Logik das Fundament der philosophischen und methodologischen Untersuchung darstellt." (Carnap, Formalization of Logic,[15E] p. XIII).

[3] ±±±) Cf. the programmatic brochure "Wissenschaftliche Weltauffassung",[16E] p. 16-17.

[3] ±±±±) E.g. Kaila, Den maenskliga kunskapen ⟨sic⟩,[17E] pp. 209ff.

[3] ±±±±±) In The Rise of Scientific Philosophy ⟨sic⟩,[18E] p. 303.

[4] ±) Cf. the critique of Norman Malcolm in The Philosophical Review, Oct. 1951, pp. 582ff.[19E]

[4] ±±) Cf. A. Pap, Elements of Analytic Philosophy,[20E] p. IX.

[4] ±±±) Cf. Quine's essay, the two dogmas of Empiricism ⟨sic⟩, The Philosophical Review, Jan. 1951, pp. 30-44 ⟨sic⟩.[21E]

[5] ±) Essay Concerning Human Understanding, II/21/7ff (Ed. Calcins).[22E]

[6] ±) The paradox of Grelling-Nelson, see J. Jørgensen, *Treatise on Formal Logic III*,[23E] p. 164 (and further literature there) and Reichenbach, *Elements of Symbolic Logic*.[24E]

[6] ±±) Cf. the discussions in the philosophical seminar of the Alpbacher Hochschulwochen, 1949.[25E]

[7] ±) Cf. Carnap, Logische Syntax der Sprache,[26E] p. 44 and the literature there.

[7] ±±) We find here, in the domain of logic, the same one-dimensionality of reasoning that Berkeley for example represents in the domain of empirical knowledge. Ayer, his loyal successor, assumed objects about which it is not allowed to say that they contained a part that has not been perceived yet. (*Foundations of Emp. Knowledge*,[27E] p. 130).
[8] ±) Meno 82b-86a.
[8] ±±) Tractatus Logico-Philosophicus,[28E] sentence 4.0031.
[8] ±±±) Principia Ethica,[29E] p. IX.
[8] ±±±±) Quoted by Popper, Logik der Forschung, p. III.[30E]
[8] ±±±±±) Tractatus, sentence 4,002 ⟨sic⟩.[31E]
[8] ±±±±±±) Cf. F. Waismann, Analytic-Synthetic II in Analysis, Vol 11/2 and F. Kaufmann, Basic Issues in Logical Positivism, in Philosophic Thought in France & the United States by M. Farber, pp. 565-588.[32E]
[9] ±) Tractatus,[33E] sentence 6,54.
[9] ±±) Inquiry ⟨sic⟩ IV/1, Edition Meiner (transl. Richter) p.35.[34E]
[9] ±±±) Richter, p. 193.[35E]
[10] ±) ⟨B.⟩ Russell, Our Knowledge of the External World,[36E] p. 89.
[10] ±±) E. Kaila, Den Maenskliga Kunskapen ⟨sic⟩,[37E] p. 216.
[10] ±±±) Kaila, op. cit.,[38E] p. 209ff, Juhos, Die Erk. U. ihre Leistung,[39E] p.5; Reichenbach, the Rise of ...,[40E] p. 304.
[12] ±) Quine, Mathematical Logic,[41E] p. 1ff; id. Two Dogmas, etc.[42E] p. 22.
[12] ±±) Quine, two Dogmas etc. ⟨sic⟩[43E] p. 22; Bolzano (Wissenschaftslehre[44E] § 148) provides a definition for A1 and the contradiction together that wholly resembles the one presented. Cf.. Jehoshua Bar-Hillel in Theoria XVI/pp. 91-117.[45E]
[12] ±±±) from p. 12 Cf. Quine, Two Dogmas etc., as well as Benson Mates, Analytical Sentences; ⟨sic⟩ The Philosophical Review 1951, p. 525ff.[46E]
[13] ±) E.g. by M. G. White, The Analytic and the Synthetic, an untenable dualism; in J. Dewey, Philosopher of Science and Freedom,[47E] p. 320. Cf. ⟨UNREADABLE⟩ Carnap, Introduction to Semantics, p. 11ff.[48E]
[13] ±±) Cf. Black's criticism of the semantical definition of truth, in M. Black, Language and Philosophy,[49E] p. 911ff.
[13] ±±±) The extreme one can be found in N. Goodman, On Likeness of Meaning, Analysis vol. 10/1 pp. 1-7 ("No non-repetitive statement will be analytic") and R. Rudner, A note on Likeness of Meaning, Anal. 10/5 pp. 115-118, who denies that a word has the same meaning in different parts.[50E]
[14] ±) R. Rudner, loc. cit.[51E]
[14] ±±) A. Tarski, der Wahrheitsbegriff in den formalisierten Sprachen, Studia Philosophica I.[52E]
[14] ±±±) Note by Quine, Two Dogmas etc.[53E]
[14] ±±±±) Cf. footnote +) on p. 7. In Vienna the conventionalist thesis is represented by Prof. Kraft. Carnap (Introd. to Semantics,[54E] p. 215, 247) has already modified the conventionalist thesis.
[15] 1) Cf. A. Pap, Synonymity and Logical Equivalence, Analysis IX/ 4.[55E]
[15] ±) Reichenbach, Elements of Symbolic Logic,[56E] §55.
[15] ±±) A. Pap, Logic and the Concept of Entailment, Journal of Philos. XLVII, p. 378-387 ⟨sic⟩.[57E]

[15] ±±±) G. ⟨sic⟩ Hempel, On the Nature of Mathematical Truth, American Mathematical Monthly, 52, 1945. Carnap, Foundations of Logic and Math.[58E]

[16] ±) Here, the material a priori is important again. Cf. Scheler: Der Formalismus in der Ethik und die materiale Wertethik.[59E]

[16] ±±) A. Pap, loc. cit.[60E]

[16] ±±±) E.g. Psychologie und Gesetzeserkenntnis in Jahrbuch der Hochschulwochen Alpbach 1948, p. 111ff.[61E]

[17] ±) Three Dialogues between Hylas and Philonous.

[18] ±) The differentiation phenomenal-functional stems from Kaila, Det fraemmande sjaelslivets kunskapsteoretiska, Theoria II,[62E] p. 114f. ⟨sic⟩.

[19] ±) Cf. E. v. Aster, Plato,[63E] p. 96.

[20] ±) Ayer, Foundation of Empirical Knowledge; Schlick, Das Fundament der Erkenntnis, ⟨sic⟩ Erk. Bd III; Juhos, Die Erkenntnis und ihre Leistung.[64E]

[20] ±±) The latter case can be found in Kaila, Den maenskliga Kunskapen.[65E]

[21] ±) Cf. Waismann, The Relevance of Psychology to Logic, Arist. Soc. Suppl. Vol. 15, 1936 ⟨sic⟩.[66E]

[21] ±±) Jørgensen, Den logiske Empirismens udvikling,[67E] p. 85. Further literature on physicalism can be found there. Cf. Kraft, The Vienna Circle.[68E]

[22] ±) Cf. St. H. Britt, Social Psychology od ⟨sic⟩ modern Life ⟨sic⟩,[69E] Ch. II.

[23] ±) Cf. Katz, Gestaltpsychologie;[70E] Nic. Hartmann, Ethik,[71E] p. 57.

[23] ±±) E. Tranekjaer-Rasmussen has undertaken an attempt of this kind in some unpublished works.[72E]

[24] ±) Friedrich Waismann, Analytic-Synthetic II, Analysis Vol. 11; No. 2, 1950,[73E] p. 28.

[24] ±±) Wittgenstein, Tractatus,[74E] sentence 4, 112. Jaspers, Einf. in die Phil.,[75E] p. 14.

Editors' Notes

EFA (24 ss., included within an untitled bounded volume, T+M). Handwritten passages are reported in italics.

[1E] Latin for "Everything is true, that is clearly and distinctively perceived."
[2E] See Schlick 1931: [171-172].
[3E] See Bacon 1620: [6].
[4E] See Ryle 1949.
[5E] See Descartes 1637: [114-115].
[6E] See Kaila 1939.
[7E] See Kirchmann 1871.
[8E] See Lewin 1931.
[9E] The correct title of Descartes' unfinished writing is "Regulae ad directionem ingenii".
[10E] See Carnap 1931.
[11E] See Carnap 1943.
[12E] See Neurath-Carnap-Hahn 1929.
[13E] See Kaila 1939.
[14E] See Reichenbach 1951.
[15E] See Malcolm 1951.
[16E] See Pap, A. 1949a.
[17E] See Quine 1951a.

[18E] It is not clear to which edition of Locke's work Feyerabend refers here (this is irrelevant, however, to identify the sections of it to which Feyerabend refers).
[19E] See JØRGENSEN 1931.
[20E] See REICHENBACH 1947.
[21E] On the 1949 European Forum Alpbach, see Vol. 1, Part II, 1948-2, n4.
[22E] See CARNAP 1934.
[23E] See AYER 1940.
[24E] See WITTGENSTEIN 1921.
[25E] See MOORE 1903.
[26E] See POPPER 1935: [xv].
[27E] See WITTGENSTEIN 1921: 4.003.
[28E] See WAISMANN 1950; KAUFMANN 1950.
[29E] See WITTGENSTEIN 1921.
[30E] See HUME 1748: [108].
[31E] See HUME 1748: [211].
[32E] See RUSSELL 1914.
[33E] See KAILA 1939.
[34E] See KAILA 1939.
[35E] See JUHOS 1950.
[36E] See REICHENBACH 1951.
[37E] See QUINE 1951b
[38E] See QUINE 1951a.
[39E] See QUINE 1951a.
[40E] See BOLZANO 1837.
[41E] See BAR-HILLEL 1950.
[42E] See QUINE 1951a; MATES 1951.
[43E] See WHITE 1950.
[44E] See CARNAP 1942.
[45E] See BLACK 1949.
[46E] See GOODMAN 1949; RUDNER 1950.
[47E] See RUDNER 1950.
[48E] See TARSKI 1936.
[49E] See QUINE 1951a.
[50E] See CARNAP 1942.
[51E] See PAP, A. 1949b.
[52E] See REICHENBACH 1947.
[53E] See PAP, A. 1950.
[54E] See HEMPEL 1945; CARNAP 1939.
[55E] See SCHELER 1916.
[56E] See PAP, A. 1950.
[57E] See TRANEKJÆR RASMUSSEN 1949.
[58E] See KAILA 1936.
[59E] See ASTER 1925.
[60E] See AYER 1940; SCHLICK 1934; JUHOS 1950.
[61E] See KAILA 1939.
[62E] See WAISMANN 1938.
[63E] See JØRGENSEN 1948.
[64E] See KRAFT 1950.
[65E] See BRITT 1941.
[66E] See KATZ 1944.
[67E] See HARTMANN 1926.
[68E] On Feyerabend and Tranekjær Rasmussen, see Vol. 1, Part I, Ch. 2, §1, n12.
[69E] See WAISMANN 1950.
[70E] See WITTGENSTEIN 1921.
[71E] See JASPERS 1950.

14 December 1951

Moderator: Prof. Dr. V. Kraft[2E]

Discussion of Paul Feyerabend's Lecture:
"The Dogmas of Logical Empiricism"

Prof. Kraft: To begin with, I would like to say that in his lecture Dr. Feyerabend has provided an excellent example of how to apply the new method. We will follow his outline:

1) Questions about method
2) Questions about both of the specific problems raised:
   a) the problem of the division of sentences into analytic and synthetic
   b) the problem of a foundation for knowledge, basic sentences, etc.

Dr. Smital: In my opinion, the problem of the division of sentences into a. and s. and its apparent inadequacy can be solved very simply: One interprets all sentences that belong neither to the group of analytic nor to the group of synthetic as definitions.

Feyerabend: This leads to the question: What is a definition? The concept of definition is problematic. I am confronted with a certain language that I need to analyze, a set of sentences. Which sentences are definitions? How do I recognize a definition?

Pichl: Schlick distinguished between statements [*Aussage*] and instructions [*Anweisung*]. The group of sentences in question, those that are neither analytic nor synthetic, would be equated with language instructions [*Sprachanweisungen*] accordingly. A solution to several paradoxes (the sentence on the backside is wrong …) is only possible on the basis of this separation.

Feyerabend: This is a problem of another dimension. This differentiation leads to a partition of object-language and meta-language. But I did not go into that at all. In his "Introduction to Semantics", Carnap replaced language instructions with semantic rules. But what are semantic rules? The problem of semantic rules really is the problem of definitions translated into the meta-language.

Pichl: You mentioned phenomenology. Maybe meta-language needed to draw on the phenomenological method.

Feyerabend: Is there a phenomenological method in a strict, established sense? What does it consist of?

Pichl: I am at a loss here.

Smital: I think the difficulty is that one cannot just start *ab ovo* in philosophy: Nothing arises out of nothing. There needs to be some things at the beginning. [1|2]3E

Feyerabend: Of course, in practice a language is always presupposed. But here it is not about that. Here it is about the logical structure [*Aufbau*] of a language-system.

Romig:[4E] You defined the logical conjunction, the 'and' connection between two elementary sentences, as if it were possible to determine the truth-value of the conjunction from knowledge about the truth-value of the elementary sentences. In my opinion this leads to the false and dangerous conclusion that a conjunction is really just the sum of its elementary sentences. Its autonomous character is not emphasized and it is not shown that both sentences merge to a new entity that is not just the sum of its parts. Primarily, it is a sentence-connector, the elementary sentences are only secondary. If you claim that the truth-value of a conjunction does not presuppose that the single elementary sentences make sense, then I think I have to point out that the entire sentence only makes sense with regard to the sense [*Sinnbezug*] of the single sentences and thus can be neither true nor false.

Feyerabend: I did not claim that elementary sentences could be senseless, only that we do not have to presuppose knowledge of the meaning of the elementary sentences to decide whether the conjunction is true or false. The assignment of truth-values is possible in a purely formal manner. In such a purely formal treatment, the elements of a conjunction are variables that have no fixed sense, but that must, however, satisfy the condition of being sentences that make sense and that can be either true or false.

X: There is only concrete sense [*Sinn*], thus, there are actually no purely formal sentences.

Pichl: May I actually talk about sense with regard to the lecturer's example of a sentence that has a truth-value that can be determined *a priori* without having determined the meaning of the words it uses such as "all", "everybody", etc.?

Feyerabend: That happens within the meta-language. In the meta-language, we use everyday-language to help determine the meaning of logical constants in an object-language. I may construct a second order meta-language for a meta-language that analyzes the meta-language I, etc.

Pichl: But we master the meta-language as an everyday-language in a purely intuitive way.

Feyerabend: A phsysicist is also a theoretical machine who has the theories intuivitely within it.

By the way, I agree with Dr. [2|3] Romig's above mentioned opinion that we first come across the conjunction in its entirety and then proceed to the analysis of its sub-sentences.

Zdarzil:[5E] In the lecture, it was observed that, today, one can no longer defend the physicalist restriction on the observational foundations with the help of which I can verify a scientific system to circumstances that can be described in a physicalist way; and that the range of observations on which science can be founded needs to be widened. Thus, not only physicalist statements are permissible, but also statements about impressions of an expression [*Ausdruckseindrücke*], etc. Now however I am convinced that this physicalist restriction was not introduced for no reason.

Because, through the extension of the observational foundations permitted in science, intersubjectivity is apparently gone. From this necessary extension of the foundational domain of the sciences, it can be concluded that not all sciences are constructed in the image, and according to the model, of physics and that not all sciences can reach the same degree of intersubjectivity. I believe that this extension of the domain of permissible observations is of tremendous significance, since it makes the humanities possible, but also biology, at least to the extent that biology would otherwise not have an object of its own. Question: Where are the new boundaries of the domain of scientifically relevant observations?

Feyerabend: With regard to the problem of physicalism: I tried to show that it is necessary to go beyond sense-data to more complex domains of experience.

With regard to the loss of intersubjectivity: What kind of sentence is it that claims physics is intersubjective? Here, I want to claim, though I do not believe it, that physics only refers to meter-readings and that physicalist statements can always be traced back to the establishement of coincidences [*Koinzidenzen*] that can be checked by any number of observers. In biology and the other sciences, the basic states of affairs [*Grundsachverhalte*] are not quite so simple. There are psychological pathologies such that the patient may retrace every single line of a picture and say, "here is a red line and there is a green one…" but he does not recognize the whole as a house; rather, he only notices coincidences. The states of affairs in the other sciences have a much more complex nature. Therefore, here it is necessary to have well-trained observers and to have a criterion to determine when an observer is well trained. If a physicist ⟨says⟩ here and here is an electric field, this statement can be verified relatively easily. But if you make a statement about the *Weltgeist*, the situation becomes much more complicated: the basic states of affairs need to be disclosed, etc… **[3|4]**

Neider: Human beings are an object of physics as well. Therefore, any non-physicalist statement must be translatable into physicalist statements, even if this leads to complicated formulations. Of course, in everyday life it is not necessary to determine whether or not a person is angry in a purely physicalist manner, but this does not alter the fact that it is possible in principle.

Feyerabend: It is a mere accident that a topologically determined facial expression of human beings does not express different feelings in different situations. For example, I might define anger by the psi-function, but who says that the psi-function always needs to refer to anger?

Romig: I claim that sense-data cannot be used as a verification foundation for the humanities since the mind [*der Geist*] cannot be traced back to sense-data. Therefore, neopositivism only applies to the natural sciences.

Feyerabend: Neopositivism does not only consist in the claim that all sentences can be traced back to sense-data. Therefore, the fact that this thesis is wrong does not entail that it cannot be applied to the humanities.

Zdarzil: I would like to draw a different conclusion from the criticism of the physicalist notion of the justification of knowledge [*Erkenntnisbegründungsansatz*]. It was shown that the physicalist basis is too narrow and that further observation domains need to be included to make the humanities, and even biology, possible. I do not mean to say, contrary to Dr. Neider's remark, that statements about anger, i.e. feelings, cannot be translated into physicalist statements. But, whenever I talk about anger and the like, I always refer to something more than just physicalist states of affairs. The question now arises as to whether there will still be any possibility of determining the domain of the intersubjectively testable observations whatsoever, or even more generally put, the domain of those that are scientifically permissible, after we have abandoned the physicalist point of view.

Feyerabend: I admit that there is no content-related criterion except for the following formal one: Anything about which an empirical consensus can be established by observers may be suitable as a foundation of science.

Zdarzil: I think that this is an insufficient criterion because here a norm is to be deduced from a statement of fact [*Tatsachenfeststellung*]. ⟨4|5⟩ Unfortunately, there is no time left to discuss this any further here.

Editors' Notes

EFA (5 ss., included within an untitled bounded volume, T)

[1E] On Viktor Kraft, see Vol. I, Part I, Ch. 1, n15.

[2E] This document follows the previous paper within the same untitled bounded volume, but with new page numbering.

[3E] Possibly, the Austrian economist **Friedrich Romig** (b. 1938).

[4E] Possibly, the Austrian educationist **Herbert Zdarzil** (1928-2008).

Chapter 3
The Feyerabend-Carnap Controversies (1954-1958)

Paul K. Feyerabend, Arthur Pap. Michael Scriven, Carl G. Hempel, Rudolf Carnap and William W. Rozeboom

1. The First Controversy – First part (1954-1955)

1. Paul K. Feyerabend, "Interpretation in Physics – A Criticism of Carnap's Theory and Related Theories", [1955]

Abbreviated version of an article to appear in Theoria, Lund[1]

Interpretation in Physics

A criticism of Carnap's Theory and Related Theories

1) If we want to construct a physical theory as an interpreted calculus with factual content we have, according to Carnap+) to proceed in the following way: We have first to lay down a calculus and we have then to state the intended interpretation in the form of semantical rules. The construction of the calculus may be achieved by using the one or the other of the following two methods:

1) The elementary terms++) of the calculus are taken as primitive, the abstract terms are introduced by means of so called reduction-sentences+++);

2) The abstract terms are taken as primitive and the elementary terms are defined on the basis of the abstract terms. In this case explicit definitions seem to be sufficient.

In interpreting the calculus we may also proceed in two different ways, namely, we may either

a) state semantical rules for the abstract terms only, or we may

b) state semantical rules for the elementary terms only.

It is assumed that in both cases the terms for which no direct semantical interpretation has been provided are nevertheless partially interpreted via the definitions (or the reduction-sentences) which connect the directly interpreted terms with the remaining expressions of the calculus. [1|2]

Combining the two methods of laying down with the two modes of interpretation we arrive at four different ways in which a physical theory may be constructed, viz.

M. Collodel, E. Oberheim (eds.), *Feyerabend's Formative Years. Volume 2. Feyerabend on Logical Empiricism, Bohm & Kuhn*, Vienna Circle Institute Library 10,
https://doi.org/10.1007/978-3-031-57519-8_3

1a, 1b, 2a, 2b. As for a physicist 2b seems to be the most natural procedure, we shall take it into consideration first.

2) Our example will be celestial mechanics. The abstract terms of this system are functors, such as "Mass", "Force", "Distance", "Acceleration" etc., the ranges of individuals of which contain material particles. The elementary terms are again functors, such as "Geometric coordinate", "High Ascension" etc. Now the elementary terms are to be defined on the basis of the more abstract expressions is shown in the textbooks of spherical geometry. No philosophical difficulty is connected with these definitions. How are we to interprete ⟨sic⟩ the elementary terms? The most natural procedure seems to be this: We interprete ⟨sic⟩ the geometric coordinate of a certain planet, say Jupiter, as the high ascension (α) and the declination (δ) of the "star", i.e. of the optical phenomenon in the sky which is observed by the astronomers, when they want to know about the position, velocity etc. of Jupiter. This interpretation may be expressed by the following semantical rule

R "Geoc.Coo (Jup)" design. $(\alpha, \delta)(J)$

(α, δ) being a functor, the range of individuals of which contains "stars" and values of which are pairs of members, viz. the high ascension and the declination of these "stars". J is the "Star" which is usually called "Jupiter".

But having established this interpretation we at once encounter the following difficulties:

a) In evaluating observations we usually have to take into account systematic errors and disturbances, such as athmospheric ⟨sic⟩ refraction. That means [2|3] that we do not directly conclude from the observed values of (α, δ) to the geometric coordinate of the planet which is being observed in any direct way; we rather transform the observed values according to certain rules which either follow from a theory other than the theory to be tested (theory of refraction) or which have been found to be valid by direct empirical investigation (personal equation). That shows that the relation between the system to be interpreted and the observations in terms of which it is to be interpreted is more complicated, than the simple rule R would suggest.

b) We never know all the systematic errors and all disturbances which distort our observations, but we discover them by steadily improving our means of observation. In doing so we may find irregularities of the movement of the planets which cannot be explained on the basis of celestial mechanics alone. We may discover that those irregularities are due to disturbances which have not been taken into account so far. But taking them into account means changing the relation between the observations (the position of the "stars") and the elementary terms of the system under discussion. Important consequences may be drawn from this simple fact.

Let us first state two principles which seem to be indisputable.

1) If it possible to compare all the consequences of a theory with experience, then every sentence of the theory is meaningful.

2) The meaning of the sentences of a theory is not changed by the fact that we are continually improving the instruments which we use in testing the theory. Example: Olaus Rømer and Michelson who used vastly different methods for determining the velocity of light were nevertheless concerned about the truth value of the same proposition, viz. the proposition that the velocity of light is such and such.

Applying the second principle we have to conclude that the sentences of [3|4] a theoretical system do no change their meaning when new systematic errors and disturbances are discovered and when the correlations between the results of observation a⟨n⟩d the elementary terms are changed accordingly. But for Carnap this change implies a change of the semantical rules which provide a direct interpretation for the elementary terms and, via the definitions, an indirect interpretation of the remaining descriptive terms of the system – and this change in its turn implies a change of meaning of these very descriptive terms and, thereby, of the theoretical system as a whole.

Applying the first principle we have to conclude that a theoretical system may have factual content independently of any <u>specific</u> correlation between its elementary terms and results of observation. For it happens very often that we determine the quantitative effect of certain disturbances, the causes of which we do not yet know, by assuming the truth of the theory under consideration and by evaluating the difference between the predictions and the observations.

It follows that the relation between the elementary terms of a theory and "experience" is much less rigid, than Carnap's scheme of interpretation would suggest.

c) Some philosophers+) have therefore tried to construe this relation in terms of probabilities rather than in terms of the relation of designation. Instead of R, from which we may derive that

$$I_1\ (x)\ (\text{Geoc.Coo}(\mathfrak{J}\text{up}) = x \mathrel{\rlap{\equiv}L} (\alpha, \delta)\ (\mathfrak{J}) = x)^2$$

they use semantical rules of lesser strength, i.e. rules R_1, …R_k from which we may only derive that

$$I_2\ (x)\ (\text{Geoc.Coo}(\mathfrak{J}\text{up}) = x \underset{p \leq M}{\Rightarrow} (\alpha, \delta)\ (\mathfrak{J}) = x)^3$$

(M being a constant <1, I_2 being the strongest sentence which follows from R_1, …..R_k and which contains elementary terms of the system under [4|5] consideration as well as observational terms.

Against this mode of interpretation two objections can be raised: 1st: It is always possible to imagine circumstances which make it highly improbable that the difference between the values of the elementary functor which is to be defined by R_1, …, R_k and the values of the observed coordinated is smaller than a certain interval, which may be as large as we like. Under such circumstances one cannot even determine the most probable values of the elementary functor on the basis of R_1, …, R_k. It is therefore not possible to evade changing the semantical rules by assuming that those rules are stated in terms of probability only. – 2nd: It very often happens that the expressions on the basis of which the elementary terms of the system are to be interpreted cannot be meaningfully applied in contexts which allow for the application of the elementary terms themselves. Thus it is under certain circumstances quite meaningful to say that the ashtray in front of me consists of atoms and that it is in the eigenstate 01 of a certain observable, say O, the circumstances being, that "ashtray" is an elementary term of a theoretical system, which contains more abstract terms, such as "eigenvalue", "atom" etc. But an ashtray which is an element of the (everyday-) thing world, on the basis of which Carnap wants to interprete ⟨sic⟩, cannot be meaningfully said to be or not to be in an eigenstate of an observable O. It is nevertheless assumed by Carnap that the semantical rules provide a complete interpretation of the elementary terms of a theoretical system what would suggest that everything that can be said of the objects which are designated by those terms can also be said of the objects which are designated by the corresponding terms of the thing-language. [5|6]

3) It has been shown

a) that a theory is to be meaningful independently of any specific correlation which has been established between its elementary terms and elements of observation.

b) that the meaning of the statements of the theory is not necessarily changed by a change of that correlation

c) that the elementary expressions may be used in contexts in which expressions of the thing-language – in terms of which the elementary expressions are to be interpreted – cannot be used at all. Nevertheless we say that the truth-values of the elementary statements i.e. of the statements which do not contain any descriptive expressions other than elementary terms – is determined "by observation". That would indicate, that there is no specific observational language apart from the language of a theoretical system the terms of which are meaningful and can be used to ascribe meaning to any theoretical system. For as we have seen the meaning of a term – whether elementary, i.e. relatable to experience, or not – is not constituted by its relation to a specific content of experience, but by the role which it plays within the theory of which it is part. It would even imply nonsense (item c!) to assume that its meaning is constituted in the way mentioned above. Saying, that the truth value of a sentence is determined "by observation" does therefore not imply anything what regards the meaning of the terms of the sentence, it indicates only a specific relation between the sentence and a certain class of observers. "Observability" is a pragmatic concept (in Morris' sense). Adding, that such and such experiences pertain to the sentence does not constitute its meaning, but tells the observer under which circumstances he is supposed to utter the sentence. Whether the sentence is true or not depends on the truth of the consequences which may be derived from it by using the theory of which it is part as well as other theories. It is, therefore, somewhat [6|7] misleading if one says that its truth value is determined by observation. By observation the sentence is created and thrown into discussion. We usually assume that it is true, but that is only a methodological device by means of which we want to find out whether or not our observations were correct. In this process the theory to which the sentence belong plays an important part. And it would be clearly impossible to proceed in the way described if the Carnapian view about the way in which we ascribe meaning to a theory were correct.

Within philosophy it is quite common to assume that elements, such as perceptions, sensations, simple impressions fulfill ⟨sic⟩ a semantical function when related to the terms of an abstract system. That prejudice derives from the idea that our theories are only instruments, which allow us to predict observational results (or classical states, as is assumed by the Copenhagen-School of quantum-mechanics) e.g. the occurrence of certain impressions. We must admit that Carnap is not prejudiced in this way – he admits, what Berkeley would have denied, that the abstract terms of our theories are meaningful. But he still assumes that the meaning of these expressions is, at least partly, determined by the meaning of the expressions of an observational language, which is the common basis for any interpretation of any theoretical system whatever. We have shown some of the untenable consequences of this view.

4) We have still to take into consideration the remaining modes of setting up a physical theory as an interpreted calculus. We shall only discuss mode 1b. In this case the use of so called reduction-sentences turns out to be necessary.[+)] It is by means of such sentences that abstract terms [7|8] are introduced into theoretical systems which are built up "from the bottom". A reduction-sentences is a sentence of the form

Red $E_1(x) \supset .E_2(x) \supset A(x)$

meaning that, if the elementary quality E_1 is present at the space-time point x, then the abstract quality A is present at x if the elementary quality E_2 is present at x. Example: If white light is directed against a material object (E_1) then this object is green (A) if it looks green (E_2) to the observer. In this way the meaning of "green" as applied to physical objects is supposed to be determined.

Two objections are evident:

a) It is quite possible that because of disturbing influences, such as e.g. partial absorption in front of the green object, shadows, dispersion, unknown effects a green object which has been put into white light does not look green to any observer. In order to exclude such a case we have to include the disturbing circumstances, viz. S_1, S_2, ...S_n into the reduction-sentence, i.e. we have to use a more complicated sentence, viz.

Red_1 $E_1(x) \& \sim S_1(x) \& \sim S_2(x) \ldots\ldots \sim S_n(x) \supset .E_2(x) \supset A(x)$

Instead of Red.

Obviously argument b, sec 2 can be applied to this case as well:
We never know all disturbing circumstances, we discover them step by step as we proceed with our investigations and we have to change our reduction-sentences accordingly. It flows, that "A" must be meaningful independently of any <u>specific</u> set of reduction-sentences. Thus we cannot say that the reduction-sentences are means of ascribing meaning to the abstract terms of the system to which they belong.

Again some philosophers try to escape those consequences by reinterpreting the reduction-sentences as probability-implications. Objection c, section 2 applies also to this attempt. [8|9]

b) In our example it is of importance that aa) physical radiation is directed against the object to be investigated (it is not sufficient that the observer gets the impression that the object has been put into white light), bb) that the object does not only <u>appear</u> green to a reliable observer, but that it actually reflects radiation of a certain wavelength. Only under those circumstances we may – other things being equal – conclude that the object <u>is</u> green (in a physical sense): The reduction-sentence is useful if and only if E_1, E_2 and A are of the same degree of abstractness. But according to the idea we are discussing at present reduction-sentences are supposed to introduce abstract expressions on the basis of elementary expressions and elementary expressions are interpreted in terms of wither a perceptional ⟨sic⟩ language or of the thing-language the expressions of which is ⟨sic⟩ in any case less abstract than the abstract terms of a theoretical system. Even if we admit (what many present-day methodologists will do) that we are only allowed to draw inductive conclusions from the occurrence of certain sense data we are not justified to assume that sense data can be used for constituting the more abstract entities of science – for, as we indicated, we can always imagine circumstances which make it highly improbable that the occurrence of a sense datum S is accompagnied ⟨sic⟩ by the occurrence of a more abstract fact, which is usually observed via S.

5) To sum up: It is impossible to define the abstract terms of a scientific system in terms of observational language or to ascribe meaning to a scientific system by correlating it in one way or another with expression⟨s⟩ of a language which is more primitive than its own language. Even the meaning of the elementary terms of a system is determined by the system itself, experience being only a means which enables us to utter some of the sentences which contain elementary terms.

Paul Feyerabend
Vienna

---

[1] +) Cp. Foundations of Logic and Mathematics; International Encyclopedia of Unified Science I/3, p. 61 ff.[4]

[1] ++) Elementary terms are terms "which we can apply to concrete cases on the basis of observations in a more direct way, than others" (loc. cit.)

[1] +++) See later, sec. 4

[4] +) E.g. Arthur Pap[5]

[7] +) In Testability and Meaning, Philosophy of Science Vol II, IV[6] as well as in Logical Foundations of the Unity of Science, International Encycl. of Unified Science, I/1[7] Carnap has given some reasons, why we have to use reduction-sentences, when proceeding according to method 1b.

Editors' Notes

HF 08-28-03 (9 ss., T+H)

[1]See Feyerabend 1955.

[2]For Feyerabend's notation, cf. Pap, A. 1955:42-43.

[3] For Feyerabend's notation, cf. Pap, A. 1955:35-36.

[4]See Carnap 1939.

[5]See Pap, A. 1953; on Feyerabend and Pap, see Vol. 1, Part II, 1953-5, n1.

[6]Carnap 1936; Carnap 1937.

[7]Carnap 1938.

2. Paul K. Feyerabend and Arthur Pap, "Carnap's Theory of Interpretation of Theoretical Systems – Summary with Critical Comments", 12 July 1955

AP:[1] 12July55

<u>Paul Feyerabend: "Carnap's theory of Interpretation of Theoretical Systems"</u>
<u>(Theoria, 1955 Part I)</u>
<u>Summary with Critical Comments</u>

Carnap distinguishes two methods of <u>constructing a calculus</u>: Either a) the elementary expressions are taken as primitive, or b) the abstract terms are taken as primitive.

Introduction of abstract terms requires reduction-sentences in case a), while we can manage with explicit definitions, according to Carnap, in case b).

There are likewise two possible methods of <u>interpretation of the calculus</u>: 1) the abstract terms get interpreted, 2) the elementary terms get interpreted. Of the resulting four possible methods of constructing a physical theory, 2b (method of partial interpretation) seems to be the most natural for the physicist. It will now be examined.

Let us take as an example of a theoretical system celestial mechanics. The abstract terms are functors like "mass," "force," distance," "acceleration," etc. the individual variables range over particles. The elementary terms are likewise functors, like "geocentric coordinate," "rectascension," "declination." Assuming that the explicit definition of the latter by means of the former is unproblematic, let us look at the interpretation of the elementary terms. A simple procedure seems to be the following: to the geocentric coordinate of each planet that is observable from the earth we coordinate ("coordinating definition"!) the luminous point on the sky whose motion the astronomer observes in order to derive the motion of the planet. With respect to a particular planet, say Jupiter, we would thus lay down the following semantic rule:

R "geoc.coo(Jup)" designates(a,d)(J).

J is that luminous point which is commonly called "Jupiter," and (a, d) is a functor whose values are the measure numbers of rectascension (a) and declination (d). [1|2]

Objections to this coordination:

a) systematic errors -- e.g. atmospheric refraction--necessitates corrections before deriving geocentric coordinates from the locations of the luminous points. The semantic rules actually employed in interpreting the system are more complicated than R.

b) We never know all the systematic errors infecting our observations, nor do we know which disturbances falsify the observational results. For example: conceivably the light rays reaching the surface of the earth are deflected a little by unknown force-fields, and hence the geocentric positions that were calculated on the assumption that no such deflecting forces are present may not correspond to the "true" positions. Such disturbances will manifest themselves in small irregularities in the planetary motions for which the laws of celestial mechanics offer no explanation. If, now, we discover through experiments in a different domain of physics that fields of a certain sort influence the propagation of light, then we <u>have</u> the explanation* which we need and we can use this knowledge for the correction of our calculations. This means, however, that the discovery of new systematic errors induces us to change the coordination of the elementary terms of the theory with elements of experience. This has important consequences.

The following two principles seem beyond dispute:

1) If it is possible to compare all the consequences of a theory with experience, then the sentences of the theory are meaningful.
2) Suppose that we test a theory by making observations with the help of certain kinds of instruments. Then we replace these instruments by better ones. Surely the theory we now test is still the same. Briefly: the meaning of the sentences of a theory is not changed by improvements on the techniques of observation that are used in testing the theory.

Application of principle 2): such sentences of a theoretical system as the sentence describing geocentric coordinates of planets do not change their meanings [2|3] when the scientist corrects his predictions in terms of newly discovered systematic errors. But such correction entail a change of the rules of coordination of elementary terms of the theory and observational data. For Carnap these rules are semantic rules which first endow the elementary terms with meaning. His theory of interpretation of theoretical systems therefore entails that the meaning of the theory changes whenever these rules change owing to new discoveries of "systematic errors."

Application of principle 1): In order for a theoretical system to be meaningful, no definite correlation between elementary terms and observational data is required. We often make corrections just because we believe in the truth of the theory, and often we make these corrections before we are able to explain the postulated disturbances.* The coordination between the elementary terms of the theory and "experience" is thus much looser than is suggested by Carnap's scheme of interpretation.

This circumstance has led some philosophers (F. means Pap)[2] to replace the Carnapian semantic rules by probability-implications:

R' (x) (geoc coo(Jup) = x $\underset{p}{\Rightarrow}$ (a, d) (J) = x)

Two objections against these looser kind of semantic rules:

1) Under certain circumstances systematic errors may occur (F. means disturbances) which make a relation of the indicated sort highly improbable. If, e.g., we know that the earth is surrounded by an intense field of some kind which has a disturbing influence on light rays reaching the surface of the earth, then we can say that the probability is negligibly small that the geocentric locus and the locus of the [3|4] luminous point are within a specific distance from each other--and this would lead again to a modification of the semantic rules specifying such a "error interval." And we may make the error interval as large as we please, still we can always conceive of circumstances which might induce us to replace the stipulated probabilistic coordination by another one. And this shows that the transition to probabilities does not improve the situation at all.*

Second objection: Frequently the expressions by means of which the elementary theoretical terms are interpreted cannot be significantly used in all those contexts in which the latter are significantly used. E.g. we can transform from geocentric coordinates to heliocentric coordinates. But it does not make sense to speak of the heliocentric coordinates of a luminous point on the sky.** [4|5]

An even more striking example is the following: an ashtray, according to physics, is an aggregate of atoms. A particular ashtray in front of me may be in a definite eigenstate of a certain observable. But we cannot meaningfully say of an ashtray as an observable thing that it is in an eigenstate of a quantum-mechanical observable. And yet Carnap assumes that the semantic rules give a complete interpretation of the elementary terms of a calculus on the basis of observational terms.*

Discussion of method 2a)

If the elementary (non-abstract) terms are taken as primitive, the abstract terms must be introduced by reduction-sentences, according to Carnap:
$E_1x \supset (E_2x \supset Ax)$, where "$E_1$" and "$E_2$" are elementary terms, and "A" an abstract term. Example: if a body is exposed to white light, then it is green of it appears green.

Comments: a) it is conceivable that a green body which reflects light does not appear green because of disturbing circumstances (absorption in front of the body's surface, visual abnormality of the observer, etc.). To take account of such possibilities, the probabilistic formulation becomes necessary again, but for the reasons already given no such probabilistic reduction-sentence can be final. This shows that "A" must have a definite meaning antecedently to its introduction by a set of reduction-sentences, i.e. that we cannot regard reduction-sentences as semantic rules which first assign meaning to the abstract terms.** [5|6]

b) With regard to the above example, it is important that 1) physical light is used for the illumination of the body whose color is to be ascertained, 2) the body not only appears green but actually reflects green light. It is only on this condition that we may--having taken account of known disturbing factors--infer from the fulfillment of the conditions that the body itself is green. In other words, the reduction sentence can--waiving the difficulty of the "ceteris paribus" provision--be used only if E_1, E_2 and A have the same degree of abstractness. But this is in contradiction to the intended function of a reduction-sentence (reduction of abstract terms to observational terms). Relatively abstract states of affairs are of course inductively inferred from observational data, and reduction-sentences are significant as directives to the observer advising him to focus attention on such and such observables in order to get inductive evidence for the theoretical statements. But they do not reduce the meaning of an abstract term to the meanings of elementary terms.* [6|7]
Conclusion: It is impossible to define the abstract terms of a scientific system by means of observational terms or to give meaning to a scientific system of sentences by connecting it in some way with a language which is more primitive than its own. Even the meanings of elementary terms of a system are determined by the system itself and so-called experience is only a means producing under certain conditions utterances of sentences that contain elementary terms.*

[2] * There is something missing in F's argument here. The astronomer would also have to have independent evidence for the presence of such fields. A disturbance cannot be explained in terms of laws alone.

[3] * A good historical example of what F. means is the discovery of Neptune, which was due to faith in the law of gravitation in spite of its apparent refutation. The assumption of closure of the dynamic system was held responsible for the discrepancy between prediction and observation of a planetary orbit.
[4] * I think F.'s argument here rests on a serious misunderstanding of probabilistic meaning-specification. In fact he overlooks that two probability statements of the form "P(E/C) = x" and P(E/C.C') = y)," where x and y are different and C and C' logically independent, are perfectly compatible. We need not actually "revise" the first probability statement, we only need to write down a second probabilistic meaning rule for the theoretical term contained in "E" which takes account of the disturbing factor, and therefore by the principle of total relevant evidence is a better rule for assigning probabilities in case the presence of the disturbing factor is known. But we may still use the original rule in those cases where there is no reason to assume the presence of C'. F. may have been thinking of statistical probability-implications. IF "P(A/B) = x" is a statistical generalization inductively derived from a non-representative sample of the reference-class B, it may indeed have to be abandoned and replaced by a generalization based on a more representative sample. If C is frequently conjoined with the disturbing condition C', the long-run frequency of E relatively to C will be significantly different from the frequency of E relatively to a sample of instances of C in which C' is never or almost never present. But probabilistic semantic rules are not, of course, to be interpreted as frequency statements. (Incidentally, I would prefer the converse of E', since we want to assign a weight to the theoretical statement on the basis of the observation statement, not the other way around.)
[4] ** Surely F. must be confused here. If there is a semantic rule interpreting "geocentric coo" in the data-language, why not also have a semantic rule for "heliocentric coo"? Why should we be committed to mix theoretical terms with observational terms in this objectionable way?
[5] * Mutatis mutandis, I would say the same about this example. Besides, it appears that F. is probably confusing semantic rules in the form of explicit definition with semantic rules on the form of definitions in use. If, e.g., we consider the syntactic counterpart of R: (geoc.coo(Jup) = x,y,z) $=_{df}$ (a,d)(J) = u,v, we see rightaway that an expression like "geoc coo (J) = u,v" is not a meaningful sentence. It is like ascribing the thing-predicate "blue" to a sense-datum, where the thing-predicate is contextually defined in terms of a corresponding sense-datum predicate. It is only if the elementary terms of the theory were explicitly defined, in the sense of simple substitutability, by observational terms, that the method of partial interpretation would have the absurd consequence which F. claims it has.
[5] ** I would of course agree that insofar as the terms introduced in the theoretical system had a prior use in the natural language, this is the case; one might therefore prefer the term "explication (partial) of meaning" to the term "assignment of meaning." But the same would be true as regards explicit definition of an antecedently used term; hence this is at any rate no special argument against reduction-sentences.
[6] * As far as I can see the discussion of the example by which this general conclusion is supported, involves simply a confusion between two kinds of dispositions that can be meant by "green" as a physical predicate. It might mean the disposition

to reflect green light, i.e. light of a specific wave-length (which of course is called "green" because a body which reflects such light also has the disposition to appear green under specified "standard" conditions), or the disposition to appear green under specified "standard" conditions of perception (the latter disposition is a secondary quality in Locke's sense). Certainly condition 2) is not required if what is to be reduced is "green" in the latter sense. Point 1) is perhaps more serious. F. in effect argues that the test-operation must be described by predicates of the same degree of abstractness as the terms to be reduced. But is it not very improbable that a body should appear green unless it is exposed to physical light? If so, isn't the probability that x is green on the evidence that appears green nearly as large as the probability that x is green on the evidence that x appears green <u>and</u> is exposed to physical light? Why then do we have to include the physical condition if the reduction-sentence is probabilistic anyway? At any rate this consideration seems to me to prove that F. does not have a new argument; it is just the argument from the probabilistic character of reduction sentences all over again.

Nevertheless I agree with F. that once we probabilize the reduction sentence, and do so not just vaguely by putting in the protecting "ceteris paribus" clause but actually specify "disturbing" conditions, then a mixture of language-levels, or of terms of different degrees of "abstractness," will occur in "E_1" and "E_2," e.g. if the reduction term belongs to the thing-language, other thing-language terms will occur in the reduction basis, or if it belongs to theoretical physics (e.g. "electronic field") other terms of theoretical physics will appear in the reduction basis: [6|7] think e.g. of the proviso that "no other forces" are present! To the extent that this feature of "impurity" of the reduction basis is incompatible with the very notion of "reduction" of terms as originally conceived, one might indeed find it confusing to speak of "reduction sentences" any longer. This may well be a major reason for which Carnap now prefers the method of "partial interpretation."

⟨7⟩ * But just how is F. going to distinguish a pure calculus from an interpreted calculus?

Editors' Notes

HF 11-27-12 (7 ss., T)

[1] "AP" stands for Arthur Pap (see Vol. 1, Part II, 1953-5, n1). Pap's annotated summary of Feyerabend 1955 was composed a few months after the publication of Feyerabend's paper, when Pap was an invited visiting scholar to a six-week conference which took place at the Minnesota Center for Philosophy of Science between 27 June and 6 August 1955 (see *below*, Feyerabend to Feigl, 19 May 1957, Ch. 3, § 1; 28 June 1957, Ch. 2, <u>§ 14</u>; and Pap to Feigl, 13 January 1955, HF 03-134-29; 29 April 1955, HF 03-134-34; Feigl to Pap, 5 May 1955, HF 03-134-35; Pap to Feigl, 7 June 1955, HF 03-134-37; and Keupink 2006: 368; Pap, P. 2006: 374).

[2] See *above*, § 1, <u>n3</u>.

3. Michael Scriven, "Comments on the Feyerabend Paper", 29 August 1955

MS:[1]29August55

Comments on the Feyerabend Paper

I am so far from understanding the specific points or general message of this paper that the following comments should not be regarded as criticisms of someone who sees but disagrees with someone's else point of view.

p. 1.6[2] I simply fail to understand what on earth could motivate someone to coordinate the geocentric coordinate of the planet with a point on the sky treated as a surface. This would be surely a flagrant violation of the most primitive intention of any system of celestial mechanics. What we do surely is coordinate the term "Jupiter" with a particular moving point in the sky not regarding the sky as a surface once we are beyond the Babylonian state.

p. 2.66.[3] Whereas the converse of this principle might be true, and perhaps tautological, I am most unclear about it as formulated; how would one know what a consequence of the theory would be, what would it be to compare a consequence of a theory to experience, etc.

p. 2.86 (AP's footnote).[4] I think a disturbance could be explained in terms of laws alone provided that it was universally associated with a certain type of condition; and I am sure that this is what PF has in mind, and it meets the substance of AP's objection which I take to be the necessity for an "applicability premise."

p. 3.45.[5] One should, I think, add to this sentence the words "or else PF's example of a "theory" is an unsound one."

p. 4.25,[6] and AP's footnote.[7] It seems to me that the mistake PF makes is not the one which AP accuses him of but merely that of trying to correlate the wrong terms. This comes out particularly in AP's second footnote on this page.

p. 7.[8] If, as PF concludes, the meaning of all the terms of a system are determined by the system itself, then how could quantum mechanics be made intelligible?

Editors' Notes

HF 12-38-32 (4 ss., T; excerpt)

[1] "MS" stands for Michael Scriven (see Vol. 1, Part II, 1957-7, n3). This note was written while Scriven was a Research Associate at the Minnesota Center for Philosophy of Science (1953-56).

[2] See *above*, § 2, p. 1, "A simple procedure seems to be the following…." Scriven's references are to Pap's annotated summary and are given according to the scheme: [page number].[line, identified through a series of progressive natural numbers printed on the left hand side border of the pages to form a sort of ruler].

[3] See *above*, § 2, p. 2, "If it is possible to compare all the consequences of a theory with experience, then the sentences of the theory are meaningful."

[4] See *above*, § 2, p. 2, n*.

[5] ee *above*, § 2, p. 3, "The coordination between the elementary terms of the theory and "experience" is thus much looser than is suggested by Carnap's scheme of interpretation."

[6] ee *above*, § 2, p. 4, "And this shows that the transition to probabilities does not improve the situation at all."

[7] ee *above*, § 2, p. 4, n*.

[8] ee *above*, § 2, p. 7.

2. The Second Controversy (1956-1957)

4. Paul K. Feyerabend, "A Note on Carnap's New Criterion of Empirical significance", [1956]

⟨undated⟩[1E]

A Note on Carnap's New Criterion of Empirical Significance

P. K. Feyerabend

The following note contains a presentation and criticism of Carnap's new criterion of empirical significance.[1] It will be shown that this criterion is useless (in a sense to be explained later in this note), grossly inadequate, and that it may even involve a logical absurdity. In order to make my point as clear as possible I shall develop my criticism in two steps. I shall first discuss what may be called a simplified version of the criterion (sec. 1). This discussion will be the basis for my criticism of Carnap's own criterion (sec. 2).

(1) Carnap's criterion of empirical significance is[2E] an improved and considerable more sophisticated version of a criterion which has been suggested by A.J. Ayer in the first edition of his Language, Truth and Logic.[3E] According to Ayer a sentence A is empirically significant if and only if there are hypotheses (i.e. sentences of the form $p \supset q$) H_1, H_2, ...H_n as well as an observation-sentence O such that O follows from A& H_1&H_2, ...&H_n but does not follow from H_1&H_2&H_3 ....&H_n alone. It has been noticed very early that on the basis of such a criterion any sentence can be proved to be empirically significant. For let be A some sentence and O an observation-sentence. Then O follows from A&(A⊃O), but not from A⊃O (the sentence which plays here the role of the required hypothesis) alone, which would imply that A is empirically significant. But as we did not presuppose anything about A it follows that according to the criterion we are using any sentence is empirically significant.

A first attempt to improve the situation would consist in stipulating (principle M) that a sentence of the form p⊃q with either p empirically significant and q devoid of empirical meaning or q empirically significant and p devoid of empirical meaning is logically true if and only if either p is logically false or q is logically true.[2] This principle makes the criterion watertight in the sense that the assumption that A is devoid of empirical meaning would contradict the assumption that A satisfies all conditions of the criterion. In order to show this, we make those conditions explicit by stating that A is empirically significant if and only if there is an observation-sentence O as well a hypothesis H such that [1|2]

(a) A&H is consistent (i.e. not logically false)
(b) A&H.⊃O is logically true
(c) H⊃O is not logically true.

Now assume that A is devoid of empirical meaning. As A&H.⊃O is logically equivalent to A⊃.H⊃O, (b) and principle M imply that either A is logically false or H⊃O is logically true. The first consequence contradicts (a), the second consequence contradicts (c). It follows that a meaningless sentence A does not satisfy the emended criterion.

In spite of this improvement the emended criterion still suffers from two defects. The first defect is a logical one: In our above argument[4E] we have assumed that $H \supset O$ is empirically meaningful, although H, i.e. $A \supset O$ contains a term which is by suppositions devoid of empirical meaning. This defect can be removed by introducing a modification which will be discussed in the next section. However this modification does not remove another defect of our criterion which is sufficient to render it useless: We have shown that, in the emended version, the assumption that A satisfies the criterion contradicts the assumption that A is devoid of empirical meaning and we have done this by showing that, <u>if</u> A is empirically meaningless, then either (a), or (b) or (c) will be violated. But in every practical case we start with a sentence A of doubtful character and we want to find out whether or not it is empirically significant. This means that we have to find out whether or not it is empirically significant. This means that we have to find out whether or not (a), (b) or (c) hold <u>without</u> assuming anything about the meaning-properties of A. But this is not possible. For if we want to exclude a sentence which is devoid of empirical meaning we must use a principle which forbids us to say that $A\&(A \supset O).\supset O$ is logically true for all A in spite of the fact that the logical form of the sentence seems to suggest just this. We have introduced such a principle (principle M) when presenting the emended version of the criterion. In applying this principle we must already know whether or not A is empirically significant. Thus is follows that either (first version) the criterion makes every sentence empirically significant which would show that it is grossly inadequate, or (emended version) that the criterion does not enable us to decide whether or not a given sentence is empirically significant, which shows that it is useless. It will turn out that Carnap's new criterion is inadequate <u>as well as</u> useless because it combines, as it were, the disadvantages of what I have called the first version with the disadvantages of what I have called the emended version.

(2) Whereas Ayer formulated his criterion quite generally, Carnap explicitly restricts it to certain language-systems which contain an observational-language L_O [2|3] and a theoretical language L_T. Whether or not a sentence is empirically meaningful depends on the rules of those languages as well as on some correspondence-rules which connect them in a suitable way. All statements about significance are relative to L_O, L_T, the correspondence-rules C as well as some theoretical assumptions formulated within L_T (physical laws, laws of psychology, sociological theories and the like). The second peculiarity of Carnap's criterion is that it is formulated in three steps, first as a criterion according to which a certain term is (or is not) significant relative to a class of other terms, secondly as a criterion of significance for a given term which does not any longer refer to other terms, and thirdly as a criterion of significance for sentences. In this note we shall deal with the first two steps of the criterion only.

Step one (D_1 in Carnap's paper): A term R is empirically significant relative to a class of terms K with respect to the theoretical language L_T, the observation-language L_O, the correspondence-rules C and the theoretical assumptions T if and only if all

K and R belong to the class of descriptive terms of L_T and there are sentences S_R, S_K in L_T and S_O in L_O such that
(a) S_R contains R and R only
(b) S_K contains all K and only those terms
(c) $S_R\&S_K\&T\&C$ is not logically false
(d) S_O follows logically from $S_R\&S_K\&T\&C$ but
(e) S_O does not follow logically from $S_K\&T\&C$ alone.[5E]

Step two (D_2 in Carnap's paper): A term R is empirically significant with respect to L_T, L_O, T, C if and only if there is a sequence of terms $M_1, M_2, \ldots\ldots M_n$ of L_T such that M_i is significant relative to the class of all M_j ($j<i$) with respect to L_T, L_O, T and C and $R=M_n$.

Before starting to criticize[6E] the criterion we shall show how Carnap tries to avoid the difficulty which we called the first defect of the emended version of Ayer's criterion. Assume for that purpose that R is devoid of empirical meaning and that it is possible to split T into two parts, T_1 and T_2, T_1 containing R and possibly other meaningless terms of L_T, T_2 containing meaningful terms only. Utilizing principle M (section 1) we find that this assumption contradicts at least one the conditions a-e, step one. For according to M the sentence $S_R\&S_K\&T\&C\supset S_O$, which is logically equivalent to $S_R\&T_1\supset.S_K\&T_2\&C\supset S_O$ is logically true if and only if either S_RT_1 is logically false – which contradicts (c) – or $S_KT_2C\supset S_O$ is logically [3|4] true, which contradicts (e). Thus either (c) or (d) or (e) (i.e. the conditions corresponding to (a), (b) and (c), sec. 1) is violated.

Now this demonstration hinges upon the assumption S, that any sentence which contains meaningful as well as meaningless terms is logically equivalent to the product of two sentences, the one containing meaningful terms, the other meaningless terms only. In proving this assumption Carnap makes use of the following principle (principle Q):[7E] A sentence which cannot be split up into two sentences, the one containing only meaningful, the other only meaningless terms and which[8E] contains some meaningful terms, contains meaningful terms only. It is this principle which leads to our first criticism.

For it is quite obvious that this principle is a return to what I have called Ayer's first version as it implies that a sentence of the form $A\supset O$ (O being some observational-sentence) contains meaningful terms only from whence it follows that all sentence are empirically significant (cf. our remark in the first paragraph of sec. 1). More especially principle Q would make empirically significant[9E] such terms as "god", or "god's volitions", as those terms can appear in sentences such as "if it is god's will, then, if I lift my hand, the sun will be extinguished".[3] And as rational theology, many systems of philosophy of nature and finally all metaphysical systems which do not rest content with developing theories about the relation between purely metaphysical entities[10E] but which also relate those entities to experience in one way or another do partly consist of sentences which contain significant[11E] terms and which cannot be split up in the way suggested by assumption S, we have to conclude that according to Carnap most assertions [4|5] about metaphysical entities are empirically significant.[12E] But this shows that Carnap's criterion is grossly inadequate <u>by his own</u> standards[13E].[4] And this inadequacy is due to the adoption of principle

Q. Without this principle Carnap's criterion cannot avoid the objection that it assumed as meaningful sentences which are neither logically true, nor logically false and which contain meaningless terms, i.e. without this principle the criterion suffers from what I have called the first defect of the emended version of Ayer's criterion. But with the principle the criterion becomes inadequate in exactly the same way in which the first version of Ayer's criterion turned out to be inadequate, as the principle assumes that any sentence of the form $p \supset q$ with either p or q meaningful contains meaningful terms only. This is our first objection against the criterion.

In our second objection we shall presuppose that S can be assumed to be true without implying Q. In this case the criterion is adequate in the sense that every term which is meaningless will violate at least one of the conditions of either D_1 or D_2. But it is useless as a means for deciding whether or not a given term is meaningful and this for precisely the same reason which we have given towards the end of the last section, i.e. it is useless, as in the attempt to decide whether or[14E] not a given term satisfies (a)–(e), D_1, we have already to know whether[15E] or not this term is empirically significant (relative to some K, with respect to L_O, etc.)

The conclusion is that Carnap's criterion is useless and grossly inadequate and that it may even involve logical absurdity. In spite of the considerably more complicated logical machinery Carnap has not succeeded in overcoming the defects of what we have called the first[16E] version (Ayer's original criterion), nor could he remove the defects of the emended version, he has only succeeded in providing a criterion which exhibits all those defects at once. We are still far from being able to exclude such apparently undesirable things as metaphysics on the basis of some criterion by a strict "proof". And considering the failure of Carnap's earlier criteria as well[5] we should perhaps give up the hope that such a criterion will ever be found.

P.K. Feyerabend
Bristol

---

[1] 1) 'The Methodological Character of Theoretical Concepts'; Minnesota Studies in the Philosophy of Science Vol. I (1956), 38-76.[17E]

[1] 2) This principle is introduced by Carnap on page 54. It is interesting to note that it implies that a sentence containing empirically meaningless terms may still be logically true (or logically false).

[4] 3) A more detailed account of the matter would be this: Take the last sentences mentioned in the text, or, in symbols, take the sentence

$$(t)(G(t) \supset .(t) \supset \mathrm{Sex}(t)) \ \langle \mathrm{sic} \rangle \qquad (1)$$

Substituting S_R for $G(t_1)$, T for (1), S_K for $L(t_1)\&C_O$ (C_O expressing[18E] conditions such that $\mathrm{Sex}(t1)\&C \supset S_O$ is logically true by virtue of correspondence-rules) we find that $S_R\&S_K\&T\&C \supset S_O$ is logically true, whereas $S_K\&T\&C \supset S_O$ is not logically true and $S_R\&S_K\&T\&C$ is consistent, i.e. we find that according to D_1 $G(t)$ is empirically meaningful relatively to K (with[19E] respect to L_O, etc.). Assuming that all the terms of K can be shown to be meaningful (with respect to…etc.) without involving T (an assumption which will be true for suitably chosen correspondence-rules) we have to conclude that G(t) ("at the time t god has a volition") is empirically significant. This conclusion could only be avoided by assuming that T is meaningless – but how else could we show this than by assuming that G(t) is meaningless from the very beginning? And apart from that the empirical significance of T is a direct consequence of principle Q, mentioned in the text. Cf. also the next footnote.

[5] 4) This objection could only be met by demanding that a sentence which contains terms of L_O as well as highly abstract terms of some L_T (whether metaphysical or not) must establish a certain kind of connexion between those terms in order to be meaningful. In his paper Carnap is very cautious not to commit himself to a specific kind of connexion, such as e.g. explicit definability, reduction-sentences and the like (probabilistic connexions are admitted!) as the discussion of his earlier criteria has invariably shown that too strict a demand of this kind will exclude scientific laws from the domain of the meaningful. But as the new criterion shows, tolerance does not help either. For now even assertions about god turn out to be empirically meaningful.

[5] 5) Cf. in this connexion the excellent criticism of Carnap's earlier attempts in Popper's paper 'The Demarcation between Science and Metaphysics', to appear in the Carnap-Volume of the Library of Living Philosophers.[20E] I am greatly indebted to this paper.

Editors' Notes

RC 088-26-17 (5 ss., T+H); KP 294.19 (5 ss., T+H); HF 08-32-06 (5 ss., T; c); HF 08-33-22 (5 ss., T; c). RC 088-26-17 and KP 284.19 are two copies of the same original typescript, but they partly differ in the corrections by Feyerabend's hand, some of which are present in the former but lack in the latter and vice versa. HF 08-32-06 and HF 08-33-22 are two identical copies of RC 088-26-17, where Feyerabend's handwritten corrections in the original were taken into account and obvious mistypes were corrected (the pagination of copies is slightly different from that of the original). At the top of RC 088-26-17: 1, we read, written by Feigl's hand: "P.K. Feyerabend (Bristol)" and "R.C. You may keep this (we have copies). H.F.". It is an easy guess to conclude that RC 088-26-17 is the original paper that Feyerabend sent to Feigl and that the latter then forwarded to Carnap, who annotated it in his characteristic shorthand writing. Most likely Hempel received a copy of the paper from Feigl. The transcription here is based on the text and pagination of HF 08-32-06 = HF 08-33-22 (Feyerabend's obvious mistypes in RC 088-26-17 and KP 294.19, and Feyerabend's handwritten corrections that can be found only in RC 088-26-17 or in KP 294.19 are reported in the editorial footnotes).

[1E] In a letter to Popper, Feyerabend wrote with reference to this writing: "I am also enclosing a brief note on Carnap's new paper on meaning which has now finally been published. I went through the paper at Christmas time when I was a bit ill and could not do more serious work + I am surprised about the fact that it should possess so obvious defects." (Feyerabend to Popper, 14 January 1957, see Vol. 1, Part II, 1957-1). Within a week from this letter to Popper, Feyerabend amended his note with two additional footnotes, and submitted it to the scholarly journal *Mind* for publication: since I sent the paper to Karl I made some further changes, adding a footnote about "meaning" (i.e. about the fact that Carnap's intention is still to give a criterion of meaning, although he says he wants to give a criterion of empirical significance) without discussing the topic in full, another footnote on the L_OL_T-reconstruction (only stating dogmatically that it is inadequate but not discussing the point) and in this form it will be published in Mind. (Feyerabend to Agassi, 21 January 1957, where Feyerabend also provides further details about the context within which his note was conceived; see Vol. 3, Ch. 1, 1957-2; *below* § 8; § 14)

[2E] In KP 294.19 Feyerabend corrected by handwriting the typed "is" in "may be described as". This correction is missing both in RC 088-26-17 and in HF 08-32-06 = HF 08-33-22.

[3E] See AYER 1936.

[4E] In RC 088-25-17 and KP 294.19 we read a typed "arument".

[5E] All the occurrences of "R" and "S_R" in this as well as the following paragraph and in note 3 are Feyerabend's handwritten corrections in RC 088-25-17 for typed "M" and "S_M" respectively in the original typescript. These corrections are missing in KP 294.19.

[6E] In RC 088-25-17 and KP 294.19 we read a typed "critizise".

[7E] In KP 294.19 Feyerabend added here in handwriting a superscript note index "⊗" referring to the following note written by his hand on the verso of the same page: «⊗ Carnap nowhere states this principle explicitly. But there is the following passage in his paper (p 58): "If a sentence connecting a meaningful term with another term in an inseparable way .... is a postulate or provable on the basis of postulates, then it is stated as holding with physical necessity{;} therefore it conveys some empirical meaning on the second term." But as a metaphysician will certainly choose metaphysical postulates as his starting-point, the restriction to the consequences of the postulates of some L_T is unnecessary.» This note is missing both in RC 088-26-17 and in HF 08-32-06 = HF 08-33-22. See *above*, n1E; *below*, § 8).

[8E] In RC 088-25-17 and KP 294.19 we read a typed "ahich".

[9E] Feyerabend's handwritten correction in RC 088-25-17 for a typed "meaningful". This correction is missing in KP 294.19,

[10E] Feyerabend's handwritten correction in RC 088-25-17 for a typed "terms". This correction is missing in KP 294.19,

[11E] Feyerabend's handwritten correction in RC 088-25-17 for a typed "meaningful". This correction is missing in KP 294.19,

[12E] Feyerabend's handwritten correction in RC 088-25-17 for a typed "meaningful". This correction is missing in KP 294.19,

[13E] In RC 088-25-17 and KP 294.19 we read a typed "standarts".

[14E] In RC 088-25-17 and KP 294.19 we read a typed "ot".

[15E] In RC 088-25-17 and KP 294.19 we read a typed "ehether".

[16E] In RC 088-25-17 and KP 294.19 we read a typed "forst".

[17E] See CARNAP 1956.

[18E] In RC 088-25-17 and KP 294.19 we read a typed "ecpressing".

[19E] In RC 088-25-17 and KP 294.19 we read a typed "eith".

[20E] POPPER 1963a; this paper, written between 1953 and 1955 for the Carnap volume of the *Library of Living Philosophers* (see SCHILPP (ed.) 1963), was distributed in a stenciled version since June 1956 (see Vol. 1, Part II, 1959-5, n6; and POPPER 1963a: 183n*; POPPER 1963b: 341n).

5. Carl G. Hempel, Reply to Feyerabend's Note, 22 February 1957

C. G. Hempel

February 22, 1957

Feyerabend's criticism of Carnap's latest criterion of significance rests on a serious misstatement of Carnap's "presupposition (1)" (p.54, par 3) in Feyerabend's "principle M" (p.1, par. 3).

Carnap's presupposition states quite explicitly that if all extra-logical constants in S_1 belong to the significant theoretical vocabulary V_1 or to the observational vocabulary O, and if all extra-logical terms in S_2 belong to the non-significant vocabulary V_2, then neither of the two sentences implies the other unless the implying one is logically false or the implied one logically true.

And Carnap later (D3, p.60) defines a significant sentence as one all of whose extra-logical terms are significant; hence a sentence is non-significant if and only if it contains at least one non-significant term. Hence, the class of non-significant sentences is much more inclusive than that of the sentences containing non-signif. terms exclusively.

And Feyerabend's grave misreading, in his principle M, of Carnap's presupposition (1) consists in asserting of all non-significant sentences what Carnap explicitly stipulates just for the sentences containing meaningless terms exclusively. Principle M is of course totally unacceptable but was never proposed by Carnap, and it is only with the help of that principle that F. establishes what he so strongly calls the uselessness and gross inadequacy of Carnap's criterion.

But I must add that I share the hunch expressed in F's last sentence, that a satisfactory meaning criterion cannot be found.[36]

Editors' Notes

HF 09-38-02 (1 s., T); RC 088-26-16 (1 s., T+H). These documents are two versions of the same text differing in line-spacing only. Feigl had these typescripts composed for distribution out of a letter of Hempel (see Hempel to Feigl, 22 February 1957, HF 02-202-18) by omitting the salutation and the whole second part of the letter (which dealt with completely different matters) as well as by doctoring the last reported paragraph (see *below*, n1). RC 088-26-16 is annotated in Carnap's characteristic shorthand writing.

[1] In Hempel's letter to Feigl this paragraph reads: "I confess that I am glad at this dénouement because of F's strong and almost aggressive language and his – to my mind a bit excessive – tribute to Popper's essay for Schilpp. But I must add that I share the hunch expressed in F's last sentence, that a satisfactory meaning criterion cannot be found. In fact, Hilary some weeks ago showed me an argument the details of which now escape me entirely, but which seemed to me to reveal a very serious defect in Carnap's latest criterion; I myself had had some minor misgivings upon reading C's ms, some of which he acknowledged. The story about Hilary's criticism should be kept confidential, in my judgment; it will be up to Hilary to tell you or others about it." (Hempel to Feigl, 22 February 1957, HF 02-202-18). The "Hilary" referred to here by Hempel is Hilary Putnam (see Vol. 1, Part II, 1957-12, n3), who was then Hempel's colleague as Associate Professor at Princeton University (1953-1960). In composing the version of Hempel's reply for distribution Feigl doctored most of this paragraph out, keeping only one sentence (see *below*, §7, n1).

6. Paul K. Feyerabend, Rejoinder to Hempel's Reply, 3 March 1957

Paul Feyerabend
March 3, 1957

I had a look at Professor Hempel's objection and I answer it as follows:

Substitution of Carnap's "presupposition (1)" (p. 54 of Carnap's paper) for what I have called "principle M" invalidates only the argument in section (1) of my paper *where I do not criticize Carnap at all* but where I make some introductory remarks about Ayer's criterion of significance. He ⟨sic⟩ does *not* invalidate my criticism of Carnap's own criterion in section (2) of the paper and therefore Professor Hempel is completely wrong when saying that "it is only with the help of that principle" (viz. that I called principle M) "that F. establishes what he so strongly calls the uselessness and gross inadequacy of Carnap's criterion." Let me explain that in detail:

The *inadequacy* of Carnap's criterion I derive from what I have called "principle A", viz. from the principle which is clearly implied in Carnap's paper [cf. pp. 57-58 – it can be shown that the restriction there, to the consequence of the postulates, is unnecessary. For, after all, a metaphysician will choose metaphysical postulates!] *that a sentence which cannot be split into two sentences, the one containing only meaningful terms, the other only meaningless terms and which contains some meaningful terms, contains meaningful terms only, i.e. is meaningful*.* From this principle it follows at once that a sentence of the form $A \supset O$ [O being some observation-sentence, A being a sentence of the form "P(b)"] contains meaningful terms only as soon as it occurs within a certain theoretical language, *whatever A may be*. But A may be a typically metaphysical statement, the theoretical language in question the language of some metaphysical system which establishes some connection (as, e.g. natural theology does) between metaphysical entities and experience. Special examples are contained in my paper, this whole argument does not once refer to what I called "principle M."

The *uselessness* follows if we consider Carnap's "presupposition (1)." But as, in deciding whether or not a term is significant, it is assumed (see above) that sentences can be separated which contain meaningless terms *only*, it does not make any difference whether or not M or Carnap's (1) is used. In neither case Carnap's criterion can be used as a *criterion* for deciding whether or not a given [1|2] term is meaningful. For assume that R is the term at stake and that $T = T' \cdot T''$ (T' contains meaningless terms *only*). Then condition (d) may be written:

$$S_R \cdot T' \supset S_K \cdot T'' \cdot C \supset S_O \text{ is logically true}$$

Now *if* (Carnap (1)) $S_R \cdot T'$ contained meaningless terms *only*, i.e. if R is meaningless, this condition contradicts (C). But how do we know that (d) is fulfilled* if we do not know whether or not R is meaningful? The logical form of the statement involved does not tell it and the recourse to either Carnap's (1) or my M make the procedure circular and therewith useless.

Thus it seems to me that Professor Hempel, though pointing out some careless formulation of mine (for in my paper I attributed "M" to Carnap) has not shown that my arguments against Carnap are invalid, as they are valid whether one uses now my M or Carnap's own "presupposition (1)."

Of course, there might be still some point which I may have overlooked, but it seems unlikely. Especially the inadequacy of Carnap's criterion on the basis of the principle I formulated above as principle Q is a most surprising thing, but I do not think it can be denied that this inadequacy exists.[1]

---

[1] *I am surprised that Professor Hempel did not at all refer to this principle, which is completely independent of either Carnap's (1) or my M and the adoption of which by Carnap leads at once into the inadequacy of his criterion.

[2] *together with (a), (b), and (c)

Editors' Notes

HF 08-33-05 (2 ss., T); RC 088-26-15 (2 ss., T+H). These documents are two identical copies composed by Feigl for distribution out of a missing letter to him by Feyerabend. RC 088-26-15 is annotated in Carnap's characteristic shorthand writing.

[1] In a letter to Popper, Feyerabend briefly commented on the reactions to his note on Carnap with the following words: "Hempel sent a violent criticism of my Carnap-paper which is, however, wrong" (Feyerabend to Popper, 30 March 1957, see Vol. 1, Part II, 1957-5).

7. Carl G. Hempel, Reply to Feyerabend's Rejoinder, 1 April 1957

C. G. Hempel April 1, 1957[38]

C.G.H. on Dr. Feyerabend's reply of 3-3-57 to C.G.H.'s earlier criticism (2-22-57, to H. Feigl) of Feyerabend's critique of Carnap.

1. It is true that Feyerabend's critique of Carnap, as presented in section (2) of his paper, does not depend on his principle M, which mis-states ⟨sic⟩ one of Carnap's assumptions. Feyerabend's critique does, however, rest on a misconstrual ⟨sic⟩ of another point made by Carnap. This occurs in attributing to Carnap the following "principle Q": "A sentence which cannot be split up into two sentences, the one containing only meaningful terms, the other only meaningless terms and which contains some meaningful terms, contains meaningful terms only." (p. 4, par. 2)

   That this principle is false can be shown by one example: Let 'O' be an observational predicate and 'Y' a metaphysical one; then the sentence

   (S) (x)(Ox v Yx)

   cannot be split in the manner provided by rule Q and thus proves Q false.
   But what Carnap asserts on pp. 57/58 is not the rule Q but rather the following: If the sentence S above were a postulate of the theory at hand, then 'Y' could not be a meaningless term since it would be connected by physical law – the sentence S – with a meaningful term and would thus have a partial empirical interpretation.
   And Carnap goes to considerable length to preclude Feyerabend's reading of this point by arguing, on pp. 56/57, that if a sentence of the kind illustrated by S is not a postulate but is merely used as the additional assumption S_K in deriving an observational sentence from the theory, then S is meaningless, and D2 excludes it from the set of meaningful sentences.
   Thus, for example, the illustration given in footnote 3 of Feyerabend's paper falls short of its purpose unless the sentence (1) there is made into a postulate of the theory under consideration. (In this case, it seems to me quite reasonable that 'G' should turn out to have empirical meaning; but this might require fuller argument.)
2. One other puzzling aspect of Feyerabend's reasoning re principle Q: He says on p. 4, line 3 that the preceding argument (which is a restatement of Carnap's) hinges on the assumption S (which entails principle Q), whereas that assumption is nowhere invoked or needed in the argument in question. In fact, it would [1|2] seem to me that the last par. on p. 3 is not needed for Feyerabend's argument unless I miss some subtlety.
3. Two minor points: (a) on p. 3 middle, in (b), Feyerabend stipulates that S_K contain all and only the terms of K, whereas Carnap merely requires that SK contain only terms of K. (b) In par. 2, p. 5, Feyerabend presupposes that "S can be assumed true without implying Q". But as his S is stated on p. 4, par. 2, it implies Q, and the presupposition is impossible.

Editors' Notes

HF 02-202-19 (1 s., T); HF 09-38-03 (2 ss., T; c); RC 088-26-13 (2 ss., T+H; c). HF 02-202-19 is a typescript either composed by Feigl for distribution out of an apparently missing letter by Hempel, or attached to such a letter; HF 09-38-03 and RC 088-26-13 are identical copies of HF 02-202-19, differing from the original only in their layout. RC 088-26-13 is annotated in Carnap's characteristic shorthand writing. The transcription here is based on the text and pagination of HF 09-38-03 = RC 088-26-13.

[1] Upon receiving Hempel's second reply to Feyerabend, Feigl wrote to the former: "I did send to Feyerabend your critical comments, but of course without the personal remarks. I am sending Carnap and Feyerabend copies of your latest reply for which many thanks. Perhaps you have sent a copy to Carnap already?" (Feigl to Hempel, 5 April 1957, HF 02-202-20 = RC 102-07-25) About the personal remarks omitted by Feigl in the version of Hempel's critical comments composed for distribution, see *above*, §5, n1.

8. Paul K. Feyerabend, Rejoinder to Hempel's Second Reply, 8 April 1957

Paul Feyerabend
April 8, 1957

Second Rejoinder to Hempel on the Meaning Criterion

Regarding Hempel's criticism for the communication of which I am very grateful, I have to say this: in an improved version of my paper which will be published in Mind* I have the following footnote:

> "The sentence which I have called principle Q is nowhere stated explicitly but it is clearly implied in the arguments on pp. 57-58 of Carnap's paper. Cf. e.g. the following passage: "If a sentence connecting a meaningful term with another term in an inseparable way … is a postulate or provable on the basis of postulates, then it is stated as holding with physical necessity; therefore it conveys some empirical meaning on the second term." But as a metaphysician will certainly choose metaphysical postulates as his starting-point, the restriction to the consequences of the postulates of some L_T is unnecessary."

This is my footnote. The point of this footnote is this: Restricting Q to postulates does not help. For will not a metaphysician at once claim that all his more general statements are postulates or provable on the basis of postulates and hence meaningful if they are indivisible and contain at least one empirical term? As long as there is no criterion which forbids us to accept as a postulate whatever sentence we like (and no such criterion of contained in Carnap's paper!!) the restriction of what I have called rule Q to postulates is empty, i.e. it is no restriction at all and may as well be omitted. And this is also the reason why I did omit it.

"Thus for example," Professor Hempel writes (second par. from bottom), "the illustration given in footnote 3 … falls short of its purpose unless the sentence (1) there is made into a postulate of the theory under construction." Now, as nothing in Carnap's paper forbids me to make it into a postulate, I need only assume it to be a postulate in order to show the inadequacy of Carnap's criterion. I conclude that Carnap's criterion is by no means better than a criterion which simply asserts that a sentence which cannot be represented in the form T & T′, T containing meaningless terms, T′ containing meaningful terms only and which contains some meaningful terms contains meaningful terms only.[1] For if Carnap adds that this is valid for postulates only and allows me at the same time to adopt as a postulate whatever sentence I like -- then he shows by this that the restriction to postulates is redundant. Hence Hempel's criticism in point 1 of his note must be rejected. [1|2] The fact which he mentions in point 1, viz. that principle Q can be easily refuted is in fact fatal for Carnap, not for me. For it is Carnap who holds Q (with the redundant addition "for postulates only" --) and not I.

Pt. 1 }
} of Hempel's criticism -- these points have been improved in the
Pt. 2 } version I sent to Ryle in January, and I agree that here my arguments have been somewhat careless.

However, so far, I think, Professor Hempel has not succeeded to show that the main thesis of my note, viz. that Carnap's criterion "is useless…, grossly inadequate, and that it may even involve some logical absurdity" -- that this thesis is false. On the contrary -- he has only helped me to defend this thesis in a better and more efficient form.[2]

---

[1] *and which I sent to Ryle in January[3]

Editors' Notes

HF 08-32-13a (2 ss., T+H); RC 088-26-12 (2 ss., T+H). These documents are two identical copies of a typescript composed by Feigl for distribution out of a missing handwritten letter by Feyerabend, with identical handwritten corrections and differing only in the handwritten side-notes. More specifically, HF 08-32-13a is a transcription of an excerpt from Feyerabend's letter that Feigl sent back to Feyerabend for completion with the following handwritten note on the top of page 1: "Please correct & fill in! And return it to me. (Your handwriting was difficult to decipher – & the sentence below I makes no sense) HF" (see *below*, n1). RC 088-26-12 is annotated in Carnap's characteristic shorthand writing.

[1] The lines starting with the first occurrence of "meaningless terms" and ending with "meaningful terms only" are those referred to by Feigl's note in HF 08-32-13a as containing a sentence that makes no sense (see *above*, introductory note).

[2] Hempel did not reply to Feyerabend's second rejoinder. In a letter to Feigl dating a month later, after he had consulted with Carnap, Hempel clarified his final position in these terms: "As for Feyerabend, I think he should now make up his own mind as to what to publish re Carnap's essay; I feel I have done my share in discussing his views, and Carnap wrote me in rather strong agreement with my reactions to F's ideas. Is he in Minnesota now or in England? He seems bright, but his eagerness to find fault with Carnap seems to get the better of his scholarly caution and judgment." (Hempel to Feigl, 4 May 1957, HF 02-202-21; see *below*, § 11, n1).

[3] See *above*, § 4, n1E; *below*, § 14.

9. Rudolf Carnap, Comments on Feyerabend's Note, 9 March - 5 April 1957

R. Carnap
March 9, 1957[1]

Comments on Feyerabend's ms "A Note on Carnap's New Criterion of Empirical Significance"

On the first objection. Both what F. calls "assumption S" and "assumption Q" (pp. 3,4), which F. ascribes to me, I should reject. My own assumptions (in the article on theoretical concepts p. 56) are very much weaker, since they do not refer to sentences in general but only to postulates of the theory T. In the first paragraph of p. 58 I have explained in detail the reason why it is essential in this context to distinguish between reference to a postulate and reference to a non-provable sentence. (Incidentally, also the principle M p. 2 is formulated too strongly.)

In his example in footnote 3 F. takes actually a postulate. Thus this example is in accordance with my criterion. But this example does not at all show an inadequacy of the criterion. My explicandum is "meaningful" in the sense of "having a minimum measure of empirical meaning namely some effect with respect to an observable prediction." If I examine F.'s predicate 'G' with respect to my explicandum, before any attempt towards a formal explanation[2] is made, then I should indeed regard this predicate as meaningful in the sense just indicated. If someone makes the assertion '$G(t_1)$' then we can find out by an experiment made at the time t_1 whether the assertion does or does not hold. Therefore the assertion[3] 'G' is certainly meaningful. The English translation with "god's volition" is misleading because it suggests meaninglessness. However, some theological terms are meaningful and some are meaningless; and the same term may be meaningful in one usage and meaningless in another depending upon the nature of the theory by which it is introduced. When I read F.'s sentence: "Carnap's criterion is grossly inadequate by his own standards," I thought: "certainly not by my standards but by the standard of Popper (in his contribution to the Schilpp volume)," and thus I was not surprised to find in the footnote 5 of F.'s remark that he was greatly indebted to Popper's article. Popper intends to refute in his article the view of logical empiricism that metaphysics is meaningless. He takes as example a sentence like: "There is a being that is omniscient and ..." and he shows that empirical criteria for "omniscient" and the other terms occurring can be specified. My reply would simply be this that Popper entirely distorts the antimetaphysical thesis of logical empiricism. We asserted merely: "If the metaphysician does not give us any empirical criteria for his terms, then his sentences are meaningless." In the Vienna Circle we would have regarded Popper's example sentence, in view of the empirical interpretation specified by himself, neither as metaphysical nor as meaningless.[4] [1|2]

On the second objection. This objection is not clear to me. F. refers here to the end of his section 1. But what he has said there on the "amended version" of Ayer's criterion does certainly not hold for my criterion. I am in agreement with F.'s requirement (p. 2), first in application to terms, viz. that the criterion of significance must be such that we can ascertain whether it is fulfilled by a given term without assuming anything about the significance of this term. My criterion certainly fulfills this requirement because my definition D1 does not refer to significance. (My assumptions on pp. 54 and 56 are not part of my criterion but belong to an informal discussion using the word "meaningful" in the informal sense of the explicandum. Since the assumptions, using the informal word "meaningful," are clearly of an informal nature, they must not, of course, be used in any formal application of my criterion in order to determine whether a given term in a given system is significant.) If F. were able to show that my criterion violates the requirement mentioned above (or the corresponding requirement for sentences), then I would certainly admit that the criterion should be modified. It seems to me that a simple inspection of my formal definitions shows that these requirements are fulfilled.

An incidental remark on footnote 2. It seems to me convenient to regard any sentence S_i as L-true if it is a substitution instance of a logically valid sentential formula, even if S_i contains non-significant terms. In the latter case the terms are either not interpreted or at least not sufficiently interpreted by the system. Now it seems not unnatural to regard sentence S_i as L-true if S_i turns out to be true irrespective of the facts on the basis of every interpretation of the descriptive terms.

Editors' Notes

RC 088-26-14 (3 ss., T+H); RC 088-26-21 (3 ss., T+H); HF 07-57-01 (3 ss., T+H); HF 07-45-04; HF 07-53-02 (2 ss., T; c). RC 088-26-14, RC 088-26-21 and HF 07-57-01 are three copies of the same typescript with identical handwritten corrections. HF 07-45-04 and HF 07-53-02 are two copies of the former typescript which Feigl had made for distribution, where, however, not all the handwritten corrections were taken into account. Those corrections that were not reported in Feigl's copies were communicated to him by Carnap in a note to his "Reply to Feyerabend's Rejoinder of May 19" (see *below*, § 11) only after these copies had already been prepared and made circulate. This is confirmed by the fact that these corrections are written in Carnap's handwriting in RC 088-26-14 = RC 088-26-21, and in Feigl's handwriting in HF 07-57-01. RC 088-26-14 is annotated in Carnap's characteristic shorthand writing. The transcription here is based on the text and pagination of HF 07-45-04 = HF 07-53-02; corrections that can be found only in RC 088-26-14 = RC 088-26-21 = HF 07-57-01 are reported in the editorial footnotes.

[1] Carnap's reply to Feyerabend's note was already announced to Feigl by Ina Carnap at the beginning of March 1957: "He ⟨Carnap⟩[E] will send you his remarks to Feyerabend soon; Carnap does not at all agree with him." (Ina Carnap to Feigl, 8 March 1957, HF 02-69-26). However, Carnap's reply was actually sent to Feigl only about a month later, accompanied by the following remarks:

> I am sending you either herewith or in the next days the following things: ⟨...⟩. My reply to Feyerabend (if you have it copied, would you please send a copy to Hempel). ⟨...⟩ What is the matter with Feyerabend? Did he fall under the spell of Popper? I find his reply to Hempel rather slippery and evasive. In his first lines he implies that his principle M is discussed only in his first section "where I do not criticize Carnap at all". But this is not true. Hempel is quite right that F. makes use of this principle in his criticism of my view (in paragraph 4 of p.3 of his original note).[X)] ⟨...⟩ [X)] In the first sentence of his note of April 1, Hempel seems to have overlooked this case! (Carnap to Feigl, 5 April 1957, HF 02-69-27)

Only a week later, most likely before receiving Carnap's reply to Feyerabend from Feigl, Hempel wrote to Carnap asking his opinion on Hempel's own replies to Feyerabend:

> Feigl mentioned that he had sent you copies of my reactions to Feyerabend's criticism; did you think I was right? I could not help feeling that he was writing in an unnecessarily involved and somehow obfuscating manner, and that, most seriously, he misconstrued what you had said in your essay in the first Minnesota volume. (Hempel to Carnap, 12 April 1957, RC 102-13-60)

As soon as they received Hempel's letter, both Ina and Rudolf Carnap sent him a joint answer. In the first part of the letter, Ina lingered at length on Carnap's reaction to Feyerabend's note and especially on the apparent influence of Popper's personality on Feyerabend's style:

> Carnap also wrote up his reactions to Feyerabend for Feigl and asked him to send you a copy if he makes additional ones. C., of course, shares your impressions about F.'s paper, he even permitted himself a mild degree of indignation. Everyone who has met F. is impressed with his ability and intelligence; for some reason he seems to have fallen under Popper's spell. Popper, by the way, is in Stanford, but we know about it only from other people. I am pretty sure that we will survive it if he does not come to Los Angeles before he returns to England. These people who are so intent on establishing their priorities for everything and anything (Tarski seems to be another one of that sort), are so strenuous. And Popper has the especially annoying habit of twisting quotations or giving them in an incomplete form... if one is therapeutically minded, one might concede that he does so perhaps unintentionally, but the result fits his contentions so well that the therapeutic attitude is not easily maintained. Even C. nowadays can give it up and be just plain put out like ordinary men would be!

On the other hand, Carnap limited himself to copying his previous remarks on Feyerabend from his letter to Feigl, 5 April 1957 (see *above*, the paragraph from "What is the matter with Feyerabend?" to the right bracket) and adding the following comment:

> Dear Lante: Am I right with the above statement? If so, did you not concede too much in the first sentence of your Note of April 1st (which I received after writing the above)) (Ina and Rudolf Carnap to Hempel, 15 April 1957, RC 102-13-59; whereas, as is well known, "Peter" was Hempel's public nickname, "Lante" was his private nickname used in his correspondence with Carnap).

[2] Carnap corrected the typed "explanation" in "explication". This correction is missing in HF 07-45-04 = HF 07-53-02. See *below*, § 11.
[3] Carnap corrected the typed "assertion" in "term". This correction is missing in HF 07-45-04 = HF 07-53-02. See *below*, § 11.
[4] Feyerabend transcribed this paragraph and sent the excerpt to Popper, highlighting the passage from "Popper intends to refute" to the end of the paragraph by circling it and adding the note "reference to your Carnap-article". See "Excerpt from Carnap's comments on my paper", KP 423.5 (1 s., T) In the letter to Popper to which Feyerabend attached this excerpt, Feyerabend briefly commented on the development of the dispute with the following words:

> I have sent a copy of my note on Carnap's new criterion of meaning to Feigl in February. Since then I have had three rejoinders, two from Hempel, contradicting each other and apart from that criticising some very small points, and one from Carnap himself. I have also copied that part of Carnap's rejoinder which refers to you and which may be interesting to you. (Feyerabend to Popper, 2 May 1957, see Vol. 1, Part II, 1957-8)

10. Paul K. Feyerabend, Rejoinder to Carnap's Comments, 19 May 1957

P.K. Feyerabend
May 19, 1957

Rejoinder to Carnap re: Meaning Criterion

Carnap's criticism, page 1, paragraph 1: Carnap says that he would reject both "assumption M" and "assumption Q". His reason for rejecting the first is that it is stronger than his "presupposition (1)" (p. 54 of his paper). To this I reply what I have replied to Hempel's first rejoinder of Feb. 22: It is true that M is stronger than Carnap's "presupposition (1)" -- but this does not make any difference to my argument as, within the context of my paper, both principles lead to the same result. I have argued this a bit more in detail in my answer to Hempel's first rejoinder and in his second rejoinder (April 1) Hempel has admitted that I was right (see the first three lines of this rejoinder).– Carnap's reason for rejecting my "principle Q" is that I have omitted the restriction to postulates. Here I repeat what I have said in my reply to Hempel's second rejoinder: It is true that Q must be restricted to postulates. But this is a purely verbal restriction. If I can make a sentence like my "if it is god's will, then, if I lift my hand, the sun will be extinguished" a postulate (which is admitted by Carnap in the first line of the second paragraph), then I can make any sentence which connects metaphysical terms with empirical terms in an inseparable way a postulate and, hence, any term a significant term. This is why the additional phrase "restricted to postulates" seems to me to be completely superfluous. This brings me to the second part of Carnap's criticism, viz. his remarks in page 1, paragraph 2. I start by admitting that I was mistaken when writing: "Carnap's criterion is grossly inadequate by his own standards" – but this mistake can be immediately rectified by omitting the underlined words. True, in this form my criticism has lost some of its force – but it is still formidable enough – or at least so it seems to me. For it shows that the boundary-line between nonsense and sense does not coincide with the line which separates metaphysics (in the historical sense and not in the technical sense given to the word by Carnap in his paper and already earlier) from the rest. Now I do not think that the idea of the Vienna Circle was merely that people do sometimes talk nonsense and that they are doing so when using terms for which no empirical criteria (in the form in which they are now provided by Carnap) are available. There was the additional idea, and it was most strongly expressed (I am sorry that at the present moment I have no source available for quotation) that metaphysics (in the historical sense, i.e. Hegel, Spinoza, the Bible, Heidegger etc.) is nearly always nonsense. Carnap's last paper implies the implicit renunciation of this additional idea, a renunciation which is hidden beneath the fact [1|2] that the word "metaphysics" is still used – but now in a new, and technical sense. For according to this criterion – and this is the main point of the criticism which still remains when I have omitted the words "according to his own standards" – Hegel's theory of the revelation of the world-spirit in history, the new testament (Christ's fate in its Theological interpretation (the suffering of god's son) and not in its historical interpretation (the suffering of a person claiming to be god's son) ⟨sic⟩ and all those things are meaningful because connected with experience, or, to be more specific, because containing statements which contain empirical terms and specifically metaphysical terms in an inseparable way. Ayer's first criterion was criticised because it made any sentence meaningful. My point is that Carnap's criterion can be criticised for exactly the same reason: According to this criterion any non-logical term can be made an empirically significant term because it can be incorporated into a sentence which can be connected in an inseparable way with a sentence which contains only observable terms. And that metaphysics can be made meaningful in such a simple way is too good to be true.

This brings me to Carnap's comments on my second objection. Here again I admit some inaccuracy in my paper which, however, can be easily remedied in the following way. Instead of saying, in the beginning paragraph, that "I try to show that Carnap's criterion is useless (in a sense to be explained later…) grossly inadequate, and that is may even involve a logical absurdity" I say now that I try to show that "this criterion is either useless…or grossly inadequate (but perhaps not both) and it may…" And the paragraph in which I describe my second objection would start like this: "In our second objection we shall presuppose that the above inadequacy has been removed. In this case the criterion is inadequate in the sense that it cannot be shown that every term which is meaningless will violate at least one of the conditions of either… But it is also useless as a means…" which is now correct as meaningfulness is now no longer decided by the criterion. One could of course say that the criterion is the only measure for meaning -- but this would lead to the inadequacy outlined above. It would also assume that (but here I may be wrong) we have a syntactical criterion for deciding logical derivability. And it seems to be to be very difficult to find such a criterion.

Editors' Note

HF 08-32-13 (2 ss., T); HF 08-33-06 (2 ss., T); RC 088-26-11 (2 ss., T+H). These documents are three identical copies of a typescript that Feigl composed for distribution as an excerpt from a letter of Feyerabend to Feigl, 19 May 1957 (reproduced *below*, Ch. 3, § 1), corrected in its obvious misspellings. RC 088-26-11 is annotated with Carnap's characteristic shorthand writing.

11. Rudolf Carnap, Reply to Feyerabend's Rejoinder, 25 May 1957

Rudolf Carnap
May 25, 1957

Reply to Feyerabend's Rejoinder of May 19

1. F. now at last admits that he was wrong in attributing to me the principle Q which did not refer to postulates: "It is true that Q must be restricted to postulates." (He adds that he said this already in his second reply to Hempel of April 8; but this is not true. On the contrary, there he still asserts in the new footnote, which he intends to insert in his paper, that the principle Q "is clearly implied in...Carnap's paper.") But he persists in basing his criticism on his principles M and Q instead of my weaker principles, because in his opinion it does not make any difference for the purposes of his discussion whether my principles are used or his principles, which are stronger and which I had rejected. If it does not make any difference, why does he then not take my principles as a basis of his criticisms? It seems to me that this would be the only relevant and fair procedure.
2. The new formulation of F's second objection is again not clear to me. In his quotation of the new formulation intended for his paper he says: "In our second objection we shall presuppose that the above inadequacy has been removed." It is not clear to me how this is to be interpreted in such a way as the next sentence makes sense. The next sentence begins: "In this case the criterion is inadequate..." In the original version he had here the word "adequate"; I do not understand his reason for the change (or is it perhaps a typing error?). The later statement "Meaningfulness is now no longer decided by the criterion" is not clear to me without more specific explanation. Furthermore, in my note of March 9, I had pointed out that in any formal application of my criterion no use must be made of the two principles; I do not see that F has taken this point into account in his discussion.
3. The intended changes in his paper indicated so far by F. seem to me entirely insufficient in order to assure that the reader receive an accurate picture of my views. I do no see any purpose in continuing this discussion any further. If two notes by Hempel and two by me are not sufficient to make it clear to a man that he should base his criticisms of an author upon an undistorted representation of his views and arguments, then nothing further will help either.
4. I should appreciate obtaining a copy of the final version of F.'s paper. If I should still find that his discussion is based on distortion, then I shall publish only a very short reply pointing out this fact. In this case I shall prefer not to enter into a discussion with F. on the problems themselves either in publication or in the planned Minneapolis Conference.[1]

Corrections to my "Comments..." of March 9.[2]
p. 1, line 5 f.b.; read "explication" instead of "explanation."
p. 1, last line; read "term 'G'" instead of "assertion 'G'."

Editors' Notes

RC 088-26-10 (1 s., T+H); RC 088-26-20 (1 s., T+H); HF 07-45-03 (1 s., T+H); HF 07-57-02 (1 s., T+H). RC 088-26-10, RC 088-26-20 and HF 07-57-02 are three carbon copies of the same typescript and present the same handwritten corrections by Carnap. More specifically, RC 088-26-10 is Carnap's personal copy annotated in his characteristic shorthand writing; HF 07-57-02 is the copy that Carnap sent to Feigl with the following handwritten note on the top-right corner: "(I have sent a copy to Hempel)"; while RC 088-26-20 is a spare copy. HF 07-45-02 is the copy of RC 088-26-10 = RC 088-26-20 = HF 07-45-03 made by Feigl for distribution, on which the transcription here is based.

[1] Carnap refers here to the "Conference on Philosophical Foundations of Physics", Minnesota Center for Philosophy of Science (MCPS), 12 August - 14 September 1957, where Feyerabend was an invited speaker (see Vol. 1, Part II, 1956-20, n4).

It should be noted that Feyerabend's stubbornly critical attitude disturbed both Hempel and Carnap. At about the time when Carnap's second response to Feyerabend's attacks was circulated, Hempel commented on it both with Feigl and with Carnap:

> To Feyerabend I think I had better not to reply further; this is a too drawn-out business; but I would hope, of course, that his objections to Carnap would come up for discussion at our meetings; in this case, Hilary might perhaps present his own objections. My feeling is that probably Carnap's latest criterion will run into difficulties, too; but Feyerabend's hasty manner – he has now conceded I think 4 or 5 misrepresentations of Carnap's statement – bothers me; and indeed he again, in his statement of May 19, seems to shrug off the fact that he had misconstrued Carnap. – He wrote me for a copy of "The theoretician's dilemma", which I promptly mailed to him. (Hempel to Feigl, 27 May 1957, HF 02-202-22; see *above*, § 8, n1; *below*, § 14; Vol. 1, Part II, Feyerabend to Popper, 26 August 1958, 1958-10, n1 in particular).

> Re Feyerabend: Just a few days ago, I wrote Feigl again that I did not think it made sense to continue commenting on F's series of modifications, and that I felt unhappy at all the misrepresentations of your position to which he has by now admitted. It is all very well to say that basically this does not affect his criticisms, I think he should write his piece after carefully studying what you say and should then present his criticisms without any distortion, and in direct reference to your views. – With an eye on point 4. in your "Reply to Feyerabend's Rejoinder of May 19", let me plead with you very urgently not to cancel your Minnesota plans (and I hope you have decided to come!) in view of this unrewarding and irritating exchange; I can see poor Herbert having beads of apprehensive perspiration on his worried brow over this. A Center Director's lot is not a happy one! And you too, dear Ina, do come along – I haven't seen you for such a long time! (Hempel to Carnap, 1 June 1957, RC 082-07-02)

Due to Carnap's and Hempel's reactions and to the possibility that Carnap could decide to decline the invitation to participate in the then forthcoming MCPS conference because of Feyerabend's presence, Feigl felt the need to apologize with Carnap in these terms:

> P.S. ⟨...⟩ Very sorry that Feyerabend's comments irritated you. My impression of F. (in Vienna, '54) were excellent – but perhaps his attachment to Popper explains at least partly his attitude. I am sure that personal discussion would be both pleasant & worthwhile. (Feigl to Carnap, 17 June 1957, RC 102-07-23; on the 1954 meeting between Feigl and Feyerabend, see Vol. 1, Part II, 1954-19, n3)

In his reply to Feigl, Carnap confirmed and further motivated his distress with the situation

> Yes, I lost my patience somewhat with F. since he costs me so much time and was unwilling to learn from Hempel. I had understood from you at first that his ms. was not going to be published in the near future, and then it turned out that he had already submitted it.

> I would mind less if his paper were published in an American journal, but the people in England have not read my publications and therefore will believe what people like Popper or Feyerabend report to them as my assertions. Well, well. (Carnap to Feigl, 25 June 1957, RC 102-07-22; see *below*, § 13, introductory note)

Eventually, Carnap did not participate in the MCPS conference, though apparently for reasons totally independent of Feyerabend's presence (see Carnap to Feigl, 5 April 1957, HF 02-69-27; Feigl to Carnap, 17 June 1957, RC 102-07-23; Feigl to Carnap, 24 July 1957, RC 102-07-21; Hempel to Feigl, 19 August 1957, HF 02-202-23; Hempel to Carnap, 26 August 1957, RC 102-13-56).

Although the "Feyerabend-Carnap controversies" left a lasting impression on Hempel and Carnap, their relation with Feyerabend recomposed after they had the occasion to become personally acquainted with him. In particular, Feyerabend met Hempel for the first time only a few months after their intense mail exchange, at the 1957 MCPS conference (see *below*, Ch. 3, § 2, n14; and Vol. 1, Part II, 1956-20, n4); and again about a year later at the follow-up 1958 MCPS Princeton conference (see Vol. 1, Part II, 1958-10, n1). The change in Hempel's opinion on Feyerabend that these meetings brought about is testified by a letter of Hempel to Dennes, 27 February 1959, CH 16-06 (1 s., T+H; cc) written in early 1959 in behalf of Feyerabend's appointment as Associate Professor at the University of California, Berkeley, and here reproduced:

re Feyerabend

February 27, 1959

Professor Williams R. Dennes
Chairman, Department of Philosophy
University of California
Berkely ⟨sic⟩ 4, California

Dear Professor Dennes:

I am very happy to give you my appraisal of the qualifications of Dr. Paul Feyerabend.

I have read a good deal of Mr. Feyerabend's published work as well as many unpublished notes and essays which were circulated by the Minnesota Center for the Philosophy of Science by way of an intensive exchange of ideas among the contributors to a forthcoming new volume in the series, Minnesota Studeis in the Philosophy of Science. In addition, I had an opportunity to discuss with Feyerabend and to see him in action during two one-week workshops, limited to some 15 participants, which were held, under the sponsorship of the Minnesota Center, at Minneapolis and at Princeton.

In my judgment, Mr. Feyerabend combines a forceful and penetrating analytic mind with a remarkably thorough training and high competence in theoretical physics and mathematics; he thus possesses to a rare extent the qualifications that are nee⟨d⟩ed for significant original work in the philosophy of science. Mr. Feyerabend has an enviable ability for lucid exposition and analysis, which manifests itself both in his publications and in oral discussion. I was much impressed, for example, by a lecture he gave at the Princeton workshop on certain problems concerning quantum mechanics, and I formed an equally favourably impression of his oral and subsequent written discussion with members of the Minnesota group, of conventionalism in physical geometry.

I have had no opportunity to talk with Dr. Feyerabend about problems on other areas of philosophy and thus I lack a basis for judging his competence here; however, his long essay on Wittgenstein's Philosophical Investigations, which appeared in the Philosophical Review, ⟨1|2⟩ is strong evidence, to my mind, of considerable power and originality in analytic approach to problems of logic and epistemology.

Some brief notes by Feyerabend that I had seen before meeting him made me wonder whether he might not be somewhat opinionated. I was happy to discover, when I made his acquaintance, that this was not at all the case: I found him open-minded and quite receptive to critical comments and suggestions concerning his ideas. I also found him a most likeable and engaging person and animated and stimulating company, full of a contagious enthusiasm for the pursuit of scientific and philosophical questions.

I think that both as a teacher and as a scholar, Dr. Feyerabend is a man of high accomplishment and promise, and I believe him therefore fully qualified for an associate professorship at so distinguished a university as Berkeley.

Sincerely yours,
⟨unsigned⟩
Carl G. Hempel
Stuart Professor of Philosophy

Hempel met Feyerabend again several times in the following years (see Vol. 1, Part II, 1966-1, n1) and expressed similar judgements. For example, after a Philosophy of Science Workshop at the University of Pittsburgh, 7-10 May 1965, in which both Feyerabend and Hempel participated (see "To all Participants in the May 7-10, 1965 Philosophy of Science Workshop at the University of Pittsburgh", 3 May 1965, RC 102-13-12; and *below*, Ch. 4, 1968-1, n1; Feyerabend to Hempel, 20 May 1968, 1968-3), the latter wrote to the organizer:

> I found the experience professionally very rewarding and personally most enjoyable. (On the latter score, I was especially delighted with Paul F. whom one would expect to be much more angular, judging by his writings, and I whom I found extremely pleasant. All the other people I knew antecedently to be highly congenial). (Hempel to Grünbaum, 11 May 1965, CH 19-08)

Hempel was requested to provide his estimate of Feyerabend's qualifications on at least two other occasions: Feyerabend's appointment to Yale University in 1969 (see Schrader to Hempel, 5 November 1968, CH 16-06) and Feyerabend's advancement within his professorship at the University of California, Berkeley in the mid-1970s (see Dreyfus to Hempel, 30 January 1974, CH 16-06). Given Feyerabend's success on both these circumstances, most likely Hempel's opinion did not undergo any substantial change during the 1960s or the early 1970s.

As for Carnap, Feyerabend became personally acquainted with him in the late winter of 1959 in the course of a lecture tour which took him through southern California (see Vol. 1, Part II, Feyerabend to Popper, 22 March 1959, 1959-5; and FEYERABEND 1995: 118-119). Carnap's reaction to their meeting was enthusiastic:

> Feyerabend was just here. He has given a lecture on "Hidden Variables in Quantumtheory", and nect ⟨sic⟩ day I had a nice private conversation with him. I was very glad to see that (as you had told me) he is a very likeable person, and easy to get along with. His lecture was very skilfull ⟨sic⟩ and clear, and discussing with him is very stimulating. On the basis of this good impression I would have recommended to the dept. inviting him at least temporarily; but he has already accepted a professorship at Berkeley. I talked with him also about the debate between him and Grünbaum. I told him that I regretted that he had repeatedly misrepresented G.'s position, as earlier mine in the discussion with Hempel and me. He said that he might have been sometimes "too radical"; later when we talked about Popper, he hinted that he had apparently taken over some of Popper's habit of overpolemic reactions and now intends to rid himself of that (but he also said good things about Popper). We had too little time to go thoroughly into the subject matter of the question of conventionality of geometry. I told him only briefly my reasons for agreeing with Grünbaum, but he still believes that in the content he is right. However, he told me that he would write a detailed and careful reply to the last long memorandum of G. and also be more cautious in representing G.'s position. F. will come again. (Carnap to Feigl, 17 March 1959, HF 02-69-45 = RC 102-06-22)

(Carnap refers here to a debate between Feyerabend and Grünbaum on the conventionality of geometry that took place during and shortly after the 1958 MCPS Princeton conference (see below, Ch. 3, § 2, n14), as testified by the following writings: Adolf Grünbaum, "Memorandum [to Princeton 1958 Conference]", HF 09-17-10 (4 ss.); Paul K. Feyerabend, "Reply to Adolf

Grünbaum's Memorandum Concerning Time Metrization", 7 October 1958, HF 08-32-09a (3 ss.) ≈ HF 08-33-13 (5 ss.); Paul K. Feyerabend, "Second Memo on Conventionalism in Geometry "Some Further Comments on Conventionality in Geometry with Special Reference to Reichenbach's Views in Sections 3-8 of his Philosophy of Space Time)", HF 08-33-17 (8 ss.); "Some Further Comments on Conventionality in Geometry with Special Reference to Reichenbach's Views in Sections 3-8 of his Philosophy of Space Time (revised version)", 15 October 1958, HF 08-33-14 (8 ss.); Adolf Grünbaum, "Further Memo on Time Metrization and Classical Mechanics" 27 Ocotber 1958, HF 09-14-01 (3 ss.)"; Paul K. Feyerabend, "Excerpt from a Letter to Herbert Feigl", 27 October 1958, HF 08-33-12 = HF 08-33-15 (7 ss., T), RC 088-26-07 (7 ss., T+H); "Further Notes on Conventionalism (Excerpts from a letter to Adolf Grünbaum Nov. 8, 1958)", 8 November 1958, HF, 08-33-19 (3 ss., T), RC 088-26-05 (3 ss., T+H); see Feigl to Carnap, 24 October 1958, RC 102-07-02 ≈ HF 02-69-34; 24 October 1958, RC 102-07-03; Carnap to Feigl, 4 December 1958, HF 02-69-35 = RC 102-07-01).

Following this episode, attempts were intensified to establish a collaboration between Carnap and Feyerabend, with Feigl's mediation, through an exchange of their respective working papers on the philosophy of physics for critical comments – in particular, drafts of FEYERABEND 1962a and of CARNAP 1977, both then planned to be published in a forthcoming volume of the *Minnesota Studies in the Philosophy of Science*. Apprently, however, these attempts did not lead to further significant developments. (See Herbert Feigl, "Report on the Conference on Philosophical Foundations of Physics, August 12 – September 14, 1957", 27 September 1957, HF 01-40-06 = HF 01-46-01; Feigl to Carnap, 10 December 1957, RC 102-07-19; Carnap to Feigl, 16 December 1957, RC 102-07-18; 4 August 1958, RC 102-07-05; Feigl to Carnap, 17 October 1958, RC 102-07-03 ≈ HF 02-69-33; 30 March 1959, RC 102-06-20 = HF 02-69-46; 1 April 1959, RC 102-06-19; Carnap to Feigl, 14 April 1959, RC 102-06-17; Feigl to Carnap, 15 October 1959, RC 102-06-10 = HF 02-69-50; Carnap to Feigl, 27 October 1959, RC 102-06-09; Feigl to Carnap, 13 January 1960, RC 102-07-03 ≈ HF 02-69-51).

Feyerabend and Carnap met again on at least three other occasions: in mid-February 1961 (11-14), during a four-day informal meeting sponsored by MCPS among Carnap, Feigl, Maxwell, Meehl, and Feyerabend, which took place at the residence of Carnap in Santa Monica (Feigl to Carnap, 13 January 1960, RC 102-07-03 ≈ HF 02-69-51; Carnap to Feigl, 1 Febraury 1960, RC 102-06-06; Feigl to Carnap, 4 February 1960, RC 102-06-05; 18 February 1960, RC 102-06-05; Herbert Feigl, "Interim Report 1959-1960", December 1960, HF 01-39-02 = HF 01-39-03); only a few months later, in the summer of 1960, at the first congress of Logic, Methodology and Philosophy of Science held at Stanford University, where Feyerabend accompanied Popper (see Vol. 1, Part II, 1960-7, n4; and CARNAP 1962); and, most likely for the last time, at the 1964 European Forum Alpbach (on Feyerabend and the 1964 European Forum Alpbach, see Vol. 1, Part II, 1963-3, n4; and FEYERABEND 1995: 119).

The good relation between Feyerabend and Carnap as well as the generally high opinion the latter had of the former can be gathered by the letter of Carnap to Taubes, 2 December 1966 (1 s., T+H, RC 088-26-02, c;) in behalf of Feyerabend's appointment at the Free University of Berlin in 1967 (see *below*, Ch. 4, 1968-1, n4; and Taubes to Carnap, 13 December 1966, RC 027-39-10), here reproduced (in English translation):

2 Dec. 1966

Professor Jacob Taubes
Philosophical Seminar
Free University of Berlin

Dear Mr. Taubes,

I understand from your letter of 24 Nov. that the Free University of Berlin is planning a new professorship for the discipline that is here called *philosophy of science* ⟨in English in the original⟩[E]. I concur in your view that such a chair is very desirable. I have also noticed with regret that this discipline has not yet found the attention it deserves in Germany.

In my judgment, Paul Feyerabend is an excellent candidate for such a chair. I have had philosophical conversations with him quite often and I also know a good part of his publications. I have also attended a seminar led by Feyerabend on the foundations of modern physics at the 1964 Alpbach conference. He knows both the physics and the philosophy thoroughly, and he is particularly well versed in the fundamental logical and epistemological problems of physics. He has a keen interest in the problems he deals with, and he is very effective in waking this interest also in the students. His vivacity leads him sometimes to become heavily polemical in discussions, both orally and in print. But he is always fair to his opponents, and in personal relations he is friendly, sociable and helpful. I am convinced that you would find in him not only a competent philosopher, but also a good colleague.

With best wishes and kind regards

Yours
⟨unsigned⟩
Rudolf Carnap

[2] See *above*, § 9, introductory note in particular.

12. William W. Rozeboom, Note on the Feyerabend-Carnap Controversy, 10 June 1957

W. Rozeboom
June 10, '57

The Feyerabend-Carnap Controversy -- A Further Comment

In a recent flurry of notes, Feyerabend has challenged the adequacy of Carnap's most recent attempt to formulate a meaning criterion. Irrespective of the intrinsic merit of Carnap's new formula (it is my suspicion that it is too narrow), it seems to me, if I understand Carnap correctly, that Feyerabend's criticisms rest on a misunderstanding that goes much deeper than merely a misstatement of Carnap's views.

1. In his first note, F. attributes to Carnap the principle M (p. 1): "a sentence of the forma $p \supset q$ with either p empirically significant and q devoid of empirical meaning or q empirically significant and q devoid of empirical meaning is logically true if and only if either p is logically false or q is logically true." As Hempel has pointed out, this is stronger than the presupposition that Carnap actually made (Assumption 1, p. 54), which was "If S_1 and S_2 are any sentences of L such that all descriptive terms of S_1 belong to V_1 or to the observational vocabulary V_O and those of S_2 to V_2, then neither of the two sentences logically implies the other unless the implying sentence is logically false or the implied sentence is logically true."

F. has asserted (May 19) that the difference between principle M and Assumption 1 makes no difference for his purpose. Be this as it may, F.'s interpretation of Assumption 1 reveals a gross misconception of Carnap's intent. It is clear from F.'s Principle M that he regards Assumption 1 as a meaning criterion which a syntactically analytic formula must also meet if it is to be regarded as logically true. The fact of the matter is, however, that Assumption 1 has nothing to do with meaningfulness. It is simply a statement of fact about conventional syntax which could have been proved, had Carnap wished to do so.

Le $S_1(a_1,\ldots,a_m)$ and $S_2(b_1,\ldots,b_n)$ be two sentences in which the a's and b's are all the non-logical constants and such that there is no term common to both the a's and the b's. If S_1 logically entails S_2, then the formula $S_1(a_1,\ldots,a_m) \supset S_2(b_1,\ldots,b_n)$ must be true in all interpretations of the formal systems. But since assignments of designata to the b's are independent of assignments to the a's, this obtains if and only if $S_1(a_1,\ldots,a_m) \supset (x_1)\ldots(x_n)S_2(x_1,\ldots,x_n)$ is also true in all interpretations -- i.e., is analytic. But $(x_1)\ldots(x_n)S_2(x_1,\ldots,x_n)$ (which I shall abbreviate 'S^*_2') contains logical constants only, and thus has the same truth value in all interpretations. Hence if $S_1 \supset S^*_2$ is analytically true, either S_1 must be analytically false, or S^*_2 must be (analytically) true. In the latter case, S_2, which is entailed by S^*_2, must also be analytically true. Hence if S_1 and S_2 contain no non-logical constants in common and S_1 logically entails S_2, either S_1 must be logically false or S_2 must be [1|2] logically true. And this, of course, proves Carnap's Assumption 1.

If it is true that F.'s use of Principle M leads to conclusions which could not also have been reached from Assumption 1, then it is indeed difficult to see how these conclusions have any bearing on the adequacy of Carnap's meaning criterion. I have not seen the latest version of F.'s critique, but in his first note, he argued that Principle

M failed to provide a meaning criterion which could actually be applied. But since Assumption 1 is not a meaning criterion, but only a fact about analytical derivability, F.'s arguments here must be wholly irrelevant.

2. In a criticism somewhat more to the point, F. has challenged Carnap with the further principle Q: "A sentence which cannot be split up into two sentences, the one containing only meaningful, the other only meaningless terms and which contains some meaningful terms, contains meaningful terms only." (p. 4) Carnap (March 9) has protested that Q is too strong as it stands, and that it should apply only to postulates of the theory. F., in turn, retorts (April 8, May 19) that restriction to postulates of the theory is superfluous, since in either case, Q admits metaphysical sentences as meaningful. I agree with F. that for his purpose the restriction to postulates is irrelevant, and it seems to me that Carnap's last note (May 25) was a bit overly harsh in this respect. But what does F.'s criticism of Principle Q amount to? Nothing more than that it admits as meaningful expressions which F. does not feel are meaningful. This hardly strikes me as a valid line of criticism. The mere fact that a sentence is believed by F. to be meaningless is scarcely sufficient grounds for asserting that it is meaningless. Carnap has attempted to explicate the conditions under which scientific hypotheses are held to be meaningful. If this criterion cannot be strengthened without excluding theories which we would like to retain as meaningful, and it can be shown that the criterion is also satisfied by certain "metaphysical" sentences, then the latter would seem to have meaning in the same way that scientific theories have meaning. The fact that a meaning criterion admits a sentence with emotionally disturbing connotations is a poor reason for rejecting a criterion. If F. holds, e.g., that theological assertions are meaningless, then presumably he does so for reasons over and above a dislike for theology. Such reasons, if tenable, would constitute a valid objection to Principle Q. But until these reasons are actually brought forth, I fail to see that F.'s argument has much force.

3. I must confess, however, that I remain dubious about the adequacy of Carnap's meaning criterion as formulated by D1 and D2 (p. 51). Let H be the theoretical hypothesis '$(x)[T_1(x) \cdot T_2(x) \supset P(x)]$,' where '$T_1$' and '$T_2$' are theoretical terms and 'P' is an observational predicate. H has empirical significance in the pre-formal sense of Carnap's criterion, since for any entity, a, '$T_1(x) \cdot T_2(x)$' [2|3] and H entail the observational consequence 'P(a).' Yet I do not see how to construct a series of terms in conformance with D_1 and D_2 which would admit either 'T_1' or 'T_2' as meaningful under the criterion. This particular example might be declared illegitimate since strictly speaking, H is a C-rule in Carnap's sense, but the same point may be made more generally. Let T_1 be a theory-cum-C-rules in which μ is a theoretical predicate which satisfies D_1 and D_2, and let T_2 be derived from T_1 by replacing μ throughout with the conjunction of theoretical predicates T_1 and T_2. T_1 and T_2 do not qualify under the formal criterion as meaningful. In general, I have very grave doubts whether for all theories that we would wish to regard as meaningful, we can demonstrate their meaningfulness in the term-by-term stepwise fashion of D_1 and D_2.

Editors' Note

HF 12-19-03 (3 ss., T); RC 088-26-09 (3 ss., T). HF 12-19-03 and RC 088-26-09 are two identical copies of the same typescript. RC 088-26-09 is annotated in Carnap's characteristic shorthand writing.

13. Rudolf Carnap, Comments on Rozeboom's Note, 24 June 1957

June 24, 1957
Rudolf Carnap

Comments on Rozeboom's Note of June 10, 1957

1a. I agree, of course, with R. in rejecting Feyerabend's conception of my Assumption 1 as part of my criterion of significance. (I made essentially the same point, but less explicitly, in my Comments of March 9, p. 3).

1b. R.'s derivation of Ass. 1 is interesting. However, Ass. 1 does not hold generally, but only under the condition that no logical relations hold between the descriptive constants of S_1 and those of S_2. (R. used the equivalent presupposition that assignments of designata to the a's are independent of those to the b's.) In my article, I stated Ass. 1, not as a general assertion, but as a more precise formulation of the presupposition characterizing the classes V_1 and V_2, replacing the previous formulation in vague terms like "meaningful."

2a. (First half of the paragraph.) The point of R.'s argument here is not quite clear to me. But I suppose that R. would agree with the following points. (1) The principle Q, which Feyerabend previously attributed to me, was never asserted by me and is obviously false (as Hempel showed, April 1, (1)). (2) Obviously, when new postulates are added to a system, the status of the primitives may be changed; in particular, a meaningless (i.e. not interpreted) term may become meaningful (i.e. interpreted). (It is for this reason that the criterion must be made relative to, among other things, the system of postulates, as was done in my D_1 and D_2.)

2b. (Second half.) I agree with R.'s remarks, which make the situation quite clear.

3a. R. shows correctly that, on the basis of my criterion, the following case can occur: each of two predicates is non-significant but their conjunction is significant (or, more exactly: if the conjunction were replaced by one new primitive term, then this term would be significant). The same case and other related cases were pointed out by Hempel ("Some Remarks on Carnap's ms.," June 1956, par. 1(ii)). He suggested that my criterion D1 be generalized so as to apply not only to predicates but also to sentential functions. In my reply ("Comments...," October 11, 1956) I agreed with this suggestion. (The Center has copies of Hempel's and my notes.)[1] I would prefer making the generalization in such a manner that the case referred to by R. would still be permitted to occur. For it might be that the rules of a system give an interpretation for the conjunction (possibly even a complete interpretation, e.g. by yielding the sentence "$t_1(x) \cdot t_2(x) \equiv$ x is a dog"), while they fail to give an interpretation for either term separately.

3b. The doubt expressed by R. in his last sentence concerns a serious problem. I am not sure myself on this point. My discussion (pp.58 f.) gives only some plausibility considerations. I have tried to construct counter-examples, in which the term-by-term stepwise procedure would not work; but I have not found any. This question must be further investigated.

Editors' Notes

RC 088-26-18 (2 ss., T+H); RC 088-26-19 (2 ss., T+H); HF 07-57-03 (2 ss., T+H); HF 07-53-01 (1 s., T). RC 088-26-18, RC 088-26-19 and HF 07-53-03 are three carbon copies of the same typescript and present the same handwritten corrections by Carnap. More specifically, RC 088-26-19 is Carnap's personal copy, annotated with some underlining; HF 07-57-03 is the copy enclosed in a letter of Carnap to Feigl, 25 June 1957, RC 102-07-22 (see *above*, § 11, n1); while RC 088-26-18 is a spare copy. HF 07-53-01 is the copy of RC 088-26-18 = RC 088-26-19 = HF 07-53-03 made by Feigl for distribution, on which the transcription here is based.

[1] See Carl G. Hempel, "Some Remarks on Carnap's Manuscript 'The Methodological Character of Theoretical Concepts'", 25 June 1956, HF 09-38-06; Rudolf Carnap, "Comments on Hempel's Remarks on Carnap's ms. 'Theoretical Concepts' (June 1956)", 11 October 1956, RC 089-04-03.

3. The First Controversy – First part (1957-1958)

14. Paul K. Feyerabend, Remarks on the Interpretation of Theories, 28 June – 4 July 1957

P. K. Feyerabend
June 28, 1957

Dear Professor Feigl:

I have been meaning to write to you for a long time, but about three weeks ago I was put out of action by the most agonising nerve-pains which I have now had for about three weeks and which I had to fight off by very strong medicines and those medicines put me into a daze most of the time. Now only a few words as to the summer-session (which I sincerely hope I shall be able to attend).[1]

1st. The Theoretician's Dilemma which I received about a month ago is a most interesting attempt to justify the use of theoretical terms.[2] I would not say on that account, however, that Hempel's position is the position of a realist. For one can admit theoretical terms into science without at the same time being a realist. This is the point of a note on Hempel's paper which I was preparing when I was caught by those nerve-pains and had to give up. But I hope that I shall be able to continue and to finish the note before the steer-session starts. This note is also intended to contain an answer to criticisms raised against my THEORIA paper in 1955 in your institute (Pap sent me a summary of those criticisms).[3]

2nd. I got Rozeboom's note,[4] but I was not yet able to have a look at it.
[... 1|2|3|4 ...]5

Yours sincerely
Paul Feyerabend (signed) [4|5]

Remarks by Feyerabend on Interpretation of Theories with an Eye Toward Quantum Mechanics

There are three theses which I hope to defend in the institute: Consider a formalism T which is interpreted by means of an interpretation-system J. Let T′ be the (interpreted) theory which results from connecting T and J, i.e. let be [T,J]=T′. Then my first thesis is that very often [T,J] = [T,J′] although J≠J′. (Cf. for this arguments on the enclosed typescript.)[6] Second thesis: Assume that L_O is an observation-language used in the interpretation of a theoretical language L_T. Then LO may be affected by the fact that it is so used. More especially: Certain statements of L_O as well as parts of the grammar of L_O will have to be reformulated. Or to be even more specific: within a theory T we cannot distinguish different levels with respect to grammar, etc. We can only draw the pragmatic distinction between more or less easily testable statements: Once the special theory of relativity has been introduced, any language used for interpreting this theory uses a concept of length as a relation between extension measured and coordinate-system used. I.e. any L_O used for interpreting $L_{spec.rel.}$ must use the relativistic concept of length once it has been connected by interpretation rules with $L_{spec.rel.}$ I think I have good argument for this thesis. Third thesis: The distinction, within an interpreted theory, of L_O and L_T is a pragmatic distinction.

The marked passage on p. 7 of the enclosed typescript[7] seems to furnish another argument for the adoption of theoretical terms for it shows that in many cases the empirical generalizations which are functionally equivalent (in Hempel's sense) to a given theory contain individual constants (names of individual telescopes in my example). Hence, if one wants one's description of nature to be free of individual constants, one has to use laws containing theoretical terms. This is an argument in favour of the use of theoretical terms which Hempel (The Theoretician's Dilemma) does not use and I think it to be a very good argument. [5|6]

Excerpt from a Typescript by Feyerabend on Interpretation of Theories

Consider a formalism T which is interpreted by means of an interpretation-system J. Let T′ be the (interpreted) theory which results from linking together T and J, i.e. let be (T,J) T′. Then it seems to me that very often
(T,J) = (T,J′) = T′ in spite of the fact that $J \neq J'$ (i.e. either J&J′ is contradictory, or J contains observation-terms which J′ does not contain or vice versa):
What a theory says is to a certain extent independent of the interpreting-system used. (I)
This is my first thesis.

In the next lines I shall first produce some arguments in favour of this thesis, and I shall then draw some conclusion from it.
ARGUMENTS

The arguments (which I have tried to develop in my note in Theoria) are by way of example.

1st example: T is the formula W const. J^2Rt, T′ is Joule's Law and J a set of interpretation-rules involving a.o. reference to mercury-thermometers. With the specification of J the class V_B is determined as well: V_B is the class of all the non-theoretical terms which occur in J. Now consider the following changes:

a) Improvement of the precision of reading the thermometer (e.g. use of a magnifying-glass). This implies a change of V_B and, hence, of J (into J′).

b) Correction of the reading of the thermometer (1) because of the non-uniform inner width of the thermometer (this would change J into J″) and (2) because of the extension of the thermometer itself ($J \rightarrow J'''$).

c) Correction of the reading of the thermometers in order to make them agree with absolute scale ($J \rightarrow J''''$).

Now it seems to be evident that
(T,J) = (T,J′) = (T,J″) = (T,J‴) = (T,J⁗) and even = (T,J̲), where J̲ is the interpretation-system resulting from (a) plus (b) plus (c). I.e. it seems to be evident that the correction of an instrument used for the test of a theory need not amount to a reinterpretation of the theory, although it amounts to a change of the interpretative system of the theory. In other words – the above equations are criteria of adequacy for any analysis of the "meaning" of T′.

Let C(T,J) be the class of sentences following from the axiomatized theoretical system $(T,J)_B$, V_B being the class of observational terms in J (for the notation see Hempel's paper p. 51).[8] Obviously $C(T,J) \neq C(T,J') \neq C(T,J'')$ etc. I.e. the class of empirical generalization functionally equivalent with (T,J) is different from the class of the empirical generalizations functionally equivalent to (T,J′) etc. – and yet (T,J), (TJ′) etc. are representing the same theory viz. Joule's law. It [6|7] follows that theories are not only to a certain extent independent of the interpretative-systems used for interpreting their formalism, but they are also to a certain extent independent of the class of empirical generalizations which follows from them. To express the same fact in a less technical way: A law does not change its meaning on account of the fact that all observers start to wear eyeglasses.

Another remark may be added here for later reference: Assume that the observation-language used contains only terms whose applicability can be decided by everyday procedures (i.e. that it is something like Carnap's "physical language" in Testability and Meaning).[9] In this case J" will contain individual terms referring to the special properties of the individual thermometer used and J"" will contain terms referring to the special physical properties of mercury. This difficulty (it will turn out that this is a difficulty) can be removed by including the term "temperature" into the observational language. But first of all this procedure could never guarantee that individual terms would not enter somewhere else and secondly I am not sure whether Carnap e.g. would allow the term "temperature" into the observational vocabulary.

To sum up: Our example shows that a change of the interpretative system of a formalism need not imply a change of the theory which is based upon that formalism. The same applies to the change of the class of empirical generalizations supposed to follow from the theory.

2nd example: (This is a modified form of my example in Theoria)

T is a (e.g. the Hamiltonian) formalization of celestial mechanics. The descriptive terms of that formalism are functors such as "mass", "force", "distance", "acceleration", "geocentric coordinate" etc. whose variables range over particles of matter. In the first part of my example I shall assume that the observation-terms are functors such as "declination", "azimuth", whose variables range over luminous points on the sky. The interpretation-system J is fairly complicated as it must take into account refraction, aberration, etc. It also changes as soon as one of the relevant constants change (J contains the constants of aberration and refraction, assertions about the change of the refracting-index with the height, the temperature, etc.). Yet, if J, J′, J″, J‴, etc. are interpretative systems based upon a series of more and more refined constants of refraction and aberration, then we would have to assume as before that (T,J) = (T,J′) = (T,J″), etc. One could even maintain that the discovery of athmospheric ⟨sic⟩ refraction (a hypothetical historical situation) and its incorporation into the interpretative-system of the formalism left the theory unchanged. This change did not amount to a reinterpretation of celestial mechanics.

So far it has been assumed that "declination" and "azimuth" belong to the observation language. According to Carnap (p. 49 of his paper – end of Ch. V.) [7|8] those terms would not be "'observable' in the sense customary in philosophical discussions of methodology and therefore belong to the theoretical concepts …"[10] Hence, it would seem advisable to admit into the observation-language only such terms whose applicability can be decided with the help of our senses only, perhaps involving some uncomplicated operations. [If so, the interpretative system of the formalism of celestial mechanics would have to contain the names of all individual telescopes which are in use, as well as of meridian-instruments and the like. It would change as soon as new telescopes are erected and with every improvement of the telescopes already in use. Yet one would not say that on that account the theory itself is constantly changing.]

These two examples are sufficient for the proof of thesis (I).

ADDENDUM

July 4, 1957

Dear Professor Feigl,

Here a very brief argument in order to support my thesis 2, last letter (The "grammar" of the observation-language used for interpreting a theoretical language cannot be different from the grammar of the latter):

(1) Assume that L_O is the language of classical physics (containing the commuting ring of physical magnitudes a, b, c, ….) and that L_T is the language of quantum mechanics (containing the non-commuting ring of linear Hermitian Operators A, B, C, ……)

(2) Assume furthermore the rules of correspondence (which are a weakened form of von Neumann's rules of correspondence)

$$aa \rightarrow AA; \lambda a \rightarrow \lambda A; (\lambda \text{ a natural number}); a + b \rightarrow A + B.$$

then for any two A and B, AB = BA, which contradicts the basic principle of quantum mechanics (for the demonstration see Temple*, Nature 135, 957).[11]

The contradiction can be avoided (α) by assuming that the ring of the a, b, c, …. is isomorphous with the ring of the A, B, C, …. which means that we ascribe to L_O the same logical structure (with respect to the use of its descriptive terms) as to L_T. This is the procedure adopted e.g. by Ludwig, Die Grundlagen der Quantenmechanik Berlin 1954. He writes "Wir werden die Observablen A mit den Hermiteschen Operatoren A identifizieren und nur den einen Buchstaben A benutzen."[12] Or we may try to avoid the contradiction (β) by replacing (2) by different rules of correspondence (Choosing a different, or, "more basic" L_O, e.g. everyday-language – would not help). Now, if (α) is not adopted there seems to be no possibility (this is a hunch borne out by reference to the procedures of symmetrization used [8|9] within QM) to find general rules instead of (2). All we can do is to give some rather pragmatic advice as to how to choose one's operators on the basis of knowledge of how the system would behave in the classical limit. (This is connected with the fact that the correspondence-principle is a pragmatic rule). I.e. in this case it is not possible to adopt an unambiguous interpretation-system. This seems to be very undesirable, hence, the only procedure left is (α), viz. admitting that any L_O used for the interpretation of quantum mechanics has the same logical structure as the language of quantum mechanics itself (although it will contain a special subclass of the descriptive terms of QM, viz. the "observable" ones).

Result: Let L_T be the language of quantum-mechanics
L_O some observation-language
(L_O, L_T, J) the set of L_O, L_T, and some rules of linking both together, i.e. this set is quantum theory itself, T,

then, if L_O has not the same logical structure as L_T (or as T), then either J is a set of rules of thumb (and therefore not further specifiable) or T is inconsistent.

I would be glad to learn whether or not this argument (which can be extended to cases where L_T is not the language of quantum mechanics, although in these latter cases the argument will presumably be less strict and more 'philosophical') is correct and whether it is relevant as a criticism of (or a remark with respect to) the double-language-reconstruction-method which seems to me highly suspicious (and I am always trying to think of arguments in order to show its inadequacy). I am also sending a copy of this letter to Grünbaum.

Yours sincerely
Paul Feyerabend (signed)

P.S. I have asked Ryle to withdraw my note on Carnap's criterion of significance from publication (I had sent it to MIND) as I realize the problem needs more consideration than I thought at first.[13]

---

[7] * Cf. also the discussion following Temple's note.[14]

Editors' Notes

HF 02-133-02 (9 ss., T), excerpt (for the omitted parts here, see *below*, Ch. 3, § 2). HF 02-133-02 is a typescript composed by Feigl for distribution, but most likely never actually distributed per mail, which collates four different documents whose originals are missing: a letter of Feyerabend to Feigl, 24 June 1957; a short note, entitled by Feigl "Remarks by Feyerabend on Interpretation of Theories with an Eye Toward Quantum Mechanics", most likely sent by Feyerabend to Feigl between 24 June and 4 July 1957; an excerpt from a "Typescript by Feyerabend on Interpretation of Theories", which was included in the same invoice as the previous document; a letter of Feyerabend to Feigl, 4 July 1957.

[1] Feyerabend refers here to the "Conference on Philosophical Foundations of Physics", Minnesota Center for Philosophy of Science (MCPS), 12 August - 14 September 1957 (see *below*, Ch. 3, §2, n14).

[2] Feyerabend refers here to a draft of Hempel 1958 (see *above*, e 11, n1).

[3] See *above*, § 1, § 2; Feyerabend to Feigl, 19 May 1957, *below*, Ch. 3, § 1; Ch. 3, § 2; and Feyerabend 1955. The note to which Feyerabend refers here might be the missing "Typescript by Feyerabend on Interpretation of Theories", an excerpt of which is included on ss. 6-8 of this document, and whose content was elaborated in the paper that Feyerabend eventually presented as his contribution to the *12th International Congress of Philosophy*, Venice-Padua, 12-18 September 1958 (see FEYERABEND 1960a), after Popper's criticism made him withdraw the paper he had initially submitted (see Paul K. Feyerabend, "Observationally Complete Theories. Some observations on quantum theory", reproduced in Vol. 1, Part III, Ch. 4; Feyerabend to Popper, 1 May 1958, 1958-5; 7 June 1958, 1958-8, n7 in particular; Feyerabend to Feigl, 19 May 1957). A draft of FEYERABEND 1960a was sent by Feyerabend to Feigl, who most likely made it circulate among the MCPS members (see Paul K. Feyerabend, "On the Interpretation of Scientific Theories" HF 08-32-01 (8 ss., T+H; d) and HF 08-32-08 (8 ss., T): HF 08-32-01 is the original paper that Feyerabend sent to Feigl; HF 08-32-08 is the copy composed by Feigl for distribution; it should be noted that while Feyerabend signed the draft of his paper as a member of the University of Bristol, in the circulated version as well as in the published paper Feyerabend is indicated as a member of the University of California, which suggests that the circular copy was prepared after September 1958).

[4] See *above*, § 13.
[5] See *below*, Ch. 3, § 2.
[6] See *below*, s. 6.
[7] See *below*, s. 8.
[8] See HEMPEL 1958: 75.
[9] See CARNAP 1936: § 15; CARNAP 1937: §§ 18-21.
[10] See CARNAP 1956: 49.
[11] See TEMPLE 1935a.
[12] See LUDWIG 1954. English translation: "We will identify the observable A with the Hermitian operator A and only use the single letter A".
[13] See *above*, § 4, n1E; § 8.
[14] See FRÖHLICH-GUTH 1935; TEMPLE 1935b; PEIERLS 1935.

15. William W. Rozeboom, Comments on Feyerabend's Remarks, 6 November 1958

November 6, 1958

ON THE INTERPRETATION OF FEYERABEND'S
"ON THE INTERPRETATION OF SCIENTIFIC THEORIES"
William W. Rozeboom

More and more I find myself impressed by the frequency with which the Meaning vs. Reference distinction keeps turning up, in one guise or another, in the professional problems of philosophers and psychologists. The importance of this distinction for semantical theory is obvious, but it may also be found, to name a few areas which come to mind at the moment, in the problems of sense-data and the ontological status of abstract entities for philosophy, and those of perception and purposes, wants and goals for psychology.

The reason that I raise this distinction here is not to discuss it – I have already given substantial space to this end in my articles "Intentionality and 'Existence'" [I&E][1] and "The Factual Content of Theoretical Concepts" [FCTC][2] – but because Paul Feyerabend has in past occasions, the latest being his note, "On the Interpretation of Scientific Theories" [IST],[3] brandished a semantical thesis which, it seems to me, is gratuitous, wholly untenable, and owes its proposal entirely to an equivocation arising from insufficient attention to the difference between Meaning and Reference*. This thesis in question, called by F. the principle of Semantic Independence, is that

(A) The interpretation of a statement describing a situation which is causally independent of the situation described by another statement should be independent of the interpretation of this latter statement. (IST, p. 4)[4]

Before much can be said, either pro or con, about this Principle, it is necessary to know what F. means by the term "interpretation," his use of which F. has not so far made very clear. A plausible suggestion, and one which F. seems to suggest on the last line of IST, p. 7,[5] is that the "interpretation" of a statement is the state of affairs, or situation, which the sentence describes. But if by "interpretation of" a statement we mean the "situation described by" that statement, then F. has an excellent case for his Principle of Semantical Independence simply because it then merely reflects a tautology at least so long as the causal independence of a [1|2] situation S_1 on a situation S_2 entails the (unqualified) independence of S_1 on S_2. For substituting definiens for definiendum, in (A), we arrive at

(B) The situation described by a statement describing a situation which is causally independent of the situation described by another statement, should be independent of the situation described by this latter statement,

or, more succinctly,

(C) For any two (statementally describable) situations S_1 and S_2, if S_1 is causally independent of S_2, then S_1 should be independent of S_2.

But to advocate this – i.e., to propose that if S_1 and S_2 are causally independent, then S_1 and S_2 should be independent – is either to make a proposal about the use of the phrase "causally independent" and "independent," or, if these expressions are being used with previously assigned meanings, to opt for either a tautology, an empirical truth, or an empirical falsehood. In any case, if the "interpretation" of a statement is the situation described – i.e., the factual referent – of that statement, then F.'s Principle of Semantic Independence has nothing to do with semantics at all (except to the extent that the parenthetical qualification in (C) is an essential part of the thesis), but concerns at best only an empirical fact about causal relations and more likely, depending on the intended meaning of "interpretations," a tautology.

But although construing the "interpretation" of a statement to be its factual referent is what gives F.'s thesis its (only) prima facie plausibility, it is apparent from the use that F. makes of his Principle that this is not the force he has given to the term in question; rather, the "interpretation" of a statement must be taken to be its meaning. For example, F. wields his Principle to smite "Positivistic" interpretations of theories. But "Positivism" in F.'s broad sense (which includes the view that theoretical terms are implicitly definable on the basis of observational terms, a position which is compatible with – in fact, calls for – a realistic interpretation of theoretical concepts) is a claim about the meaning of theoretical terms, not about their referents (see my FCTC). Again, F.'s insistence that a change in "meaning" necessitates a change in "interpretation" (p. 4, lines 13 ff.)[6] makes clear that by "interpretation" F. primarily means "meaning," for no one who distinguishes between Meaning and Reference would hold that two expressions with different meanings must have different referents. Finally, that for F., "interpretation"$=_{def}$ "meaning" is made overwhelming certain by the parenthetical remark on p. 2, lines 24 f.[7] [2|3]

(I doubt that F. would disagree that what he means by the "interpretation" of a statement is its meaning. However, it is highly likely that he would also hold that the interpretation of a statement is what the statement describes, for as the concluding lines on p. 7, IST,[8] make all clear, he uses these three expressions interchangeably. And it is precisely because of this confounding of Meaning and Reference that he is able to move so blithly ⟨sic⟩ from the tautology, that the situations described by statements describing independent situations must be independent of one another, to the falsehood that the meanings of statements describing independent situations must also be independent of one another.)

Once the "interpretation" of a statement becomes identified as its meaning, however, and the distinction between Meaning and Reference is held firmly in mind, then F.'s Principle of Semantic Independence stands clearly revealed as false. For example, a given unobserved event U may indeed be "causally independent" (i.e., is not causally determined by – see IST, p. 2, lines 1-3)[9] of the observed effects O for which it is (in part) responsible, yet it may well be that our only knowledge of U is obtained through knowledge of its effects O, and that the way in which we refer to U is by use of an expression whose construction draws upon the fact that O occurred and whose meaning is therefore not necessarily independent of the meaning of an expression which refers to O. More generally, if $E_1,\dots,E_n$ are a set of expressions which refer, respectively, to entities $\varepsilon_1,\dots\varepsilon_n$, then the meanings of $E_1,\dots,E_n$ can be mobilized to construct an expression $F(E_1,\dots,E_n)$ which refers to a further entity ε_{n+1}, irrespective of whatever causal relations are sustained by $\varepsilon_1,\dots,\varepsilon_{n+1}$, so long as a predicate which applies to ε_{n+1} can be constructed from $E_1,\ldots,E_n$. I have gone into this matter in some detail in Sections III and IV of FCTC, so for present illustrations, I merely point to the use of descriptions. Let us agree that definite descriptions are genuine referring expressions (see FCTC, version dated Sept. 1958, p. 90 ff),[10] rather than syntactically misleading abbreviations as Russell claimed. Then

"The current President of the U.S.A. is bald"

refers to the fact that Eisenhower is bald; and the fact that Eisenhower is bald is causally independent (presumably, since he was already bald before taking office) of the fact that Eisenhower is currently President of the U.S.A. Yet the meaning of

"The current President of the U.S.A. is bald"

is not independent of the meaning of

"Eisenhower is currently President of the U.S.A." [3|4]

nor is there any reason why it should be – the facts about the Presidency make it possible for the meaning of "being currently President of the U.S.A." to be used in discussing Eisenhower's tonsorial needs. Similarly it is because as a matter of fact certain unobserved entities have certain observable effects that we are able to use the meaning-components involved in describing these effects also to refer to the unobserved causes. It is not the referent of the theoretical statement which is dependent upon the referent of the observational statement, but only the meanings – i.e., certain linguistic properties in virtue of which expressions have referents – of these statements which are so connected.

While proper heeding of the Meaning vs. Reference distinction should therefore suffice to sink F.'s current arguments against "Positivism" (in F.'s sense, which includes that form of Realism which I have Empirical Realism ⟨sic⟩, as well as Positivism in the narrow sense) without ripple or visible wreckage, I would still like to comment briefly on a couple of F.'s more specific arguments. On p. 3, IST,[11] F. draws upon his earlier trichotomy of non-logical terms into theoretical, observational, and mediational, to argue that participation of mediating terms in explication of a theory poses an insurmountable hurdle for "Positivistic" interpretations of theories. But on what ground does F. draw a distinction between theoretical terms and "mediating" terms? While any philosophical reconstruction of scientific practice must necessarily be somewhat arbitrary in its selection of what is to count as "observational," all extra-logical terms which are not observational are, by definition, theoretical. What F. seems to be disturbed by is that when we single out a particular theoretical statement we frequently have to adduce additional non-observational statements in order to arrive at an observational consequence. But a sophisticated "Positivism" (in F.'s sense) does not attempt to give a list of separate interpretations of individual theoretical statements; rather, it provides a simultaneous implicit definition of all theoretical terms in force, including the "mediational" ones (see my FCTC).

Another specific objection raised to "Positivism" by F. is that "given two situations S′ and S″, which are causally independent of each other, a change of our theories about S′ which preserves this causal independence and leaves unchanged the character of S″ may yet imply a change of the interpretation of S″." (IST, p. 3f)[12] There is indeed a problem here, but not the one that F. sees. For what worries F. is that we can change that part of our theory which concerns proximal events without changing what we want to say about distal events; yet this would not be possible, F. believes, under the [4|5] "Positivistic" interpretation of theories since the observational consequences of the distal theory are mediated by the proximal theory and modifying the latter would hence necessitate a change in the interpretation of the former. But even if changing our theory about proximal events alters the meanings of theoretical statements about distal events, it by no means follows that what the statements of the distal theory are about has been altered, for we can change the meaning of an expression without affecting its referent – "The current President of the U.S.A. is bald" and "The man who was Commander-in-Chief of NATO in 1948 is bald" both describe the same fact.

Nonetheless, the issue that F. touches upon here does pose a problem. For if the meaning of the terms by which we refer to distal events are dependent upon our theory about proximal events, and the proximal theory is in fact incorrect, do the statements of our distal theory have factual reference at all? For example (to move to a simpler level which exemplifies the difficulty more vividly), suppose we believe that Eisenhower was the senior U.S. senator from Alaska in 1950. We might think that

> "The man who was the senior U.S. senator from Alaska in 1950 is bald"

refers to the fact that Eisenhower is bald; however, since there were no U.S. senators from Alaska in 1950, this sentence actually has no factual referent at all. But if an error in one segment of the theoretical enterprise can contaminate the referential potency of theory in its entirety (as the reconstruction I propose in FCTC entails), is it ever justifiable (since presumably we never in practice have theories which are entirely correct) to think that theories actually in use have factual reference? I think the resolution of this difficulty lies along a line which I will only sketch here. Philosophical reconstructions of actual linguistic practices describe "pure" semantical states to which actual practices are but imperfect approximations. Where the "pure" reconstruction sees all semantical situations as either black or white (e.g., either statement S is true of S is not true; either expression E designates entity ε or E does not designate entity ε), these situations are regarded in de facto languages as varying shades of grey (e.g., S is almost but not quite true, since it is the conjunction of 100 logically independent and equally important statements only one of which is false). It seems to me, then, that in the de facto use of theories we would want to say that theoretical statements about distal unobservables, if otherwise correct, do in practice describe distal events even though the meanings of these statements are built upon a somewhat incorrect theory about proximal events and hence in a strict, all-or-none [5|6] reconstruction would have to be said to be without factual reference. For example (to return to the simpler level of analysis), the statement

> "The man who is currently President of the U.S.A., and was Commander-in-Chief of the NATO in 1948, and [etc.], and was the senior U.S. senator from Alaska in 1950, is bald,"

where the "[etc.]" appends a great deal of true biographical data about Eisenhower, does not, strictly speaking, describe any state of reality because there is no entity which completely satisfies the description. In practice, however, we would undoubtedly want to say that this statement does indeed refer to the fact that Eisenhower is bald, but that while doing so, it incorporates a slight error. If this line of thought is sound, it is then possible for theoretical statements whose meanings are based on a somewhat incorrect theory of proximal events nonetheless to make essentially true statements about distal events, and improvement of the proximal theory would (perhaps) only decontaminate the meaning of our distal theory while preserving its factual reference, just as dropping the clause about Alaskan senatorship ⟨sic⟩ in the statement above would purify its meaning but not affect its factual reference as seen from the point of view of ordinary language.

Finally, I want to raise an eyebrow or two at F.'s unabashed and carefree espousal of metaphysics. I do not feel that the term "metaphysics" deserves to be treated as the dirty word it has become in the mouths of contemporary empiricists. But F.'s brand of "metaphysics" seems to me to be a kind of high-society or let-the-other-fellow-do-the-dirty-work-and-be-damned-to-him Realism. It is all very well to bask Realistically ⟨sic⟩ in the knowledge that one has avoided many ontological and syntactical difficulties which beset alternative interpretations of theories, but this does not free one from the intellectual responsibility of coming to grips with the semantics-epistemological problem of <u>how</u> it is possible for us to talk about entities which we have never observed. To be sure, if one believes that human minds have some transcendental access to everything there is – a sort of apprentice omniscience – then there is little more to discuss here. But if a person's knowledge about the world is given to him only <u>through</u> his observations, as any consistent empiricist must agree, then the constructive labors of the Realist begin only when he leaves off criticizing other interpretations of theories and tries to explain and delimit the manner in which the data of experience can be smelted and recast in the form of concepts which referentially ignore the prison of our senses.

---

* This charge looks harsher in type than I intended. What I wish to say is that Paul's thesis, which seems reasonable so long as Meaning and Reference are confounded, falls apart immediately as the distinction is heeded.

Editors' Notes

HF 12-19-01 (6 ss., T); RC 088-26-06 (6 ss., T). RC 088-26-06 is annotated in Carnap's characteristic shorthand writing.

[1] See Rozeboom 1962a.

[2] See Rozeboom 1962b.

[3] Rozeboom refers here to the preprint version of Feyerabend 1960a prepared by Feigl for distribution (see *above*, § 14, n3).

[4] Page references are to the circulated copy of the preprint version of Feyerabend 1960a (HF 08-32-08; see *above*, n3). See Feyerabend 1960a: [40].

[5] See Feyerabend 1960a: 158, end of the first paragraph [42, end of the second paragraph].

[6] See Feyerabend 1960a: 154-155 [39-40].

[7] See Feyerabend 1960a: 153, end of the first paragraph [38, end of the second paragraph].

[8] See Feyerabend 1960a: 158, end of the first paragraph [42, end of the second paragraph].

[9] See Feyerabend 1960a: 152, middle of the first paragraph [38, top].

[10] See Rozeboom 1962: 349-351.

[11] See Feyerabend 1960a: 153-154 [38-39].

[12] See Feyerabend 1960a: 154 [39].

16. Paul K. Feyerabend, Reply to Rozeboom's Comments, [1958]

P.K. Feyerabend

NOTE ON ROZEBOOM'S MEMO OF NOV. 6, 1958.

(1) In his memo of Nov. 6 (to be referred to as RM),[75] Bill Rozeboom severely critizises ⟨sic⟩ my note (FM)[76] on the interpretation of scientific theories. As his criticism is partly connected with the fact that I never made explicit the purpose of my note, and as RM interprets this purpose in a way slightly different from the way in which I should like to see it interpreted, I shall start my reply by explaining this purpose a little more in detail.

What I attempted in FM was to find, or to make some steps towards, an adequate explication of the terms "interpretation of a theory", or "interpretation of a scientific statement", or, if you like, "meaning of a scientific statement". With respect to this attempt, the sentence called "A" by RM, and 'the principle of semantic independence' (PS) by FM plays the role of a criterion of adequacy. Being such it cannot be attacked by showing that in some sense of 'interpretation' it is incorrect (for such an attack cf. the first four lines, par. 2, page 3 of RM). Nor can it be attacked by saying that "it owes its proposal entirely to an equivocation arising from insufficient attention to the difference between meaning and reference" (RM, p1). Such an attack would be admissible only, if I had intended, with my principle of semantic independence, to formulate a theorem about some types of meaning already in existence (e.g. about what is termed the 'connotation' of a term, or a sentence) rather than a necessary condition of a perhaps not yet existing explication. In order to see this point more clearly consider the following example. Assume that a philosopher attempts to find an explication of the concept of necessity and that, as a criterion of adequacy, he introduces the principle that any assertion to the effect that a certain statement is necessary must itself be necessary. It would be unjustified to critizise ⟨sic⟩ this principle by pointing out that, necessity being a property of sentences, any statement to the effect that a sentence is necessary must itself be contingent. For the rejoinder would simply consist in the remark that necessity interpreted as a property of linguistic entities was not an adequate explication of the concept of necessity as determined by the above principle. It is of course correct that criteria of adequacy may themselves be subject to criticism and I think that some of Rozeboom's remarks (to be dealt with in section 3 of the present note) may be interpreted as attacks against the PS considered as a criterion of adequacy. Yet the explanations of the present section seemed to be necessary in order to see the subsequent [1|2] discussion in better perspective.

(2) Before turning to those of Bill's remarks which may be interpreted as arguments against the plausibility of the PS as a criterion of adequacy I would first like to point out a slight inaccuracy in his presentation. On p3 he asserts that the PS obtains an appearance of truth from the fact that sometimes I interpret 'interpretation' to mean 'meaning', and sometimes I interpret it to mean 'reference'; and that in the latter case the PS is true, but trivially so: "It is precisely because of this confounding of Meaning and Reference that he is able to move so blithly ⟨sic⟩ from the

tautology that the situation described by statements describing independent situations must be independent of one another, to the falsehood that the meanings of statements describing independent situations must also be independent of one another" (RM, p3, top paragraph). Three points ought to be made. First that Rozeboom, proceeding from already existing notions of meaning sees in the PS an ambiguity which in fact does not exist. For, as was said in sec. 1, it is not the case that I want to defend the PS as a theorem about some already existent meanings of 'meaning', and that I am therefore intent on showing that it is true (by using one sense of 'meaning') and interesting (by switching to another sense of 'meaning'). Any such attempt, whether based upon ambiguity or not, would be utterly pointless as all that I am doing is to put forward a criterion of adequacy and to look for an explanation which satisfies this criterion. – My second point is this: Rozeboom charges that sometimes I identify meaning with reference, that this makes the PS true, but trivially so. My answer is that if this were the case it would be an argument in my favour and not against me. For as I am looking for an explication of 'interpretation', as any such explication will consist of a set of statements all of them analytical, as the explication is adequate only if its criterion of adequacy can be shown to be derivable from it, it would plainly follow that an explication is satisfactory only if after due substitution its criterion of adequacy can be shown to be trivially true. Hence, Rozeboom's charge of triviality, far from being an objection, turns out to be an argument in my favour, if at any stage of my paper I had identified 'meaning' and 'interpretation' with 'reference'. But this – and therewith I turn to my third point – is plainly not the case. Quite on the contrary: on the very place to which Rozeboom refers (p3 of RM) in order to establish his contention that I [2|3] "would also hold that the interpretation of a statement is what a statement describes", viz. on p7 of FM (last two lines) I explicitly say that "the interpretation of a scientific statement depends upon nothing but the state of affairs it describes" – which is a very different matter indeed. I.e. I hold interpretations to be linguistic entities uniquely determined by (but not identical with) the reference of the corresponding sentence. This disposes of R's criticism on p1, last paragraph, page 2 first and second paragraph ⟨sic⟩, page 3, top paragraph, and page 5 top paragraph of his paper.

(3) However the most important question is now whether the PS is acceptable as a criterion of adequacy for an explication of 'interpretation of a scientific theory'. I now turn to those of R′ ⟨sic⟩ arguments which seem to imply that it is not. R points out that we may refer to an unobservable U "by use of an expression whose construction draws upon the fact that" an observed effect O of which U is causally independent "occurred and whose meaning is therefore not necessarily independent of the meaning of an expression which refers to O" (RM, p3). As a very simple example Rozeboom uses the sentence

The current president of the U.S.A. is bald (A)

whose 'meaning' (in the common-and-garden-sense of 'meaning') depends upon the meaning of 'being a president' although what it describes (the baldness of Mr. Eisenhower) is causally independent of Mr. Eisenhower's being a president. A similar example which is nearer to the example used in my own paper would proceed from the sentence

The greatest source of energy for the motion of the athmosphere ⟨sic⟩ of the earth has a surface temperature of 6000 degrees centigrade. (B)

Here one could argue that the fact that the sun has 6000 degrees surface temperature is causally independent of the motion of the athmosphere ⟨sic⟩ of the earth; and yet the meaning of a sentence expressing that fact, viz ⟨sic⟩ the meaning of (B) depends upon the meaning of the expression 'motion of the athmosphere ⟨sic⟩ of the earth'. These two examples make it appear unreasonable to adopt a principle (viz. the PS) which rules them out from the very beginning. Two points ought to be made in connexion with this criticism. First we may ask the question whether it is indeed the case that what (A) asserts, expresses, or the fact to which (A) refers is Eisenhower's baldness. Let us assume that it is. Then if (A) is true this fact must be said to contain a further component, viz. Eisenhower's presidency. I.e. it would then be incorrect to [3|4] assert that the fact referred to by (A) is nothing but Eisenhower's baldness, this fact would have to be described by a phrase like 'Eisenhower's baldness-presidency'. Now if this is the case then the existence of (A), (B) and of similar sentences cannot any longer be used as an argument against PS. I am aware that I have here entered the very slippery ground of descriptive phrases, but the point is that I am not convinced that the account usually given of them is either correct, or the only correct account possible. Formally my argument could be said to consist in the consideration that if BALD[(ιx)PRES(x)], then, provided we abbreviate '(ιx)PRES(x)' as 'X', it follows that BALD(X)&PRES(X).(*) – My second point is this: assume Rozeboom's interpretation of (A) and (B) is correct. Does it follow that every empirical sentence is of that type, i.e. does it follow that every empirical sentence contains a descriptive phrase involving states of affairs of which the reference of the whole sentence ('reference' now in the sense of Rozeboom, and not in the sense suggested in my point 1 above) is causally independent? (B) is plainly such a sentence. It refers to the temperature of the sun with the help of a descriptive phrase whose referent does not influence it causally. But is

The sun has a surface temperature of 6000 degrees Kelvin (C)

also such a sentence? It is not. But now the Carnap-Hempel model suggests that even (C) which does not contain any descriptive phrase involving a state of affairs of which the temperature of the sun is causally independent, implicitly uses such descriptive phrases in which reference is made to effects which the sun has upon the earth. "It is because as a matter of fact certain unobserved entities have certain observable effects that we are able to use the meaning component involved in describing these effects also to refer to the unobservable cause" says Rozeboom (RM, p 4, top paragraph) which seems to mean that in the last analysis we always talk about theoretical entities by either explicitly, or, in the great majority of cases, implicitly using descriptive phrases which describe some of their observable effects. Now before I turn to a more detailed discussion of this last thesis which I think to be implicit in the Carnap-Hempel approach, it is worth pointing out that even if Rozeboom should have shown the PS to be untenable for statements such as (A) and (B) which explicitly use a descriptive phrase⟨,⟩ he has thereby not shown that the PS will be untenable for (C) as well, i.e. he has not shown that the PS will be untenable also [4|5] for those statements with which I was concerned in my paper (Scientific theories, or singular statements of science), which never contain a descriptive phrase explicitly. The remark that they may contain it implicitly cannot any longer be regarded as constituting an argument against me for this is just my charge: that I find it very implausible that a statement about the sun, which is not of type (B) should change its interpretation when a statement about the athmosphere ⟨sic⟩ of the earth changes its interpretation, although the phrase 'athmosphere ⟨sic⟩ of the earth' does not appear anywhere in the statement.

(4) In this section I want to deal with various disconnected points made in Bill's paper. On page 4 he criticizes my trichotomy theoretical-observational-mediational terms. His criticism is a wellcome ⟨sic⟩ opportunity for formulating a question which has puzzled me for some time. Carnap and Hempel both distinguish theoretical terms and observational terms and Carnap stresses (cf. the quotation on page 2 of FM) that the theoretical terms possess <u>no interpretation over and above what is conferred upon them by the fact that they are connected with the observational terms in a certain way</u>. This means that the interpretation of say, the term 'mass' is completely determined by its connexion (via a suitable interpretative system) with some L_O as well as by the postulates of classical mechanics. The same applies to every other term of classical mechanics. Now these terms will also occur in interpretative systems for theoretical languages different from the theoretical language of mechanics. Here the question arises: is their interpretation when used in this way to be determined by their occurrence and their function within mechanics <u>only</u>, or does the fact that they now occur as parts of (viz. in the interpretative system for) a different theoretical language further contribute to their interpretation? It seems to me that Carnap, although not explicit on that point, would take the first position, whereas Rozeboom explicitly adopts the second. However this second position has peculiar difficulties of its own. In order to demonstrate these difficulties consider the case of two theories, T′ and T″ which are formally contradictory, the interpretative system of $L_{T''}$ containing terms belonging to $L_{T'}$. As a concrete example take T′=classical mechanics; T″=special relativity; and assume that we test T″ with the help of instruments involving simple mechanical devices (the example can be worked out in detail). Then the interpretative system of T″ will contain the terms 'mass', 'force' etc. Now if, as Rozeboom assumes, the connexion, via an interpretative system, of a theoretical language with an observational language provides an interpretation not [5|6] only for the descriptive terms of the former, but also of the theoretical terms contained in the interpretative system itself, then we are faced with the following two possibilities: (a) the interpretation of 'Mass', 'force' as used in the interpretative system of T″ is different from the interpretation of the very same terms as used within $L_{T'}$, the first one being uniquely determined by $L_{T''}$ and the structure of the interpretative system itself, the second one being uniquely determined by the structure (including the axioms) of $L_{T'}$, and of the interpretative system used for the interpretation of $L_{T'}$. In this case the mechanical properties of mass and force as well as the methods of their observational determination could not be used

in the interpretation of T″, i.e. the interpretative system would not serve its purpose; the second possibility (b), is that the interpretation of 'mass', 'force' in T′ and T″ is the same and depending upon both T′ and T″, which would make the terms inconsistent. The only way out of these difficulties seems to me to consist in the assumption (if one wants to retain the Carnap-Hempel model) that the interpretation of the theoretical terms of an interpretative system is independent of the theory resulting from the application of some L_O + this interpretative system to some theoretical language. But if that is the case then the distinction of the terms of a theory into theoretical terms (i.e. descriptive terms of the L_T of the theory under review); mediating terms (i.e. descriptive terms of the L_T of some other theory); and observational terms seems to me to be perfectly justified.

Another thing that makes Rozeboom feel uncomfortable is my use of the word 'positivism'. I would be the last one to quibble about words. However I have always felt that realism asserts that existence and observability are two different things and that therefore the meaning of a statement (which is uniquely determined by what the statement is about) cannot be exhaustively characterized by the way in which it is related to some L_O + the meaning of the terms of this L_O itself. Now the fact that the principle of semantic independence which I still think to be a very plausible principle and a good starting point for any theory of meaning, leads to the same result seems to me to be strong independent evidence for the plausibility of realism. If we assume that the meaning of a scientific statement cannot depend upon the meaning of statements describing states of affairs of which the state of affairs described by the first statement is causally independent, then a change of the interpretative system of a theory cannot influence the meaning of its primitive terms either. It then [6|7]
follows that this meaning is not wholly empirical in the sense that it is not uniquely determined by experience and the formal features of the LT of the theory. Insofar as the theories listed in sec. 1 of FM imply the opposite, viz. insofar as they imply that the meaning of every theoretical term is uniquely determined by experience and the formal features of the L_T of the theory (there exists "no independent interpretation for them" says Carnap), they still have an important property in common with all sorts of other positivistic theories, a property moreover, which is not shared by any honest-to-god realism (see above!). This is my reason for drawing the distinction between positivism and realism in a way which is different from the way in which Rozeboom wants to draw it.

Now with respect of the underlined statement in the above paragraph it is important to realize that it is not the starting point of my analysis, but rather the result (the same is true of my Theoria-paper.[77] On page 59 of that paper I say: "So far I have shown (a) that a theory possesses an interpretation independently of any particular correlation between its elementary expressions and simple facts of observation; and (b) that its interpretation is not necessarily changed by a change of this relation" – and this assertion is the result of my analysis, and not my point of departure.) And as what is not uniquely determined by experience + the formal structure of the theory concerned has frequently been called 'metaphysical', this result may also be expressed by saying that the content, or the interpretation, or the 'meaning' of a scientific term is partly metaphysical.

This brings me to the third point I want to make in the present section. The occurrence of the word 'metaphysical' in my paper is described by Rozeboom as an "unabashed and carefree espousal of metaphysics", as a "kind of high-society or let-the-other-fellow-do-the-dirty-work-and-be-damned-to-him-Realism" (p 6 – an excellent demonstration, to a foreigner like myself, of the ample opportunities the English tongue provides for the philosopher). Now what I have said above should have made it clear that this presentation of may ⟨sic⟩ way of bringing in metaphysics is somewhat distorted – to say the least. For however unacceptable my point of departure may be, it most certainly does not consist in the invitation to exercise some kind of transcendental ability, or to join a high-society-club of epistemological beatniks, or anything of that sort. My point of departure is a semantical principle (the PS) which I hold to be fairly plausible and which, if applied, has the somewhat unexpected result that scientific terms, if meaningful, are partly metaphysical, a result which I [7|8] then tried to make plausible from the point of view of realism (see sec. 7 of FM). Hence the only presuppositions of my paper are (1) the PS; and (2) the invitation to use what science implies about the relation between things and their observational effects. And what I asserted was that on the background of these two assertions the Carnap-Hempel model makes a curious impression indeed! Of course, when denying that in my paper I started from the assumption of transcendental abilities (abilities, that is, which transcend the domain of the observational results available at a certain time) I did not mean to deny, as Rozeboom does, that such abilities exist and play an important role in the development of science. Quite on the contrary it seems to me that Bill's principle that "a person's knowledge about the world is given to him only through his observations" (p 6) makes it completely impossible to explain the growth of science. It is unclear to me how the law of inertia (involving the occurrence of motion into infinity; by the way one can show, by simple calculation that an inertial movement will never be accessible to observation!), or how Newton's physics (involving infinite space); or how Maxwell's electrodynamics (which involves the continuity of the electromagnetic field) could ever have been given to the scientist "through his observations". All these theories transcend the observations known at their time to such an extent that the ability to invent them must indeed be regarded as a transcendental ability. Of course, this "does not free us from the intellectual responsibility of coming to grips with the semantical-epistemological problem of how it is possible for us to talk about entities which we have never observed" (P 6 of RM) – but the invention of an untenable model of scientific procedure which behaves as if the reference to experience were sufficient does not free us from this responsibility either. For if it is true that without some transcendental ability and some kind of metaphysical insight the progress of science is impossible, and the occurrence of this progress inexplicable, then it is the empiricist (for whom 'metaphysics' frequently is a dirty word indeed and who wants to get on without it) who must be accused of retiring to a high-society club while the other fellow (i.e. the scientist) has to do the dirty work for him by exercising his disreputable transcendental abilities and inventing theories which he, the empiricist then disinfects and contemplates independently of their unmentionable origin, overlooking, as I contend, that in the process of disinfection he so thoroughly changes them as to make them altogether unrecognizable. [8|9]

(5) Let me conclude with a somewhat sketchy remark on the thesis underlined in sec. 3 viz. on the thesis that <u>in the last analysis we always talk about theoretical entities by either explicitly, or, in the great majority of cases, implicitly using descriptive phrases which describe some of their observational effects</u>. This thesis seems to me to lead to peculiar difficulties. Consider the case that a certain theoretical entity is characterized by reference to <u>all</u> the observational effects ascribed to it by a theory T′ and that we want to assert that it is identical with an entity characterized by all the observational effects ascribed to it by T″. If T′ and T″ are formally contradictory then this identification will be clearly impossible. But it will also be impossible <u>if the interpretative system of T′ is incompatible with the interpretative system of T″</u> although in this case we would be inclined to say that we were talking about the same things allright ⟨sic⟩ the difficulty being that we had made incorrect assumptions about the way in which their causal effects (which we also described correctly) were affected by disturbing influences (such as the athmosphere ⟨sic⟩ of the earth). One could attempt to remedy this difficulty by declaring that the one interpretative system contained statements which are in fact false. But this would imply that no identification can be made between the two entities mentioned above as long as we insist that all we can do is to introduce them by descriptive phrases about their observational effects: no identification can be made between Jupiter as observed through a bad telescope and Jupiter as observed through a good telescope. This absurd consequence seems to me to be a decisive argument against the underlined passage above.

(6) These are the reasons why I doubt that Rozeboom's arguments, and his final sermon about what is proper and what is not proper for a philosopher to do are strong enough to "sink my argument against 'Positivism' without ripples or visible wreckage" (p4). In giving my reasons I have used RM only and I have not taken into account R's positive theses in the other papers which he refers and which I had not yet any opportunity to consult. [9|10]

(7) Rereading the first point I made in section 3 I feel that a little more elaboration is needed. Rozeboom asserts that what

The present President of the U.S. is bald (A)

refers to or is about is Eisenhower's baldness. Now take the further assertion

A sentence is true if and only if what it is about is in fact the case (B)

Combining this with the assertion that

What (A) is about does not contain Eisenhower's presidency (C)

we arrive at the result that

The present President of the U.S. is bald and Eisenhower is not the present President of the U.S.
(D)

from which we may infer, via (B), that

(A) in not about Eisenhower's baldness (E)

which contradicts Rozeboom's assertion that it is. In order to escape this inconsistency Rozeboom might point out that he holds not (C), but the weaker assumption that

(A) is not about Eisenhower's presidency

(C′)

and that (A), (B), (C′), taken together do not lead to the result (E). But if (C′) is accepted instead of (C) then Rozeboom cannot any longer repeat his original argument. Remember this argument: (A) is about Eisenhower's baldness. Eisenhower's baldness is causally independent of his Presidency ⟨sic⟩. Yet the meaning of (A) in not independent of the meaning of 'President of the U.S.'. Now in this argument it is assumed that what (A) is about is causally independent of Eisenhower's presidency and, a fortiori, does not contain it. If we use, not the assumption that (A) does not contain Eisenhower's presidency (which, as is shown above, leads to inconsistency) but the assumption that (A) is not about Eisenhower's presidency, admitting however that a fact which is described by a sentence not being about Eisenhowers ⟨sic⟩ presidency may yet contain this fact as a constituent (this is the move from C to C′ above), then an inconsistency with the PS cannot any longer be asserted. The result seems to be that either Rozeboom's argument is insufficient for deriving the conclusion he wants to derive (viz. that the existence of sentences such as (A) contradicts my principle of semantical independence), or that it is inconsistent. In both cases it is unsatisfactory.

---

[4] (*) For a more detailed criticism turn at once to sec. 7

Editors' Notes

HF 08-33-16 (10 ss.; T+H)

[1] See *above*, § 15.

[2] See *above*, § 15, n3.

[3] See Feyerabend 1955.

Chapter 4
Feyerabend at the Minnesota Center for Philosophy of Science (1957-1963)

Paul K. Feyerabend

1957

1. Feyerabend to Feigl, 19 May [1957]

19. May

Dear Prof. Feigl,

I thank you very much for sending me Carnap's comments[1] as well as the program of the summer-session.[2] I have comments on both and I shall start with Carnap:
<u>His criticism, page 1, paragraph 1</u>: [... 1|2 ...] [3]

So far my remarks about the criterion. Now as to the program. What regards the more general topics I am most interested in part III.[4] If there is any opportunity I would like to discuss the problem of the relation between theoretical language and observation-language which I have very sketchily discussed in my note in <u>Theoria</u>.(+) This problem seems to me to be of some importance for two reasons. The first reson ⟨sic⟩ is that the double-language-methode ⟨sic⟩ of Carnap seems to me to be definitely inadequate. I hope that I shall be able to send you a brief note on this matter in a very short time. The second [2|3] reason is that some similar ideas (in fact they are nearly identical) seem to be part and parcel of the Copenhagen-interpretation. I could also point out where the fact that physicists believe in those ideas leads into difficulties. Hence the double-language-thing could be then discussed in connexion with some very concrete difficulties. I shall not be more definite about this now as I hope that I shall be able to send you a short note within the next two weeks. But I do enclose the final version of my Bristol-paper (I have changed parts of it and I think that it is now much clearer than it was before; also some inaccuracies have been removed).[5] In section (6) of this paper (starting on page 10) which I would ask you to read if you have not time to read the whole paper, I try to connect the special

M. Collodel, E. Oberheim (eds.), *Feyerabend's Formative Years. Volume 2. Feyerabend on Logical Empiricism, Bohm & Kuhn*, Vienna Circle Institute Library 10,
https://doi.org/10.1007/978-3-031-57519-8_4

results obtained with a broader philosophical background, especially with the double-language-theory. But the remarks there are only very sketchy, as the paper is essentially on a physical ⟨sic⟩ problem.[6] Please do not any longer use the paper which Scriven brought along as it contains several inaccuracies.[7]

I have got nearly finished a paper on the Einstein-Podolski-Rosen-paradox. The Bristol-Volume will be out in October.[8]

Many thanks for your communication

Yours sincerely
Paul Feyerabend

P.S. Perhaps in the session on probability the Neumann-proof can be discussed as some people have assumed that it necessitates a revision of our concept of probability.
In the next letter I shall send some more detailed suggestions with regard to the meeting.

---

[2] (+) *I still have Pap's objections which he made against this paper in 1955, I believe in your institute. I think I can reply to most of those objections.*[9]

Editors' Notes

HF 02-32-12 (3 ss., T+H)

[1] A Feyerabend had already acknowledged the receipt of Carnap's reply (see *above*, Ch. 2, § 9) with a brief comment in a previous letter to Feigl: "I have not yet had time to take a close look to Carnap's criticism. I am supposed to deliver some lecture in Birmingham + must prepare myself – but I shall write to you as soon as I have studied it more carefully. Page 1 seems easy to answer, but not so page 2" (Feyerabend to Feigl, 6 May 1957, HF 02-133-01).

[2] Feyerabend refers here to the "Conference on Philosophical Foundations of Physics", Minnesota Center for Philosophy of Science (MCPS), 12 August - 14 September 1957; on Feyerabend and the 1957 MCPS conference, see Vol. 1, Part II, 1956-20, n4;

[3] See *above*, Ch. 2 § 10.

[4] Apparently, no copy of the "Circular letter: Suggestions for the Summer Conference 1957" (see "Minnesota Center for Philosophy of Science, Activities Report for 1956-57", 25 June 1957, HF 01-36-01 = HF 01-40-05) has been preserved.

[5] Feyerabend refers here to Paul K. Feyerabend,"On the Quantum Theory of Measurement", HF 08-33-27 (11 ss., T+H); see FEYERABEND 1957a; on the history of the development of FEYERABEND 1957a, see *below*, § 2, n12.

[6] See FEYERABEND 1957a: § 6.

[7] Feyerabend refers here to Paul K. Feyerabend, "On the Quantum-Theory of Measurement", HF 08-33-26 (11 ss., T+H; d); see Vol. 1, Ch. 2, 1957-7, n3; FEYERABEND 1957a; on the history of the development of FEYERABEND 1957a, see *below*, § 2, n12.

[8] See KÖRNER-PRYCE (eds) 1957.

[9] See *above*, Ch. 2, § 2, § 14.

2. Feyerabend to Feigl, 28 June 1957

P. K. Feyerabend
June 28, 1957

Dear Professor Feigl:

I have been meaning to write to you for a long time, but about three weeks ago I was put out of action by the most agonising nerve-pains which I have now had for about three weeks and which I had to fight off by very strong medicines and those medicines put me into a daze most of the time. Now only a few words as to the summer-session (which I sincerely hope I shall be able to attend).[1]

⟨...⟩[2]

<u>3rd</u>. <u>The proofs of the papers and discussions of the Bristol-Meeting will be finished by July 20.</u>[3] <u>I shall then be able to send you a set of proofs which may be of importance for the discussions at the institute</u>.

<u>4th</u>. I considered the question which problems should be treated in the section devoted to quantum-mechanics (Section V of the preliminary program).[4] These are my suggestions:

There are essentially two large groups of problems. Group I is concerned with the analysis of quantum mechanics <u>in its present form and interpretation</u> (and by "quantum mechanics" I shall always understand elementary quantum mechanics, preferably without spin); group II is concerned with suggestions as to the possible torn of a future theory of microscopic phenomena. I would suggest to concentrate on the problems of group I and only occasionally to touch problems of group II. This, of course, excludes Bohm's new theories (with the sole exception of his theory of 1952)[5] – but I think in matters as difficult as those it is wise to be too narrow rather than too wide.

Secondly it seems to me to be of advantage to give a abort introduction into the language of the theory to be discussed, i.e., into its <u>syntax</u> as wall as into its <u>semantics</u>.

The SYNTAX of the language of QM is essentially that of the calculus of vectors and operators in Hilbert-space. An elementary explanation of the rules of this calculus would be advisable. <u>Literature</u>: Dirac Chs. I and II, von Neumann Ch. II or [1|2] Reichenbach, Part II.[6] Reichenbach does not use von Neumann's notation which at times turns out to be very useful (representation of the state of a system not by a vector, but by its "statistical operator" – for a very recent account cf. U. Fano, <u>Revs. Mod. Phys</u>., Vol. 29 (1957), 74.)[7]

The SEMANTICS of the language of QM comprises those rules which are (necessary and) sufficient for transforming the formalism into a fully-fledged physical theory. The semantical rules of the formalism of QM have become known as the <u>Born-interpretation</u>. It should be noted that the Born-interpretation connects an <u>uninterpreted</u> (in the sense of semantics) formalism with "reality".

According to the Born-interpretation the probability to obtain the value a′ of the observable A on measurement of A in the state $|\psi\rangle$ is $|\langle \psi'| \psi\rangle|^2$, $|\psi'\rangle$ being the eigenfunction of A belonging to a′.

For philosophical discussion it is advisable to use the Born-interpretation in the form Exp(R) = Tr(RW): The expectation-value of the magnitude R in the state W is the trace of the product of the operator R (which corresponds to the magnitude R) with the statistical operator W corresponding to the state W.

This interpretation raises the problem of the correspondence between operators and physical magnitudes, a problem which seems still to be unsolved (Cf. Groenewold's paper in Physica 12 (1946).)[8] The semantics of QM is still unsettled. (This would be a topic for discussion.)

ONTOLOGY However, when discussing the question which is the proper interpretation of quantum mechanics, a wave-interpretation, a particle interpretation or e.g. the Copenhagen-interpretation, physicists and philosophers are not concerned with semantical problems, i.e. trey are not concerned with the problems how an uninterpreted formalism ought to be connected "with reality". The question "particles or waves?" rather presupposes that the symbols of quantum mechanics have already been given a certain meaning, i.e. it presupposes that all syntactical and semantical problems have been settled in a satisfactory way. What is to be interpreted is not a formalism, but a physical theory. This is the reason why it seems to be advisable to distinguish between two different kinds of interpretation of a physical theory, between its semantical interpretation and its ontological interpretation. The Born-interpretation is a semantical interpretation of the formalism of elementary quantum-mechanics. The Copenhagen-interpretation (or the wave-interpretation or the particle-interpretation) is an ontological interpretation of quantum-theory. Problems connected with ontological interpretations I shall call ontological problems. This distinction between syntactical problems, semantic problems, ontological problems, seems to be very useful, especially in the case of quantum mechanics. (Reichenbach is not very clear on that issue; he introduces the Born-interpretation as an ontological interpretation [2|3] (p. 22) but discusses it later as part of the semantic interpretation (p. 82).)

Traditional philosophers have tried to solve ontological (or metaphysical) problems such as e.g. the problem of causality (or the narrower problem of determinism) by speculation on the basis of (sometimes very scarce) experience. The existence of very general scientific theories enables the philosopher to change the methods of ontological research. For it may turn out that a theorem of one of those theories either contradicts, or implies a statement of metaphysics. Such a theorem may be called "ontologically relevant". And a hypothesis as to the ontologically relevant theorems of a given theory may be called an ontological interpretation of that theory. Ontological interpretations in this sense can be tested by comparing their consequences with theorems of the theory so interpreted. It is not always easy to carry out such a test. This is the reason why there is still so much argument about the (ontological) interpretation of quantum-mechanics. On the other hand they may be introduced with the help of certain arguments which do not at all refer to theorems of the theory so interpreted and which strongly resemble the ontological arguments of traditional metaphysics. Most of Bohr's arguments are of this kind, although his results are shown to be correct by many theorems of the theory itself.

In order to show that the above presentation is correct one need only analyse Bohr's and Heisenberg's arguments against determinism (another topic for discussion) which are by no means conclusive (in fact Heisenberg's microscope-arguments are all circular) but lead to the correct result as is shown by the Neumann-proof.

The same applies to Bohr's and Heisenberg's contention that quantum-mechanical description of reality is complete (the correctness of the result of these arguments is again shown by von Neumann's proof).

There are only two major* objections which have been raised against the Copenhagen-interpretation (both raised at the Bristol congress).

1) The paradox of Einstein, Podolski and Rosen
2) The paradox of macroscopic determinatedness (in two form: Schroedinger's cat, Naturwissenschaften Vol. 23 and Einstein's Box in Scientific Papers presented to Max Born)[9]

A thorough discussion of those two objections is advisable. This discussion will have to touch upon three more philosophical problems

(a) Quantum-Mechanics and Subjectivism
(b) Quantum-Mechanics and Polyvalued Logics
(c) Are there Quantum-Jumps? [3|4]

and will also have to deal with the more technical problem of the quantum theory of measurement* which is still in a rather sad state.

In my opinion the result of the discussion of those two objections shows that the Copenhagen-interpretation is, in general, on the right track. (a) and (c) will be answered in the negative. Bohr himself has never been happy about polyvalued logics in quantum-mechanics, and it also seems to me that the need for such new logics is only felt when one makes certain mistaken assumptions about the theory.

Other ontological interpretations could be discussed such e.g. (a) Einstein's (which is still defended by Popper) (b) Bohm's first interpretation (c) similarity between quantum-mechanics and the theory of diffusion – here questions of reversibility come in. An excellent review-article exists on this: ter Haar in Revs. Mod. Phys. July 1955[10] which may be regarded as a continuation of the very important Ehrenfest paper). (d) Schroedinger's interpretation (in his "Are there Quantum-Jumps?" British Journal for the Phil. Sc. 1952).[11] – and finally – but this already belongs to Group II, – Bohm's new papers.

I have made these suggestions not, because I think that they all can be discussed in the short time available, but in order to suggest a certain order for the topics of discussion – the clear distinction between syntactical, semantical and ontological problems. In the connexion with the Born-interpretation the role of probability within quantum-mechanics could perhaps also be discussed.

I have already sent you two papers by myself on the theory of measurement[12] as well as part of my discussion with Weizsaecker on polyvalued logics (referring to his paper in Naturwissenschaft 1955).[13] I hope that I shall soon be able to send you a paper on the Einstein-Podolski-Rosen-case as well as on the more general matter of Professor Hempel's paper to the discussion of which I am really looking forwards.[14]

In the meantime I have also received Scriven's paper to Vol. II of the Proceedings[15] and I find myself in great sympathy with his views which are there set out with extreme clarity and lucidity. I am also enclosing a paper of measurement which is to be published in Zs. Physik within a few months now on (October 1957).[16]

Yours sincerely
Paul Feyerabend (signed)

---

[3] * i.e., objections which are not purely philosophical
[4] * Perhaps this should be the main topic of discussion.

Editors' Notes

HF 02-133-02 (9 ss., T), excerpt (for the omitted parts here, see *above*, Ch. 2, § 14).

[1] Feyerabend refers here to the "Conference on Philosophical Foundations of Physics", Minnesota Center for Philosophy of Science (MCPS), 12 August - 14 September 1957 (see *below*, n14).

[2] See *above*, Ch. 2, § 14.

[3] Feyerabend refers here to the ninth symposium of the Colston Research Society "Observation and Interpretation", Bristol, 1-4 April 1957 and to KÖRNER-PRYCE (eds) 1957 (see Vol. 1, Ch. 2, 1956-14, n4).

[4] See *above*, § 1, n4.

[5] See BOHM 1952a; BOHM 1952b.

[6] See DIRAC 1930: Chs I and II; NEUMANN 1932: Ch. II; and REICHENBACH 1944: Part II, respectively.

[7] See FANO 1957.

[8] See GRÖNEWOLD 1946.

[9] See SCHRÖDINGER 1935; and EINSTEIN 1953, respectively.

[10] See TER HAAR 1955.

[11] See SCHRÖDINGER 1952a; SCHRÖDINGER 1952b.

[12] Feyerabend had already sent to Feigl a longer, first draft, of FEYERABEND 1957a (see Paul K. Feyerabend,"On the Quantum Theory of Measurement", HF 08-33-24 (27 ss., T+H); Feyerabend to Feigl, 6 May 1957, HF 02-133-01), and its final version (see *above*, § 1, n5), while a second draft was taken by Michael Scriven with him from the 1957 Bristol conference (see *above*, n3) to the Minnesota Center for Philosophy of Science (see *above*, § 1, n7); on the history of the development of FEYERABEND 1957a, see *above*, § 1, n5 and n7 in particular; Vol. 1, Part II, 1957-4, n11).

[13] See Paul K. Feyerabend, "Weizsäcker's Criticism of My Note on the Logic of Quantum-Mechanics (Part of his letter to me)", HF 08-33-25 (2 ss., T+H); Feyerabend to Weizsäcker, 26 April 1957, HF 04-83-01 (8 ss., T+H); and WEIZSÄCKER 1955, respctively.

[14] See HEMPEL 1958; *above*, Ch. 2, § 14, n3 in particular.

Eventually Feyerabend succeeded in participating in the "Conference on Philosophical Foundations of Physics", Minnesota Center for Philosophy of Science (MCPS), 12 August - 14 September 1957. He left Bristol with his wife in late July and spent a few days in the East Coast – without however visiting either Adolf Grünbaum (see Vol. 1, Part II, 1957-12, n2) or Arthur Pap (see Vol. 1, Part II, 1953-5, n1) as apparently initially planned – before reaching Minneapolis for the five-week event; a visit to Grünbaum in Bethlehem (PA) was paid by Feyerabend on his way back to Bristol, in late September. On Feyerabend and the 1957 MCPS conference, see Vol. 1, Part II, 1956-20, n4; and Feigl to Grünbaum, 21 June 1957, HF 02-169-46; Grünbaum to Feigl, 25 June 1957, HF 02-169-47; Feigl to Grünbaum, 1 August 1957, HF 02-169-51; Pap to Feigl, 12 August 1957, HF 03-134-47; Feigl to Pap, 22 August 1957, HF 03-134-49; Grünbaum to Feigl, 27 September 1957, HF 02-169-54.

A description of the conference works can be found in Herbert Feigl, "Report on the Conference on Philosophical Foundations of Physics", 27 September 1957, HF 01-40-04 = HF 01-46-01 (6 ss., T), here reproduced:

September 27, 1957

Minnesota Center for the Philosophy of Science

REPORT
ON THE CONFERENCE ON PHILOSOPHICAL
FOUNDATIONS OF PHYSICS
August 12 – September 14, 1957

A conference, sponsored by the National Science Foundation, was conducted at the Minnesota Center for the Philosophy of Science from August 12 to September 14, 1957.

The participants were (in alphabetic order):

Alexander, H. Gavin; Lecturer, Philosophy of Sci., Univ. of Manchester, England (Commonwealth Fellow)

Cassirer, Eva; Ph.D. (London Univ.), Lecturer, Free Univ., Berlin, Germany

Feigl, Herbert; Director, Minn. Center, University of Minnesota

Feyerabend, Paul; Ph.D. (Physics), Lecturer in Philos. of Sci., Univ, of Bristol, Bristol, England

Hempel, C. G. ; Professor & Dir. of Grad. Studies, Philos., Princeton Univ.

Maxwell, Grover; Ph.D. (Chemistry), Research Asst., Minn. Center, Univ. of Minn.

Mehlberg, Henry; Professor of Philosophy, University of Chicago

Nagel, Ernest; Professor of Philosophy, Columbia University

Putnam, Hilary; Asst. Professor of Philosophy, Princeton University

Rozeboom, William; Ph.D. (Psychology, Univ. of Chicago) National Science Foundation Fellow, Minn. Center 1956-58

Scriven, Michael; Asst. Professor of Philosophy, Swarthmore College

Sellars, Wilfrid; Professor of Philosophy, University of Minnesota

For several of the sessions in which the discussion concerned the logic of quantum mechanics we also had the benefit of the participation of the following University of Minnesota

Professor Warren B. Cheston Professor Edward L. Hill Professor Norton M. Hintz	Physics
Professor William N. Lipscomb	Physical Chemistry
Professor Otto H. Schmitt	Biophysics

faculty members:

Professors C. G. Hempel and E. Nagel joined the conference only for the last week. During this period (September 7-14) all previous discussions were intensively reviewed, and important critical and constructive contributions were made by these two outstanding philosophers of science. ⟨1|2⟩

[Professor Adolf Grünbaum (Lehigh University) , originally scheduled to participate in the conference, was prevented from coming to Minnesota because of the illness of his wife. But ever since his earlier collaboration during his sojourn at the Center (September 1956 to January 1957) we have continued and shall continue collaboration by correspondence. Professor Rudolf Carnap (U.C.L.A.) was unable to come to the conference, but has also contributed valuably by previous consultation and by continuing correspondence.]

The procedures of the conference consisted mainly in intensive plenary discussion sessions, – four or five per week; many additional discussions in smaller subgroups; the writing of a number of memoranda all of which were circulated among the participants, and also transmitted to some of our correspondence. In accordance with the major aim of the conference we attempted to achieve a genuine meeting of minds, a thorough thrashing out of difference in points of view. It was not expected to reach full agreement on all issues. But ewe have been extremely satisfied with the measure of mutual understanding and of the degree of clarification of basic ideas which has been attained. The sentiment that the conference was eminently worthwhile was shared by all participants. This is well expressed in a personal letter I received from Professor C. G. Hempel (Princeton University). I take the liberty (with appropriate blushes) of quoting a passage from this letter:

"I have sung, and will continue to sing, the praises of the Center in general, and of this summer's meeting in particular: I was confirmed in the impression formed at my previous visit that this is an organization that works most effectively and productively, and the credit for that goes essentially to you, Herbert! On the whole, I tend to shy away when colleagues want to discuss philosophy because I do not find it sufficiently relevant to my interests or not sufficiently enlightening to be justifiable; but at Minneapolis I sensed a strong effort to get right at the heart of important problems and the discussion was of a kind that I found definitely stimulating and profitable."

The work and the topics of the conference were all well prepared by a great deal of antecedent consultation and correspondence. Fundamental problems in the logic and methodology of scientific theories in general, and of physical theories in particular, formed the main subject of the conference. The only deviation from the anticipated program consisted in a shift of emphasis: We found that we could ⟨2|3⟩ "tackle the impossible," viz., the logical analysis of quantum mechanics immediately and fairly comprehensively. We had originally planned to work on this thorny and controversial subject only by a way of a first orientation, and to reserve more intensive work in this domain until the summer of 1958. But to our pleasant surprise we found ourselves well equipped for fruitful discussion and genuinely enlightening interpretations. Thus almost half of our work concentrated in this area. Starting out with a critique of the (otherwise divergent) deterministic interpretations published by E. Nagel, H. Margenau, D. Bohm, de Broglie, etc., we reviewed the contributions of von Neumann, H. Reichenbach, et al. and analyzed in some detail the various logical problems involved in the physical meaning of Schrödinger's wave mechanics, its statistical interpretation by M. Born, the positivistic views of the Copenhagen School (w. Bohr, W. Pauli, et al.). Due attention was paid to the syntactical and semantical aspects of the basic laws of quantum mechanics, and in this connection the meaning and fruitfulness of the application of a three-valued logic was carefully scrutinized. Work on all these and closely related issues continues currently.

The various logical problems connected with the formulation and confirmation of scientific laws (deterministic or statistical) were studied intensively.

Among the more general and fundamental problems of the logic of science we also discussed fairly thoroughly the following topics:

1. The logic of scientific explanation (deductive, as well as probabilistic models); with side glances at the logic of the social sciences.
2. Pragmatic and ontological interpretations of determinism and indeterminism.
3. The meaning of probability.
4. The role of laws and definitions in scientific theories.
5. The "Theoretician's Dilemma," – i.e. the subject of an essay by C. G. Hempel for our Volume II of Minnesota Studies in the Philosophy of Science, concerning the question of the indispensability of theoretical concepts and postulates.
6. The closely related issue of syntactical positivism vs. various realistic interpretations of "existential" hypotheses in science in general, and in the atomic and quantum theory in particular. (In this connection: critique of operationism and subjectivistic interpretations of quantum mechanics.)
7. The meanings of reduction and identification in scientific theories. ⟨3|4⟩
8. Various special problems such as concerning time and irreversibility; the significance of P. F. Ramsey's and William Craig's formal analyses; the explication of scientific laws in terms of rules of inference; and related problems in logical theory were discussed to a more limited extent.» (pp. 2-4)

—

Two evenings were devoted to "extracurricular" symposia: one with the University's Philosophical Society and the Society of Pai Chi (graduate psychological fraternity, with an audience of about 150 persons, including many faculty members) in which Dr. M. Scriven presented a talk on "The Nature and the Necessity of Scientific Theories," followed by a panel discussion in which several members of the conference (Feyerabend, Mehlberg, Putnam, etc.) participated.

The other symposium was with several members of our Philosophy Department and concerned Issues of Determinism, and of Inductive Probability.

Both evenings were most profitable and enjoyable.

—

Tentative plans for publication of the results of the conference and of our continuing work have been thoroughly discussed. We intend to publish this material in Volume III of Minnesota Studies in the Philosophy of Science, possibly under the subtitle: "The Philosophical Foundations of Physics." We have asked the participants (and some correspondents) to deliver their papers, if at all feasible, by March 1st, 1958. Then these papers will be distributed among the contributors for discussion, comments, rejoinders, etc. This will no doubt be helpful in regard to the quality and the degree of integration of the entire volume. We hope to finish the editorial work on the book before Christmas 1958, and expect that publication (by the University of Minnesota Press) will be possible sometime in 1959.

—

Our plans for next summer (1958) include beside the editorial work on Volume III, the writing of discussions and rejoinders, possibly also a smaller conference of shorter duration, mainly for sake of a final "comparing of notes" and for integration of the discussion section of the book. – A tentative outline of our current ideas regarding the table of contents is as follows: ⟨4|5⟩

Tentative table of contents
of Volume III
of the Minnesota Studies in the Philosophy of Science
[Titles and sequence of articles subject to revision]

THE PHILOSOPHICAL FOUNDATIONS
OF PHYSICS

E. Nagel:	Analytic and Synthetic Statements in Physical Theory
M. Scriven:	The Meanings of Scientific Explanation & The Logic of Negative Temperatures
C. G. Hempel:	Probabilistic Explanation
W. Sellars:	The Logic of Theoretical Concepts and/or: Time and Becoming
W. Rozeboom:	The Factual Content of Theories
A. Grünbaum:	Time and Irreversibility (possibly with special attention to Quantum Mechanics)
G. Maxwell:	Observational and Theoretical Language
H. Mehelberg;	The Problem of Objectifiability in Quantum Mechanics (and/or: Formalization of Quantum Mechanics)
P. Feyerabend:	Interpretations of Quantum Mechanics – a critical analysis
H. Putnam:	The Logic of Quantum Mechanics
H. Feigl:	Interpretations of Indeterminism; or: The "Arbitrary" Elements in Scientific Theories
Various Contributors:	Discussion and Rejoinders

Possibly also:

R. Carnap: Logical Considerations regarding the Entropy Principle
H. G. Groenewold: Comments on the Philosophy of Quantum Mechanics ⟨5|6⟩

The National Science Foundation will receive a detailed account of expenditures from the University's Business Administration (Controller's Office). ⟨...⟩

—

Ma I say in conclusion that, in my personal evaluation, the conference was definitely worth the great amount of effort and work that it required. As I had anticipated and hoped, the size of the group, as well as the selection of the participants, was very nearly optimal. The intensity and the fruitfulness of our work compares, in my recollection and estimation, well with that of the Vienna Circle (1925-30) during some of its most productive periods.

—

On behalf of the participants in our conference and of the Minnesota Center for the Philosophy of Science I wish to express the most sincere appreciation to the National Science Foundation for sponsoring our project.

Herbert Feigl, Director
Minnesota Center for the Philosophy of Science

September 27, 1957

The follow-up conference mentioned in the report took place at Princeton University, 5-12 September 1958 under the auspices of the Minnesota Center for Philosophy of Science, and was actively attended by Feyerabend (on Feyerabend and the 1958 Princeton MCPS Conference, see *above*, Ch. 2, § 11, n1; and Vol. 1, Part II, 1958-10, n1).

The publication plans of the conference proceedings, where Feyerabend's contribution was to appear, underwent various changes and delays, eventually causing Feyerabend projected paper to appear, under a different title, in the first volume of the series *Pittsburgh Studies in the Philosophy of Science* (see Feyerabend 1962a; on the history of the development and publication of Feyerabend 1962a, see *below*, Ch. 4, 1961-1, n14; and Vol. 1, Part II, 1957-12, n1).

[15] See Scriven 1958.

[16] See Feyerabend 1957a.

1959

3. Paul K. Feyerabend. "The Logic of Measurement – A Comparative Study in Classical Physics, Statistical Mechanics, Elementary Quantum Mechanics, and Quantum Field Theory", [1959]

Paul K. Feyerabend. "The Logic of Measurement – A Comparative Study in Classical Physics, Statistical Mechanics, Elementary Quantum Mechanics, and Quantum Field Theory

(The following outline was written by Professor P. K. Feyerabend.)[1]

It is my intention to develop a theory of observation and measurement and to show the force of this theory by applying it to various concrete cases. My work will therefore naturally divide into two parts, the first part being more general and "philosophical," the second part more specific and "physical."

As regards the general part, I shall have to deal with two problems, which I shall call the logical and the psychological problem of measurement, respectively. The logical problem consists in the question: Under what circumstances a given experiment constitutes a test, and is in that sense relevant to, a physical hypothesis, h? The psychological problem of observation is concerned with determining which situations are accessible to direct [1|2] inspection by human beings and how such situations can be either explicitly defined, or characterized axiomatically on the basis of behavioristic terms and terms of set theory. The solution of the psychological problem of measurement must be accomplished if the distinction, characteristic for the Carnap-Hempel approach, between theoretical terms and observational terms is to have any precise meaning. At the saw time, it will have to provide a criterion for classifying terms as observational and theoretical. I have already attempted to solve the psychological problem (cf. my Theorie der Basissaetze, Vienna 1951), but I think now that the suggestions made there need modification and elaboration. It will, of course, also be necessary to defend the point of view, presupposed in the above classification of the more general problems of observation, that (a) observability is a purely psychological concept which does not involve a reference to a specific set of entities such as sense-data, aspects, perspectives (Russell) and the like; and (b) that the meaning of a term is independent of its being observable (measurable).

The more specific part of my investigations will consist of an attempt to solve what I have called the physical problem of measurement. This problem consists in the question by what physical means we may establish a correlation between a situation S to be measured, and some observable (in the psychological sense introduced above) state of affairs O such that (a) the occurrence of O allows us to infer that S possesses a certain property out of a category in which we are interested (e.g., a certain position, velocity, or a certain mass); and (b) this property obtains either immediately before, or immediately after, the occurrence of O. It is obvious that the physical problem of measurement involves a reference to physical theories, and it will therefore be necessary to discuss its solution separately for each physical theory considered. I propose to discuss the following cases: (a) observation, with the [2|3] unaided eye, of the properties of material objects of a size comparable with the size of the observed; (b) measurement in classical particle mechanics; (c) measurement in elementary quantum theory; (d) measurement in the field theories (classical, and quantum field theories).

(a) It is this case that has usually been discussed by philosophers. Their discussion is impaired by the fact that a clear theory of the interaction between material object (in the everyday sense and not in the sense of contemporary physics) has never been developed, the hidden assumption being that such a theory could, and should be replaced by an analysis of material objects in terms of some specific occurrences such as sense-data or the like. I shall attempt to develop a theory of interaction which will be adequate for the everyday level.

(b) Classical physics does not contain a fully developed theory of measurement. First of all, it is not the case, as within quantum theory, that the postulates of measurement range among the fundamental postulates of the theory. Measurement in regarded as just another special case of the interaction between two or more physical systems. Secondly, it has never been attempted to develop a classical theory of the interaction between a measuring instrument and an investigated system in sufficient detail. With respect to the first point it seems reasonable to assume that formalizations of classical particle mechanics may be found which really contain postulates of measurement among their fundamental axioms. It will be my task to discuss to what extent such formalizations even allow the application of Bohr's interpretation within classical particle mechanics. With respect to the second point, I intend to give a critical review of the investigations of Szilard, Brillouin, and others concerning the changes of entropy connected with a successful measurement. Szilard's investigations proceed from the assertion, made by Smoluchowski, that measurement plus utilization of its results may [3|4] lead to the violation of the second law of thermodynamics. He attempts to show that such a violation does not take place, as the increase of information obtained is more than balanced by a corresponding increase of entropy in parts of the measuring apparatus. I do not believe that the problem posed by Smoluchowki has been dealt with in a satisfactory way by Szilard, although it is to be admitted that thermodynamic considerations may play an essential part even if we restrict ourselves to the observation of particles satisfying the laws of classical mechanics. My task will be to show to what extent that is the case and to what extent (on the basis of what axioms) thermodynamic considerations enter even the most simple processes of measurement (such as the measurement of the position of a ball by elastic reflection of another ball).

(c) The only physical theory with respect to which the physical problem of measurement has been treated in some detail is elementary quantum theory. Moreover, the quantum theory possesses a unique position in that any specific theory of measurement which may be developed will imply an interpretation of the theory as a whole. Therefore, within quantum theory, the discussion of the physical problem of measurement is intimately connected with such general philosophical problems as the problem of the reality of the external world, the distinction between subject and object, and so on. (It is of course obvious that this connection will obtain only if the problems are reformulated in a way which makes them accessible to reasonable discussion.) It is my aim (A) to clarify this connection; and (B) to develop a satisfactory quantum theory of measurement and thereby a satisfactory interpretation of elementary quantum theory.

A solution of problem (B) was offered long ago by Niels Bohr, whose interpretation of elementary quantum theory still seems to me to be the only one which is consistent, exhaustive, and also philosophically satisfactory. [4|5] The only disadvantage of Bohr's interpretation is that, so far, it has been presented only in very general and sometimes vague terms. A formally satisfactory presentation of it has not been given. Von Neumann's theory, which was developed with the explicit purpose of formalizing some of Bohr's ideas, leads into a completely new direction and also violates Bohr's fundamental postulate to the effect that the measuring instrument must always be described in classical terms and not with the help of a statistical operator. Von Neumann's theory must therefore be regarded as an alternative theory rather than as a more precise formulation of Bohr's own point of view.

One way of developing a consistent and all-embracing quantum theory of measurement would, therefore, be to present Bohr's point of view in a formally satisfactory way or, what amounts to the same, it would consist in a formalization of the language of elementary quantum mechanics which would contain Bohr's assertions about the role of the classical concepts and about the block-character of the phenomena – either as formation rules, or as consequences of formation rules. The main problem here is the choice of the atomic sentences. The existing attempts at formalization are inadequate for one or the other of the following two reasons: (a) the rules of formation do not involve any reference to the concepts of classical physics which play such an important role in Bohr's picture (this applies to the partial formalization discussed by Birkhoff and won Neumann); (b) the logic of the language chosen is not the ordinary two-valued logic (this applies to Reichenbach's sketch as well as to the much more detailed formalization of von Weizsaecker). It will be part of my task to develop a formalization which does not suffer from these two shortcomings. I believe that such a formalization would lead to a consistent solution of the physical problem of measurement with respect to elementary quantum theory. At the same time it would be the first instance [5|6] of a language which is observationally complete in the sense that it does not allow for the formulation of sentences expressing unobservable states of affairs (classical physics is not such a language) and which yet allows for the formulation of important physical laws possessing a rich predicative content. Bohr's guess that any future theory will comply with his interpretation can now be expressed by stating that for any future theory there will exist an adequate formalization sharing certain formal properties with the formalization of elementary quantum mechanics just discussed. The characterization of these formal properties may be regarded as a formalization of Bohr's ontology which is independent of the specific theory used for explanation and prediction on the atomic level.

Bohr interprets the basic terms of the elementary quantum theory as instruments for the prediction of classical states of affairs (his so-called "phenomena"). He therefore assumes that quantum theory must not be interpreted, as earlier physical theories were, as introducing a new set of entities. It must rather be interpreted as a new way of arranging old and well known entities, i.e., classical states of affairs. Von Neumann's theory of measurement may be regarded as a first step in the direction of a realistic alternative. In von Neumann's theory of measurement, it is assumed that the macroscopic apparatus, too, may be dwelling in a certain quantum mechanical state. The idea behind this theory seems to be that once a new theory has been introduced it must be possible, not only to predict new effects with its help, but also to give a new account of macroscopic affairs in its terms. Now although von Neumann's theory is internally consistent, it possesses some very unsatisfactory features. The reason for this is that it does not proceed from a satisfactory account of the behavior of macroscopic objects and, therefore, it does not make use of the new assumptions which any such account would have to introduce. Such an account has been given by Ludwig and others. [6|7] According to Ludwig the process of measurement must be regarded as a complicated thermodynamical process which can be approximately represented by 'quantum jumps'. Although this account (which is not too well known) disposes of the difficulties of von Neumann's theory, it is still far from satisfactory. It implies that fluctuations will occur even after the value of a certain magnitude has been established and that these fluctuations will be large enough to be accessible to test. Yet all our experiences seem to indicate that they do not occur, and that the transitions, caused by measurement, of states into eigenstates, occurs with absolute precision. These and other considerations seem to show that elementary quantum theory is not powerful enough to provide a physical theory of measurement which is closed in the sense that it is formulated in its own term only, i.e., formulated without bringing in classical terms in an essential way. Basically, the difficulty is that no account of all the classical properties of macroscopic objects in terms of elementary quantum mechanics is available. Arguments will be presented to the effect that such an account can be given only within a non-linear theory: the attempt to solve the problem of measurement in the way first suggested by von Neumann forces us to look for a new quantum theory. Some very general properties of this new theory can be stated already today.

(d) Quantum electrodynamics which might well possess these properties (and in thin respect be inconsistent with Bohr's ontology) has not yet developed a full-fledged theory of measurement. The investigations of Bohr and Rosenfeld as well as Thirring's more recent suggestions will be discussed with respect to the development of such a theory. In the course of my research I hope to be able to develop such a theory in greater detail.

Editors' Notes

HF 08-33-18 (4 ss., T+H); Herbert Feigl, "Request for Support of Research" 20 January 1959, HF 01-40-08 (25 ss., T) excerpt (pp. 19-25). HF 08-33-18 is the original draft written by Feyerabend which was then included into HF 01-40-08, on which the transcription here is based.

[1]The research project outlined by Feyerabend was part of an application submitted by Feigl to the National Science Foundation with the aim of getting a grant to financially sustain the activities of the Minnesota Center for Philosophy of Science in 1959-1961. Feigl had tried to help and support Feyerabend by arranging visiting appointments for him at the MCPS since their meeting in 1954 (see Vol. 1, Part II, 1954-19, n4). After the attempts made for part of the 1955-56 session (see Vol. 1, Part II, 1954-19, n4, n27), and for the whole 1956-57 session (see Vol. 1, Part II, Feyerabend to Popper, 2 July 1955, 1955-22; 4 July 1955, 1955-23) – both failed due to Feyerabend's choice of more promising alternatives – Feigl's latest invitation, originally planned for the Spring Semester 1959 (see Feigl to Grünbaum, 12 December 1957, HF 02-169-57; Feigl to Alpert, 8 January 1958, HF 04-164-19), was later postponed to the Fall Semester of the same year, since Feyerabend met difficulties in obtaining a leave of absence for a semester while he was visiting associate professor at the University of California, Berkeley (1958-59). Eventually Feyerabend's temporary position at Berkeley was turned into a tenured one (see *above*, Ch. 2, § 11, n1; and Vol. 1), Feigl's application to the NSF was successful, and Feyerabend officially started his tenured appointment at Berkeley with a leave of absence that allowed him to spend the Fall Semester 1959 as associate visiting professor at the MCPS (see Vol. 1, Part II, 1959-2, n7; FEYERABEND 1995: 115).

1961

4. Paul K. Feyerabend, "Research Project on the Logic of Measurement, Probability and Confirmation, with Applications in Atomic and Quantum Theory", [1961]

Research Project on the Logic of Measurement, Probability and Confirmation, with Applications in Atomic and Quantum Theory.

Paul K. Feyerabend

My research project at the Minnesota Center for Philosophy of Science is intended further to advance the work I started in 1959 with the help of an NSF grant and additional support from the Minnesota Center.[1] As I pointed out in my last report to Professor Herbert Feigl (and submitted to the NSF in 1960)[2] only part of this project has been completed. You will remember that I intended to give a thorough discussion of the process of measurement. Any such discussion is confronted with three distinct kinds of problems which we may call the logical, the psychological, and the physical problems of measurement respectively. The logical problem consists in the questions as to what must be the relation between a sentence and a theory in order that we may be able to say that the former either supports, or undermines the latter. I think I have definitely solved an important part of this problem viz. the problem as to whether the meaning of the observation statement can (or should) be defined independently of the postulates of the theory to be tested. The solution is contained in some papers I referred to in my last report and in some papers I have written since, viz.:

> "How to be a good Empiricist" ⟨sic⟩ to be presented to section L of the annual meeting of the AAAS, Denver 1961 and to be published in Daedalus ⟨sic⟩ and Vol. II of the Pittsburgh Publications in the Philosophy of Science ⟨sic⟩.[3]
> "Some Comments on the Logic of Empirical Support," to appear in The Rational Approach, Festschrift for Karl Popper.[4]
> Knowledge without Foundations, the Nelly Rand ⟨sic⟩ Lectures held in November 1961 at Oberlin College, to appear in print in Summer 1962.[5]

Of course, this solution leads at once to the problem of circularity: of the meaning of an observation statement depends on the theory to be tested, will then the latter not be safe from refutation? (Kant's whole philosophy consists in an elaborate procedure leading to the answer YES). This problem I think I have solved developing, in the papers mentioned above and in other papers on which I am working currently, a theory of tests which utilizes not one physical theory, but a whole set of mutually inconsistent theories. I think it is very important to develop this point of view concerning tests in greater detail and to discuss it with those who might feel reluctant to accept it. I do not know of any better place for such s discussion than is the Minnesota Center (with Professors H. Feigl and G. Maxwell) and this especially in view of the presence, there, for several months of Professors K.R. Popper and Wesley Salmon; as well as (for shorter conferences) of Professor E. Nagel, R. Carnap, H. Mehlberg, and possibly others.

All this leaves out the question of the quantitative, or even the comparative evaluation of theories on the basis of evidence. I have the suspicion (and some preliminary arguments supporting this suspicion) that once the above test-model is adopted all of the contemporary theory of confirmation (and [1|2] Popper's theory of corroboration) will turn out to be completely and irreparably inadequate. Again, I do not see a better way to test, and further to develop these ideas, than by discussion ⟨sic⟩ them with the distinguished members and visitors of the Center. Also such a discussion would give me an opportunity more effectively to convince those in doubt of the correctness of my ideas than I could do in correspondence, or in papers. And it should also enable me much more quickly to discover the weak point of my own ideas.

The second set of problems deals with the psychological situation that must obtain in order that a sentence be observational. Roughly speaking this has something to do with the presence of sensations in the observer and implies therefore a certain amount of psychological research. A first, rather crude attempt to answer questions of this kind was made by R. Carnap in his Testability and Meaning.[6] I tried to improve upon Carnap's definition ("An Attempt at a Realistic Interpretation of Experience" Proc. Arist. Soc., 1958)[7] I have not been able to advance this particular aspect of my project but I have realized that it involves a thorough discussion of the mind-body-problem. This was the reason why I spent most of my last visit in Minnesota (November 1961) discussing the mind-body-problem with Professors Feigl and Meehl. I know that Professor Popper and Carnap have definite (though conflicting) views on this subject. The meaning dependence, asserted in the partial solutions of the first problems reappears here and this is what connects the first with the second problem. Also a solution, along the lines suggested by myself, of the first problem will be satisfactory only if a causal account can be given of the psychological aspects of observation. All this is shown in my paper:

> "The Trivialization of Epistemology" (to appear in a Volume to be published by the University of Minnesota Press)[8]

and this paper should be consulted by those who think that considering the mind-body problem in connection with measurement is going a little too far or unnecessary.

Finally there are the physical problems of measurement, i.e. the problems of the physical processes which occur during a measurement. The solution of the physical problems consists in (1) deriving theorems from the theories used and (2) in the interpretation of the theoretical primitives of these theories. The second part of this problem is also known as the problem of the existence of theoretical entities. This problem I have attempted to solve in the following three papers (some of which are listed in my last report).

"Das Problem der Existenz Theoretischer Entitaeten" Probleme der Wissenschaftstheorie Vienna 1960.[9]
"Problems of Microphysics," Pittsburgh Studies in the Philosophy of Science Vol. 1.[10]
"Realism and Instrumentalism" in The Rational Approach, and in other papers.[11]

As regards the first problem I have so far only succeeded in giving a relative evaluation of all the major theories of measurement which exist today with respect to the elementary quantum theory. The result of this evaluation is that Bohr's theory of measurement, properly understood, is the best theory available. I have already given what I think to be a satisfactory [2|3] account of Bohr's theory that is freed from all metaphysical admixtures. What is still missing is a theory of measurement for the field theories, or parts of them. It turns out that the construction of such a theory is very closely connected with the first problem, i.e. the problem of confirmation and the interpretation of probability. This is the reason why I expect important results from my discussion with Professors Popper and Salmon.[12] But I also intend to consult very fully some physicists in Europe before starting with this part of my project, especially Professors Heisenberg (Munich), Rosenfeld, Møller, Petersen (Copenhagen), Groenewold (Utrecht), Bohm (London), and Thirring (Vienna).

Editors' Notes

RC 090-63-06 (3 ss., T)

[1] See *above*, § 3, n1 in particular. The research project outlined by Feyerabend was part of a larger joint research project on "The foundations of probability, with applications to the logic of atomic physics", involving Karl R. Popper and Wesley Salomon, which Feigl submitted to the National Science Foundation with the aim of getting a grant to financially sustain the activities of the Minnesota Center for Philosophy of Science in 1962-1963. (Feigl to Spaulding, 30 November 1961, RC 090-63-03.; see Feigl to Popper, 14 March 1961, KP 294.6). On Feyerabend's 1962 visiting appointment to the MCPS, see Vol. 1, Part II, 1961-2, n5; on Popper's 1962 visiting appointment to the MCPS, see Vol. 1, Part II, 1961-2, n3.

[2] Feyerabend may be referring to Herbert Feigl, "Interim Report 1959-1960", December 1960, HF 01-39-02 = HF 01-39-03

[3] First delivered within the second (1961-62) Annual Lecture Series of the Center for Philosophy of Science, University of Pittsburgh (30 November 1961; see *below*, Ch. 4, Hempel to Feyerabend, 16 October 1961, 1961-3; "University of Pittsburgh – The Program in the Philosophy of Science – Schedule of Lectures 1961-62", HF 09-16-01), and a few weeks later presented to Session I "Empiricism and the Status of Theories" of Section L (History and Philosophy of Science) – chaired by Hanson (see Vol. 1, Part II, 1956-14, n5), and with George T. McClure and Wilfrid Sellars as discussants – of the 128th Annual Meeting of the American Association for the Advancement of Science, Denver, 26-31 December 1961 (26 December 1961), this paper did not appear either in the scholarly journal *Daedalus* – which in any case was founded in 1955 as the journal of the American Academy of Arts and Sciences, and not of the American Association for the Advancement of Science, whose scholarly journal is *Science* – or in Vol. 2 of the *Pittsburgh Series in the Philosophy of Science*. Eventually the paper was published in the proceedings of the 1962-63 University of Delaware Seminar in the Philosophy of Science (see FEYERABEND 1963a), where Feyerabend gave a talk on "The Pernicious Influence of Empiricism" about a year later (23

October 1962), with the indication: "Revised copy of paper originally appearing in *Inquiry*" (FEYERABEND 1963a: 3*); yet, no related paper by Feyerabend ever appeared in the scholarly journal *Inquiry* (see Vol. 1, Part II, 1961-12, n4). See Vol. 1, Part II, 1961-4, n7.

[4] See FEYERABEND 1964a.

[5] See FEYERABEND 1961a. The paper combines two lectures delivered by Feyerabend on the Nellie Heldt Lecture Fund (see *below*, Ch. 4).

[6] See CARNAP 1936; CARNAP 1937.

[7] See FEYERABEND 1958a.

[8] Most likely Feyerabend refers here to a long paper that he eventually completed in the summer of 1962 and which was planned to appear as Feyerabend's contribution to the *Feigl Festschrift* – a projected volume of the University of Minnesota Press under Feyerabend and Maxwell's editorship – with the title "Linguistic Philosophy and the Mind-Body Problem" (see Vol. 1, Part II, Feyerabend to Popper, 18 July 1962 undated, 1962-2, n3 in particular; Vol. 3, Ch. 2, Watkins to Feyerabend, 27 August 1962, 1962-5; FEYERABEND 1963a: 8; BAUMRIN (ed.) 1963: 531). However, probably due to the various delays in the publication process of that volume (on the history of the publication of the *Feigl Festschrift* (FEYERABEND-MAXWELL (eds) 1966), see *below* and Vol. 1, Part II, 1961-4, n8), a part of this paper was reworked as an independent article, most likely during Feyerabend's visiting appointment at the Minnesota Center for Philosophy of Science in the Fall Semester 1962 (see *above*, n1), submitted to the scholarly journal *The Review of Metaphysics* in early 1963 and there published in the late summer of that year (see Smart to Feyerabend, 25 March 1963, PF 2-7-43; FEYERABEND 1963b).

It should be noted that a further large essay by Feyerabend, whose draft was complete in the late winter of 1963, and which was at least since then planned to appear in Vol. 2 of the *Pittsburgh Series in the Philosophy of Science* under the title "Problems of Empiricism" (see *below*, Vol. 3, Ch. 2, Feyerabend to Watkins, undated, 1963-1; Watkins to Feyerabend, 3 April 1963, 1963-2; and Smart to Feyerabend, 15 March 1963, PF 2-7-44; FEYERABEND 1963a: 8*; BAUMRIN (ed.) 1963: 531; FEYERABEND 1963b: 52 and n4) – in connection with the papers delivered by Feyerabend within both the second (1961-62) and the third (1962-63) Annual Lecture Series of the Center for Philosophy of Science, University of Pittsburgh, i.e., respectively "How to Be a Good Empiricist" (see *above*, n3), and "Language and Science: Is Linguistic Analysis A Useful Philosophical Method?", 13 November 1962) – ended up including significant parts of FEYERABEND 1963b (as well as of FEYERABEND 1963b) in its revised published version (see FEYERABEND 1965a; on the history of the publication of FEYERABEND 1965a, see Vol. 1, Part II, 1964-1, n2).

A paper by Feyerabend entitled "Notes on Linguistic Philosophy and the Mind-Body Problem" (see Paul K. Feyerabend, "Problems of Empiricism", HF 08-21-01 (210 ss., T+H), pp. 11-12, 124; Feyerabend to Feigl, December 1963, HF 02-133-09) or "A Note on Linguistic Philosophy and the Mind-Body Problem" (see FEYERABEND 1965a: 153, 230-231n25) was planned to appear in the *Feigl Festschrift* at least until late 1963. Yet, Feyerabend never published a paper with a similar title, while – possibly due to the publication of FEYERABEND 1963b and FEYERABEND 1965a – his actual contribution to the *Feigl Festschrift* turned out to be a completely different paper (see Vol. 1, Part II, Feyerabend to Popper, 26 February 1964, 1964-1; and FEYERABEND 1966a).

[9] See FEYERABEND 1960c.

[10] See FEYERABEND 1962a.

[11] See FEYERABEND 1964a.

[12] See *above*, n1.

1963

5. Paul K. Feyerabend, "Notes on Scientific Method", 4 February 1963

Paul K. Feyerabend[HF]
February 4, 1963[HF 1]

Notes On ⟨sic⟩ Scientific Method.

(1) It is the purpose of methodology to design rules for the handling of statements which are supposed to constitute knowledge. Different sets of rules can be devised depending on different ideals of knowledge. The choice between these sets is a genuine choice which cannot be replaced by a stronger instance such as for example proof. Wherever proof occurs we are already moving within a chosen system and we have only failed to notice that a choice has been made. Nor does the actual existence of systems which have been built up in accordance with a certain set of rules give this set an advantage over other sets. There is method in many kinds of madness and what we want to find are methods we are able to approve of. Moreover the actually existing should not be exempt from criticism. Methodological discussions must not therefore be influenced by what does and what does not exist; and a methodologist must not be intimidated by the remark that his ideas "do not agree with", say, "actual scientific practice". Arguments between different methodologies may depend on demands not directly connected with the theory of knowledge. Thus the demand that the capabilities of human beings be developed as fully as possible, or the demand that a clear distinction be drawn between laws of nature and laws of society leads to the exclusion of certainty as an ideal of knowledge. In the person note I would like to examine the issue between a deductivist methodology and an inductivistic methodology.

(2) It has been doubted (in our discussions at the center) that there exists a difference between inductivism and deductivism (as we shall lable ⟨sic⟩ the two points of view). More especially, it has been asserted that both methodologies are doing exactly the same and that the difference [1|2] is only one of formulation. No doubt this was encouraged by Popper's occasional admission (not necessarily in these words) that a justification, or a comparative justification of scientific theories could be obtained and by an inductivistic interpretation of this phrase, or of the thought behind it, on the side of the opponents. Now even if the difference were only a verbal one one could still argue that one formulation, for example the formulation given by the deductivist, was simpler, or more concise than the formulation given by the inductivist. However before discussing this possibility it is better to find out whether there is really nothing but a verbal issue at stake and whether there are not decisive differences, at least of attitude, between the two points of view. We must also distinguish between the intentions of the inductivist (and the deductivist) and the shape which the inductivistic methodology is actually in. It may well be that all

the valid rules of induction which have so far been formulated can be accounted for by the deductivist. This does not yet invalidate the belief of the inductivist that independent rules may one day be found and that they must be found if a satisfactory account of our knowledge is to be given.

(3) It is generally admitted that a logic of invention is impossible. The invention of theory is considered, both by inductivists and by deductivists as an irrational act. Methodology starts only after this act has been completed. It therefore assumes as given (1) a set O of observation statements; and (2) a set S of theories designed to explain (at least) what happens inside O. The proper handling of observation statements is of course also a concern of methodology. There are good reasons to assume that observation statements do not always possess greater force than do theories. Nor are they as distinct from theories as will be postulated in the discussions to follow. Their content, the assumptions concerning [2|3] their validity and even their observational character may depend on theories and, more especially, these elements may depend on the elements of S itself. I do not intend in the present note to discuss the problems arising from such a dependence. I shall assume a domain O of singular statements, a class S of theories invented with the purpose of explaining what is going on inside S, and I shall also assume that there is no "interaction" whatever between O and S. These assumptions are made both by a inductivists and by deductivists. Without them the problems to be discussed could not even be formulated. Adopting them we are doing what every philosopher does – we restrict ourselves to some problems, and assume or take for granted a definite solution for others. It may of course be (and I for one think there is a great deal of truth in this supposition) that this procedure renders our discussion as unrealistic as is the inquiry concerning the number of angels capable of inhabiting the point of a needle.

(4) Our procedure is therefore this: we assume a class S of theories, a class O of singular statements. We ask: given O, what can be said about S? It is important to point out in this connexion that S contains only the theories which have been actually suggested. It does not contain all possible theories. In special cases S may contain a single theory only. Methodological rules concerning the selection of special theories from S which refer to "all possible theories", or to "all possible theories, given these and these conditions" are therefore inapplicable unless further hypotheses are introduced. This remark applies to the deductivist and to the inductivist alike. It seems however (and I might return to this point) that deductivism is not in need of such additional hypothesis. [3|4]

(5) There are certain methodological rules which are shared by the inductivist and the deductivist. I mention some of them:
The elements of S are supposed to be

(i) consistent
(ii) relevant, i.e. capable of explaining at least O
(iii) not ad hoc, i.e. capable of explaining not just O
(iv) not refuted

Rule iv is rejected by some inductivists. And this they cannot do provided they regard theories as instruments of prediction and refuse to regard them as descriptive of certain existing features of the universe. Instruments of prediction may be retained even after they have been refuted and one might find it interesting even then to ascribe to them degrees of reliability. For a realist such a procedure is clearly impossible. It is also not easy to see how a refuted theory could be used as an explanans in a satisfactory explanation. Moreover the issue between realism and instrumentalism can be formulated so as to be independent of the issue between inductivism and deductivism. I shall therefore exclude from consideration those versions of inductivism which reject iv.

Rules ii and iii are capable, as well as in need of, more precise formulation. Restrictions must be imposed upon the additional premises used in the derivation of what is regarded as the explanandum. This is a task which can be carried out by inductivists and deductivists alike.

The application of rules i to iv leads to a reduction of the number of initially suggested hypothesis. There are other rules of a comparative kind which may further reduce the number of admissible hypotheses. Of them I shall mention only three [4|5]

(v) a hypothesis which is more highly testable is to be preferred

(vi) the more coherent hypothesis is to be preferred

(vii) the more general hypothesis is to be preferred (in a given situation take always the most general theory).

The three rules just mentioned try to embrace what may be contained in the notion of simplicity insofar as this notion is dependent on possible evidence rather than on actually existing evidence. The more general theory is to be preferred because it says more. Two theories may be of an equal degree of generality, that is they may entail the "same number" of statements, and nevertheless the one may be more highly testable than the other. Thus "All ravens are black except one and that one is white" though as general as "all ravens are black" is yet less testable as there is a situation (non-existence of ravens other than black and white) where its refuting instance (at least two white ravens) is more complex than the refuting instance for "all ravens are black" (one white raven). The statement is also less coherent. I admit again that all these notions need refinement and more precise definitions are needed in order to explain their content. But this is again a problem that is common to inductivis⟨ts⟩[E] and to deductivists. Both are convinced that rules v, vi, vii are good rules as far as they go and that all that is needed is a better formulation.

(6) So far the situation is as follows. Both the inductivist and the deductivist started with a pair (S,O) and try to reduce S by applying methodological rules. We have listed some rules which they have in common. The differences between the inductivist and a deductivist start after some of the elements of S have been eliminated. Basically there are the following two possibilities: (1) inductivist and deductivist agree in their evaluation of the theories which have survived application of i-vii but disagree in their attitude towards the result of this [5|6] evaluation. (2) They already disagree in their evaluation of the remaining theories. It will soon be clear that only the first possibility merits serious consideration. However the situation is sometimes complicated by the following fact: an inductivist may pride himself to

have found rules which enable him to evaluate his theories in a more positive fashion than is permitted to the deductivist and whose application leads to a guarantee for future success (or a certain probability of futures success; or the reasonableness of the expectation of future success). He may be unaware of the fact that his rules are deducible within the deductivist scheme and therefore incapable to do more than i-vii with whose negative character he agrees; and he may therefore be nothing but a misguided deductivist. This case only apparently belongs to (2); it can be reduced to (1) by proof of the deductivistic character of the allegedly positive rules of "induction".

The first possibility, spelled out in greater detail, is however base this: inductivist and deductivist agree that application of i-vii (and of further rules which the deductivist may use but which the inductivist views with some suspicion – I am here mainly thinking of the use of metaphysical ideas in the selection of physical hypotheses) does in no way guarantee the future success of the remaining hypotheses, although we have received the assurance that they are the best hypotheses available. For the deductivist this is all that can be achieved. Assurance of future success is beyond the task and the capabilities of the theory of knowledge (and of the scientific method). For an inductivists the mentioned result of the application of i-vii shows the need for further and more positive methodological rules, for a "logic of induction" which either provides reasons for present expectations, or else gives assurance of at least partial future success. We shall soon discuss this demand in greater detail. At [6|7] the present moment I find it feasible to give a more detailed account of what can be expected from a consistent application of i-vii and other similar rules and of what cannot be expected from such an application. This discussion is a necessary element of all arguments against what we have above called "misguided deductivists". If our evaluation of one i-vii is correct then proof to the effect that an allegedly inductivistic rule follows from these rules immediately punctures any promise of success that may be connected with the application of it.

(7) It has already been indicated that the deductivist may be able to further reduce the group of theories obtained from an application, to S, of i-vii. He may make use of metaphysics, of more general hypotheses asserting the similarity between different domains, and he may even base his selection upon aesthetic prejudice (Galileo, for example, abhorred ovals, both in churches, and in the sky). Now whatever the rules used for selection, the attitudes towards the remaining theories will be this: these theories are the best theories available (given the information O, that is); it is rational to choose any single one of them; one is perhaps even justified to choose them (in the sense in which one is justified of choosing what is provided by rules one has adopted). However beyond this the deductivist is not prepared to go. More especially, he is not prepared to say (and he has not the slightest reason for saying) that the procedure of selection has established the remaining theories as faithful reflection of reality; or has shown that they can in any way be trusted to give accurate predictions in domains not yet investigated. The very next observation may refute them. Nor will he be prepared to ascribe to these theories a high probability expressing either a high chance of success in as yet unexplored domains, or a high

expectation of success (rational expectation, that is) in such domains. Such probabilities he will not ascribed except on [7|8] the basis of further theories establishing a connexion between the existing theories and O. (All these considerations are of course independent of the number of elements of O). In the sense in which we are justified to make use of the best means that our disposal the use of any one of the remaining theories may of course be regarded as justified; we are justified in preferring what is better or what is worse. However the fact that a selection has been made according to i-vii does not provide any basis whatever for the belief that these best theories will fulfill our expectations – that is that they will predict correctly, or at least with high probability. There is no possibility to argue that we are now closer to success, or that we are likely to be successful in the future. Thus far the attitude of the deductivist. This attitude is based on an analysis of the rules which the deductivist accepts (i-vii). This analysis is, I think, accepted by many inductivists. It is as follows.

(8) Speaking intuitively we might say that i-vii are more likely to lead away from success than towards success. We have excluded ad hoc hypotheses which means that we have proceeded into unknown domains about which no information is available. We have preferred statements of higher content to statements of lower content which means that we took a rather high chance and refused to play safe. We have also preferred highly testable statements to safe statements which means that we have chosen the most vulnerable theory. Considering all this it seems rather foolish to expect the theory to succeed in new domains or to assume that the predictions are now very likely to turn out right. The only possibility to achieve security would be to remain to remain within O – and this one has refused to do.

(9) It is instructive to repeat what has been said so far in the case of a concrete example. O = a is a raven in is black and b is a raven and is black and c is a raven and is black. We consider the following hypotheses: [8|9]
The group to follow the call group R:
T All ravens are black
T′ All regions but one is black
T″ All ravens but two are black

----

t No raven but two are black
t′ no ⟨sic⟩ raven but one is black
t″ no ⟨sic⟩ raven is black.
These are by no means the only hypotheses which can be devised. We might also consider
A′ T′ and one raven is white
A″ T″ and two ravens are white…
or
B All birds are black
B′ All birds but one etc. as in group R.

We shall restrict ourselves to group R. The reasons are obvious, but it is advisable to spell them out in detail. Inductivists are always been in the following dilemma. On the one side the aim is to achieve general knowledge; on the other side it is attempted to make this knowledge as safe as possible. One wants to generalise, yes, but one does not want to generalise too much. Seen from this point of view "All ravens are black" is to be preferred to "ALL BIRDS are black". It is general, but it is not too general. However is it not clear that between "All RAVENS are black" and O there is an infinity of further statements of decreasing generality which may even form a dense sequence? Just assume that "All G are black" is a statement at which we have arrived. Now consider a property, P, possessed by the ravens of O, but not contained in the definition of G. Assuming there [9|10] are infinitely many "pure" properties in many individual ("pure" properties being properties which do not contain any preference to individuals such as do some of Goodman's properties) such a property can always be found and "All PG are black" will then be "between" O and "All G are black" in order of generality. This being the case the rule "given O, use the least general generalization" is impossible to carry out. Moreover, the use of "All RAVENS are black" (rather than "All BIRDS are black") can be defended on deductivist grounds. The deductivist demands to use the most general hypothesis one can think of and which is compatible with the evidence O. "All BIRDS are black" is clearly excluded by the evidence available to a birdlover. However if O is the only evidence available, then this statement cannot be excluded. The inductivist, making surreptitious use of what he knows uses these evidence nevertheless. He therefore proceeds exactly as a deductivist would who starts from a richer evidence $O' \neq O$. He interprets his procedure has being based on an independent inductivistic rule and expects therefore "All RAVENS are black" to be successful. In this he is nothing but a misguided deductivist (cf. above, section 6).

Restricting ourselves to group R we may now argue as follows: t, t′, t″ are refuted by O; T′, T″ etc. are statements of decreasing testability. T, "All Ravens are black" remains as the most general, most highly testable, unrefuted statement of R and will therefore be accepted. Considering the manner in which it was obtained we have no guarantee whatever that the next relevant observation will be a confirmation. What could provide such a guarantee? The observations O? But T goes beyond these observations and there is no reason why the particular way it went beyond them should represent the actual state of affairs. For this particular way in which we have made T go beyond the observational information is determined not by any further knowledge we possess concerning the course of nature but by demands which are dictated by reasonable maxime ⟨sic⟩ that where nature [10|11] does not help us to make a selection among theories, we should try to make a reasonable selection ourselves. We cannot work with 10 theories at the same time. We must choose. Of course, we should choose the correct theory and we should therefore let indications of correctness be guides in our choice. Observations are such indications. But compared with what we want to achieve, – general knowledge – observation says always too little. Every theory therefore possesses a domain where observations cannot guide us any more. In such domains we might make use of theories we already possess. These theories bring in further actual observations, viz. the observations which have been collected for the purpose of testing them. But they also transcend these observations, as does the theory we are examining: let T be this theory, D its domain of transcendence, T′ the theory used for making a selection, and D′ its domain of transcendence. It is clear that inside DD′ a choice between T and T′ is a matter that cannot any longer be decided on empirical grounds. Experience has already had its say, and this say does not extend into DD′. Also experience has had its say negatively, on the basis of rule iv, so that T and T′, although being survivers ⟨sic⟩, cannot be regarded as containing information which, although going beyond observations, can be guaranteed to be successful, or reliable. This means that inside DD′ we are completely free to do whatever we see fit to do. If we should decide to adopt certain rules here then the justification cannot be that they will guarantee success – in this case they would have to be based upon empirical hypotheses and we have already assumed that we are working in a domain where all empirical hypotheses are equally conjectural, and where therefore a decision must be made on non-empirical grounds. We shall select according to purely human standards (when no one else tells us what to do then we may do whatever we like to do). For example, we shall select the simplest theory and we shall select it not because we shall in this way come closer [11|12] to the truth, or because such a procedure guarantees predictive success, but because working with simple hypotheses makes life easier. There is no reason why a scientist should wilfully ⟨sic⟩ complicate matters. And there is no reason either to try to justify the very sensible demand for simplicity in an "inductive" fashion. The same remarks apply to all the other demands i-vii (with the exception of iv) which are made the basis for selection. They are "formal" in the sense that they have no factual basis and no factual consequences. They are designed to regulate the construction of knowledge in domains where we are completely free to do whatever we see fit to do. Is it not clear, now, that a selection according to these purely human rules does not contain the slightest guarantee for future success? One might as well expect that a beautifully painted car will be swifter than a car of an ugly color. And is not also clear that our expectations should go parallell ⟨sic⟩ with this actual situation and that we should not try to build a logic of expectation which may be regarded as a guide for our feeling of security?

(10) The fact that a certain manner of choice seems to be part of our natural behavior has led to some very misleading arguments. It has been assumed that the existence of such habits justifies the choice of certain hypotheses and the trust we possess concerning the future, the success of these hypotheses. On this basis the problem of induction was rejected as a pseudo-problem. Now it is of course true that if it is customary to do a certain thing then we are justified in doing it given the custom. That is, if it is customary to generalize from the PaandQa and PbandQb and PcandQc to all P are Q then proceeding in the fashion in a specific case is justified by reference to this general custom. And if it is customary to call a sentence which has been obtained by following this habit "justified on the basis of the evidence", then this is what "justified" means and this is how we use the term. However the problem of induction, our problem starts with the question for the rationale behind the habit that leads [12|13] one from the singular to the general and that is verbalised in the phrase "justified on the basis of observation". That there was such a habit was well known to Hume and he tried his best to explain its origin. At the same time he

asked for the arguments which could be adduced in favor of proceeding according to the habit and he arrived at the result of that no such arguments can be given. According to the deductivist there are such arguments. We can give reasons as to why it is better to choose "All Ravens are Black" rather than "All ravens are black but one and that one is white" when the evidence is 10 black ravens. We can strongly support this choice of "All ravens are black" on the basis of the rules i-vii. However as soon as it is realised that the choice is supported by i-vii, in the very same moment we must give up the hope to have are also arguments in favor of the success of the chosen hypothesis "All Ravens are Black". That is, we can justify the habit, but we cannot justify the expectation that following the habit we shall arrive at statements which are more successful, more likely to lead to correct predictions than other statements. Of course, and this item deserves repetition, we can base such an expectation upon further hypotheses, such as is the hypothesis of the existence of laws of nature, or the existence of limits of relative frequencies. However the whole structure of these more general hypotheses together with the more specific ones is entirely conjectural.

(11) To sum up: according to the deductivist logic and "formal" considerations (that is considerations concerning the structure of the theories considered and the extension of the relevant evidence) are sufficient for selecting those hypotheses we intend to use. In the process of selection observation plays a negative role and it therefore cannot guarantee success to the remaining theories (unless other the theories are used – see above). The "formal" considerations are adapted to our own convenience and even increase the chance of failure (intuitively spoken). The deductivist [13|14] can in this fashion justify many generalising habits, such as the transition from the black ravens to "All Ravens are Black", especially if the intuitive knowledge of further evidence which guides such habits (and which in the present case excludes "All Birds are Black" or even "All things are black") is utilised. These habits are thereby shown to be reasonable, and not just existent. At the same time it emerges that the expectation of success, although psychologically explainable and also explainable on the basis of further conjectural beliefs does not have a led to stand on if the permitted legs are the existing observations and rules i-vii. Clearly further rules are needed in order to give it some reasonable support. It is the conviction of the inductivist that such further rules do in fact exist, that they enable the methodologist to make a further selection in cases where application of i-vii has left more than a single theory, and that they can also deal with cases which constitute a great difficulty, or alleged difficulty for the deductivist, such as the case of probabilistic laws (iv not capable to lead to the rejection of probabilistic laws). The interesting difference between the inductivist and the deductivist therefore lies in this. The former agrees that i-vii are reasonable rules. He agrees with their evaluation by the deductivist (success cannot be guaranteed on the basis). And he demands therefore the use of further rules leading to such guarantee. He also hopes that these further rules will solve difficulties for which the deductivist has at present no solution.

(12) Now at this point it is necessary to emphasize the merits of deductivism. Its merits lie in its simplicity. Basically it can do with logic (syntax, semantics) alone.

In the definition of the main terms of i-vii [14|15] there will occur only logical notions (apart from the notion of possible evidence) and only logical rules of deduction will be used. For a deductivist the solution of difficulties is much more well defined. To talk in an exaggerated fashion: if the deductivist cannot handle certain statements, he has to think and try to accommodate them within his scheme. The inductivist need only invent a new rule of induction, and everything is taken care of. Even if there should be criteria for the selection of "good" rules of induction, even then there are many more degrees of freedom in his system than are in deductivist system and alleged successes therefore do not count very much. It seems to be a good heuristic principle first to try to solve a problem on the basis of the deductivist scheme and not to give up however many difficulties might arise. Such a procedure may have quite surprising consequences. It might turn out that many so-called "inductivistic" rules are as a matter of fact consequences of i-vii and therefore incapable of guaranteeing success in as yet unexplored domains. The rule of simple generalization the (from 10 ravens to all ravens) is most certainly of this kind, and so might be Reichenbach's rule acc. to which the relative frequencies should at any moment to be identified with the limit [It seems to me that this rule could be derived from i-vii made more explicit. First, the use of just m and n would follow from the considerations of simplicity; then the use of m/n rather than of another function would seem to follow (I argue here intuitively, but I think the argument can be carried out in detail) from the demand of 0 and 1 as the extreme values and the additivity of the probabilities for disjoint domains. Again such derivation would show it to be a deductivist rule which is therefore incapable of guaranteeing success.]

But this, now, is the main objection: that deductivism cannot guarantee success; and this also is the main motive of any site search for a different kind of logic. [15|16]

(13) At this point it is imperative to distinguish between two different kinds of rules. We shall call them active and passive rules respectively. Passive rules do not pretend to be more than <u>descriptive</u> – all they want to do is to describe what is actually being done by the members of a certain community such as the community of scientists. Active rules intend to <u>interfere</u>. They do not describe scientific practice, they want to <u>guide</u> scientific practice. Rules i-vii are clearly intended (at least by myself) as active rules. If a scientist does not conform to them then all the arguments which might be used in favor of them turn against him and condemn his procedure. The question arises whether the additional rules which the inductivist wants to use are intended as active rules, or as passive rules. And if as active rules, whether they will further restrict the use of theories over and above i-vii, or whether they will also "liberalise" some of the rules i-vii. I think that the liberalization of i-vii must not be permitted. I shall not give the arguments here, but the reader might know which arguments I would propose were I to take up the matter. As regards the further restriction of the choice of theories, or such restriction combined with a liberalization of i-vii the result might indeed justify the promise held out by the inductivist: <u>he might indeed be capable to promise success, and to keep his promise</u>. However I am afraid he will be capable of doing so not because he understands nature so well, but because he has imposed restrictions upon the future replacement

of theories. And from the existence of these restrictions he can of course derive the non-occurrence of replacement and therefrom ⟨sic⟩ a certain measure of success. This is what has happened in physics were alternative theories are forbidden although it is only with their help that the difficulties of the existent theories can be discovered. Rules which are not of this kind can be criticised in a different manner. Consider the rule [16|17] which ascribes to every statement a function F(n) depending on the number n of confirming instances such that for n>n′ F(n)>F(n′). This rule invites to use among different competing theories the theory with the higher F and thereby actively guides our way of building up knowledge. Is there any rationale for the rule? Consider the following possibilities:

(A) Observations of different individuals of nature are mutually independent.
(B) There are periodicities of length M±a (a a small integer) after which properties change.
(C) There are stable complexes of properties and the properties investigated are part of such complexes.

If (A) is correct, then the rule is misleading in the same sense in which the gambler's fallacy is misleading. If (B) is correct then the rule is misleading in an even stronger sense. The rule we should use would then be that for n>n′ F(n)>F(n′) in the neighborhood of M. Only in case (C) does the rule have any sense at all. Now if we known neither (A), nor (B), nor (C) but know that they are actual possibilities, then all the possible rules should be given equal weight, that is all the hypotheses selected according to all the possible rules should be given equal weight, that is no single rule of the kind mentioned, and this include⟨s⟩[E] of course the most popular first rule should be used. That is, no active restrictive inductive rule should be used.

To be continued.

Editors' Notes

HF 08-29-01 (17 ss., T+H); HF 08-34-05 (17 ss., T+H). HF 08-29-01 and HF 08-34-05 are two identical copies of the same typescript differing only in the handwriting of the note stating the author and date on the first page of the document. The transcription here is based on HF 08-34-05.

[1] In HF 08-29-01 Feigl's handwritten note reads: "P. K. Feyerabend Feb '63".

Chapter 5
The Feyerabend-Hempel Correspondence (1961-1977)

Paul K. Feyerabend and Carl G. Hempel

1961

1. Feyerabend to Hempel, [4 October 1961]

⟨4 October 1961⟩PM CH

Dear Peter,

I wonder whether you got my letter about the Feigl-Festschrift. We (i.e. Grover Maxwell[1] and I) plan to prepare a Volume ⟨sic⟩ on the occasion of Herbert's 60th birthday. Nagel,[2] Popper,[3] Wilfrid Sellars,[4] Viktor Kraft,[5] Michael Scriven,[6] Russ Hanson,[7] Wolfgang Köhler,[8] S. S. Stevens,[9] Salmon,[10] Hilary Putnam,[11] Hill, the physicist from Minnesota[12] and others have already agreed to contribute. It would be excellent if you, too, could write an article for the volume. The article should be submitted, i.e. sent to me by Jan. 31 1962 and it can be on any topic you would care to choose. I may inform you that there are 7 contributions in the mind-body problem which topic is the one that occurs most frequently. Two contributions on induction (Nagel, Kraft), two on meaning criteria (Putnam, Salmon) and other topics. I do hope you will be able to write an article for the volume.[13]

I do not know whether I have sent you my latest on the quantum theory. In any case I am sending a copy with the same mail.[14]

With best wishes

Yours sincerely
Paul (Feyerabend)

M. Collodel, E. Oberheim (eds.), *Feyerabend's Formative Years. Volume 2. Feyerabend on Logical Empiricism, Bohm & Kuhn*, Vienna Circle Institute Library 10,
https://doi.org/10.1007/978-3-031-57519-8_5

Editors' Notes

CH 16-06 (1 s., H)

[1] See Vol. 1, Part II, 1960-1, n1.
[2] See Vol. 1, Part II, 1957-11, n7; eventually Nagel did not contribute to the *Feigl Festschift* (see FEYERABEND-MAXWELL (eds) 1966).
[3] See POPPER 1966.
[4] See Vol. 1, Part II, 1959-9, n5; 1961-4, n7; and SELLARS 1966.
[5] See Vol. 1, Part I, Ch. 1, n15; and KRAFT 1966.
[6] See Vol. 1, Part II, 1957-7, n3; and SCRIVEN 1966.
[7] See Vol. 1, Part II, 1956-14, n5; and HANSON 1966.
[8] Feyerabend refers here to the German experimental psychologist Wolfgang Köhler (1987-1967); see KÖHLER 1966.
[9] See STEVENS 1966.
[10] See Vol. 1, Part II, 1961-10, n4; and Salmon 1966.
[11] See Vol. 1, Part II, 1957-12, n3; eventually Putnam did not contribute to the *Feigl Festschift* (see FEYERABEND-MAXWELL (eds) 1966).
[12] See Vol. 1, Part II, 1957-12, n4 ; and Hill 1966.
[13] See FEYERABEND-MAXWELL (eds) 1966; eventually Hempel did not contribute to the *Feigl Festschift*. On the history of the publication of FEYERABEND-MAXWELL (eds) 1966, see *below*, Hempel to Feyerabend, 16 October 1961, 1961-2; Feyerabend to Hempel, 25 October 1961, 1961-3; 10 October 1963, 1963-1; Hempel to Feyerabend, 24 October 1963, 1963-2; Feyerabend to Hempel, 5 January 1964, 1964-1; Hempel to Feyerabend, 5 January 1964, 1964-2; and Vol. 1, Part II, 1961-4, n8.
[14] See FEYERABEND 1962a; on the history of the development and publication of Feyerabend 1962a, see *above*, Ch. 3, § 2, n14; *below*, Feyerabend to Hempel, 25 October 1961, 1961-3; and Vol. 1, Part II, 1957-12, n1.

2. Hempel to Feyerabend, 16 October 1961

October 16, 1961

Professor Paul Feyerabend
Department of Philosophy
The U of Cal
Berkeley 4, California

Dear Paul:

I am very sorry indeed not to have responded to the circular you sent me quite some time ago inviting me to write a contribution to the volume you and Grover[1] are planning to publish in honor of Herbert Feigl's 60th birthday.[2] The reason was, very simply, that I just hated to say I could not do it; but this is what I have to say after all. I have a number of earlier commitments still hanging over me, among them the presidential address for the December meeting of the APA, and several other papers, and I would rather contribute nothing than submit a hastily written and not carefully consider piece. As it is, I find that even the things I have given a lot of thought to, and that I have re-written several times before submitting them, still turn out to have shortcomings; and what I could produce on fairly short notice would be only more or less a re-hash of previously published stuff or something tentative and half-baked. I have had to declined several such invitations, but in the case of the volume for Herbert, it is very painful indeed for me to have to say No ⟨sic⟩. I have thought about it again and again, but I just can't get anything done that would be suited for the purpose. I will have to participate in a symposium in November and will have to think up some ideas for the occasion; if – contrary to my expectation – there should seem enough substance to what occurs to me to make it suitable for the Feigl volume, I will promptly let you know.

Thank you for the dittoed version of your paper, "How to be ⟨sic⟩ a Good Empiricist",[3] which arrived two days ago. I have read some parts of it so far and certainly found your discussion stimulating, but it seems to me that your basic point, the need for a critical metaphysics, is closely akin to what Einstein often expressed by saying that the formulation of scientific theories was a matter of inspired invention, of ingenious imagination, rather than of any inferential procedure leading from the "data" to an explanatory system; and this surely has been acknowledged in the writings of many empiricists – e.g. by Carnap, in his book on Probability ⟨sic⟩;[4] and the point that the "facts" by which supposedly the theories of science are tested are themselves theory-dependent is made, for example in Cohen and Nagel's [1|2] Introduction to Logic and Scientific Method,[5] ch. XI, sec. 5. This is not to say, of course, that there is nothing further to your own discussion, but only that perhaps your initial statement of your thesis is unduly extreme. I'll read your essay with much interest and will see how far I agree. – There are, by the way, a number of bibliographic references in your piece that I was not aware of, and which to be sure to look up.

Once again, please let me assure you as Grover how very sorry I am not to be able to contribute to the bouquet that Herbert deserves to ⟨sic⟩ well.

Cordially, and with best wishes,
⟨UNSIGNED⟩
Carl G. Hempel

Editors' Notes

CH 16-06 (1 s., T+H; c)

[1] Hempel refers here to Grover Maxwell (see Vol. 1, Part II, 1960-1, n1).
[2] See *above*, Feyerabend to Hempel, 4 October 1961; on the history of the publication of FEYERABEND-MAXWELL (eds) 1966, see *above*, 1961-1, n13.
[3] On the history of the publication of FEYERABEND 1963a, see *above*, Ch. 3, § 4, n3.
[4] See Carnap 1950.
[5] See COHEN-NAGEL 1934.

3. Feyerabend to Hempel, [25 October 1961]

University of California, Department of Philosophy, Berkeley[1]

Rec. 10 - 25 - 61[CH]

Dear Peter –

thank you very much for your letter. It is a pity that there is only such a small chance that you may write a contribution, but of course I understand your predicament. But I still hope that the unlikely will happen, and you will find something to contribute. After all, it need not be a full-length paper. A note of about 10 pages, or less, will also do. It may even be possible to extend the deadline. I shall write you about this after my visit to the University of Minnesota Press in November.

Thank you also for your comments concerning my 'how to be' etc.[2] By the way – my main point is not that the formulation of a scientific theory is a matter of inspiration, nor is it that facts are theory-dependent. I completely agree with you that these two items have become safe possession of contemporary empiricism. My main point is rather this: assume that there is a theory T that is highly confirmed in a domain D and whose principles have also being developed with great precision. An example is the Phenomenological thermodynamics (consider it in the form given to it by Caratheodory ⟨sic⟩[3]). Assume that somebody suggests another theory, T′, which, in the form in which he puts it forth has very little formal precision and moreover contradicts T inside D. An example would be the various attempts at the kinetic account of heat (a more concrete example is T: orthodox quantum theory. T′: Bohm). Now the question arises: [1|2] Shall we encourage the further development of T′, or shall we tell the people who introduced T′ in the first place to give up and to rest content with T which, after all, is formally very satisfactory and highly confirmed. It seems to me that many contemporary physicists would take the second position. This is, after all, the main argument against a Bohm: "what do you want. Our theory works, is successful, formally satisfactory. Your ideas are undeveloped. So better give up". There are of course some people that would allow Bohm to continue and their argument is that he might be right. What I try to show is not only that there are no reasons why one should not try out alternatives to the commonly accepted point of view but there are methodological reasons in favor of such alternatives. And the reasons are that (using the above nomenclature) the empirical content of T will be increased by introducing T′ and developing it in some detail. What I argue for is this: do not rest content with the theory you possess, even if it is highly confirmed and formally very satisfactory. Try to develop alternatives to this theory which are inconsistent with it and which therefore lead to new predictions. And you should do this because in this way you will increase the empirical content of the theory from which you started. This is my main point. I would be most grateful if you could tell me whether this point has been made before, and if so where, and by whom. I do not know the literature too well, so I may [2|3] well have overlooked something. Duhem, of course, says that because of the latitude admitted by the observation statements such alternatives are possible. But as far as I know he does not give any reason why they should be invented.

Thank you again for your letter,
and all the best

Yours friendly
Paul

P.S. Did I ever send you my opus magnum on the quantum theory (title 'Problems of Microphysics')?[4] It will be published, together with your paper on explanation in Grünbaum's Pittsburgh Volume.[5]

Editors' Notes

CH 16-06 (3 ss., H)

[1] UNIVERSITY OF CALIFORNIA
DEPARTMENT OF PHILOSOPHY
BERKELEY 4, CALIFORNIA

[2] See *above*, Hempel to Feyerabend, 16 October 1961, 1961-2.

[3] Feyerabend refers here to the Greek mathematician Constantin Carathéodory (or Constantine Karatheodori) (1873-1950).

[4] See *above*, Feyerabend to Hempel, 4 October 1961, 1961-1; and Feyerabend 1962a.

[5] See Hempel 1962; on the history of the development and publication of Feyerabend 1962a, see *above*, 1961-1, n14; Ch. 3, § 2, n14; and Vol. 1, Part II, 1957-12, n1.

1963

1. Feyerabend to Hempel, 10 October [1963]

University of California, Department of Philosophy, Berkeley

⟨10 October⟩PM CG
rec 10/16CH

Dear Peter,

the deadline for the Feigl Festschrift has been postponed until December 1963. If by any chance you have something ready now, please let me have it. I do not want to urge you and if you have neither time, nor a paper ready, please disregard this letter. However if you should find a little time, it would be most wonderful to have something by you, if only a short note, in the volume honoring Herbert.[1] I am enclosing two of my most recent literary crimes.[2]

With best wishes
Paul

Editors' Notes

CH 16-06 (1 s., H)

[1] On the history of the publication of FEYERABEND-MAXWELL (eds) 1966, see *above*, 1961-1, n13.
[2] Feyerabend refers here to FEYERABEND 1963b and FEYERABEND 1963c (see *below*, Hempel to Feyerabend, 24 October 1963, 1963-2).

2. Hempel to Feyerabend, 24 October 1963

Center for Advanced Studies in the Behavioral Sciences, Stanford[1]

October 24, 1963

Professor Paul Feyerabend
Department of philosophy
The University of California
Berkeley 4, California

Dear Paul:

Many thanks for your note, which reached me here a few days ago.[2]

I truly appreciate your offering me another chance to contribute something to the Festschrift for Hebert; as you know, I would be delighted to do so for many reasons. But I have nothing in hand now, short or long, that would be appropriate for the purpose; and I don't want just to dash off something inconsequential for the occasion. Nor do I feel that I can afford to take out the time that would be required to write even a brief piece which would be worthwhile: I am at work on two projects which I am very anxious to complete during my stay here, and that, unfortunately, will require all the time and energy I can muster. I have for the same reason also declined the various lecture invitations I have received here. I do hope you will understand![3]

It was very good of you to send me offprints ⟨sic⟩ of your review of Nicolai Hartmann's <u>Philosophie der Natur</u> and of your article on materialism and the mind-body problem.[4] I look forward to studying the latter, and I have already read your review, which had for me a special personal interest because I once took a course with Hartmann in Berlin; and when Reichenbach, my thesis adviser, left Berlin for Istanbul, Hartmann was one of the men who agreed to read my thesis and be my oral examiners. He could easily have failed me in the oral since I did not know much about his philosophy, but he did not even ask questions about it, as I recall, and treated me very decently. Still, I am afraid that I would have found his book almost impossible to read or digest, and I really admire you for giving such a lucid account of what Hartmann is after, and such a clear and pointed criticism of his work.

With many good wishes,

Cordially yours,
Peter
Carl G. Hempel

Editors' Notes

PF 2-7-81 (1 s., T); CH 16-06 (1 s., T; c)

[1] CENTER FOR ADVANCED STUDIES IN THE BEHAVIORAL SCIENCES
202 Junipero Serra Boulvard
Stanford, California

[2] See *above*, Feyerabend to Hempel, 10 October 1963, 1963-1.

[3] On the history of the publication of FEYERABEND-MAXWELL (eds) 1966, see *above*, 1961-1, <u>n13</u>.

[4] See FEYERABEND 1963c and FEYERABEND 1963b, respectively (see *above*, Feyerabend to Hempel, 10 October 1963, 1963-1, <u>n2</u> in particular).

1964

1. Feyerabend to Hempel, 5 January [1964]

University of California, Department of Philosophy, Berkeley

5. Jan. ⟨1964⟩

Dear Peter,

this is to inform you that the deadline for the Feigl-Festschrift has been extended to end of March. Perhaps it is now possible to contribute? I very much hope you will be able to write a contribution.[1]

Kindest regards plus
best wishes for the New Year
Paul

Editors' Notes

CH 16-06 (1 s., H)

[1] On the history of the publication of FEYERABEND-MAXWELL (eds) 1966, see *above*, 1961-1, n13.

2. Hempel to Feyerabend, 5 January 1964

Center for Advanced Studies in the Behavioral Sciences, Stanford

5-1-64

Dear Paul –

Many thanks for your review of Russ Hanson's book[1] and for your earlier interesting materials![2] I was particularly intrigued by your comment on the explanation-prediction issue[3] and if I understand your brief remarks right, I quite agree; & I ⟨should⟩ think that if use is made of hidden variables, then still to the extent that we can explain we could potentially have predicted. I hope to send you some things I've written, but not quite yet.

Best wishes,
Peter

Editors' Notes

PF 2-7-75 (PC, H)

[1] See Feyerabend 1964b.
[2] See *above*, Feyerabend to Hempel, 10 October 1963, 1963-1, n2 in particular.
[3] See Feyerabend 1964b: 266.

3. Feyerabend to Hempel, [September 1964]

University of California, Department of Philosophy, Berkeley

rec 10/2[CH]

Dear Peter –

this is only a brief request. Could you please send me details about the way the philosophy of science program is run at Princeton? I may have to draw up a program myself and would like to profit from your experience.[1]

Herbert Feigl and I had a wonderful time at a seminar in Alpbach/Tirol. Carnap participated also and was criticized as being too liberal (he explained the most recent version of his criterion of cognitive significance) by the younger generation of German physicists and philosophers of science. They really want to return to the early time of the Vienna Circle.[2] With best wishes, and many thanks

Paul Feyerabend

Editors' Notes

CH 16-06 (1 s., H)

[1] Feyerabend's request is connected with the plans for his appointment to the Chair of the History and Philosophy of Science at the University College London (see Vol. 1, Part II, 1964-2, n1); on Feyerabend and the University College London Chair of the History and Philosophy of Science, see Vol. 1, Part II, 1962-2, n4.

[2] On Feyerabend and the 1964 European Forum Alpbach, see Vol. 1, Part II, 1963-3, n2.

4. Hempel to Feyerabend, 2 October 1964

October 2, 1964.

Professor Paul Feyerabend
Department of Philosophy
University of California
Berkeley 4, California

Dear Paul:

It was nice to hear of your pleasant experience in Alpbach. I hope Carnap will stop over here on is way back to Los Angeles, and that I will then hear more about the proceedings.

The best way of informing you about our Graduate Program in the History and Philosophy of Science is probably to send you the enclosed folder; on page 7 ff. you will find the details. Under separate cover, I am mailing you a copy of our Graduate School catalogue; pages 121-123 and 152-156 are relevant for your purposes.

If you have any further questions, please let me know; and I will be glad to try to answer them.

With many good wishes,

Cordially yours,
⟨UNSIGNED⟩
Carl G. Hempel.

Editors' Notes

CH 16-06 (1 s., T; c)

1965

1. Feyerabend to Hempel, 23 October [1965]

⟨Vienna,⟩PM 23. Oct.

Dear Peter –

about one year ago I was asked by a gentleman of your department (whose name I have forgotten) to deliver a lecture in Princeton, in case I was coming East. Now I could deliver a lecture on March 23 or 24 or 25 1966, if there is still some interest.X Could you please inform the gentleman in question of the matter?[1] It will not be possible to reach me by mail until December when I shall be back in Berkeley.

In the meantime I have after long hesitation decided to decline the appointment with the chair in London + to stay in the U.S.[2] I am now off duty for a year and spending the time partly lecturing, partly in research.[3] In a month I start a lecture tour through Germany (both East and West) and I shall return to the U.S. in December.[4] The students are very good here, and very pleasant to talk to. In Salzburg I met Lorenzen who not content with his mathematics now branches out into the foundations of physics strictly after the fashion of Dingler. His 'protophysics' (which is jokingly called 'prototoon physics') and his 'Primary Language' (which we christened 'Primate-language') is simply Dingler all over again. But [1|2] he is excellent in argument, and very amusing to talk to. If I am not quite mistaken he is now in Texas) ⟨sic⟩.[5] The greatest philosophical ability is found here not among the philosophers but among the scientist.

With best regards
Paul

---

[1] X I would come from Pittsburgh

Editors' Notes

CH 16-06 (1 s., H)

[1] It is not clear whether Feyerabend eventually delivered a lecture in Princeton in March 1966 or not (see Vol. 1, Part II, 1966-1, n1).

[2] On Feyerabend and the University College London Chair of the History and Philosophy of Science, see Vol. 1, Part II, 1962-2, n4.

[3] Feyerabend took a leave of absence from July 1965 to June 1966 thanks to a National Science Foundation grant (see Vol. 1, Part II, 1965-3, n3).

[4] On Feyerabend's 1965 lecture tour in Germany, see Vol. 1, Part II, 1965-3, n3.

[5] Feyerabend met Lorenzen (see Vol. 1, Part I, Ch. 1, n5) at the 1965 research discussion (*Forschungsgespräch*) organized by the *Internationales Forschungszentrum für Grundfragen der Wissenschaften Salzburg* (Salzburg International Research Center on Key Issues of the Sciences); on Feyerabend and Lorenzen, see Vol. 1, Part I, Ch. 1, n5 in particular; Part II, Feyerabend to Popper, 13 October 1965, 1965-3, n6 in particular; and FEYERABEND 1995: 136.

1966

1. Feyerabend to Hempel, 1 June [1966]

University of California, Berkeley, Department of Philosophy[1]

1. June

Dear Peter,

enclosed a discussion on explanation between Smart, Putnam, Sellars, and myself.[2] Things are still pretty rogue and unsatisfactory, but you might find the one or the other remark interesting. Best wishes, and looking forward to seeing you again

Paul

Editors' Notes

CH 16-06 (1 s., H)

[1] UNIVERSITY OF CALIFORNIA, BERKELEY
DEPARTMENT OF PHILOSOPHY

[2] See See SMART 1965, PUTNAM 1965, SELLARS 1965 and FEYERABEND 1965b, respectively.

2. Hempel to Feyerabend, 16 June 1966

June 16, 1966

Professor Paul Feyerabend
Department of Philosophy
University of California, Berkeley
Berkeley, California 94720

Dear Paul,

Thank you very much for sending me reprints of the Smart-Putnam-Feyerabend discussion in the latest Boston Colloquium volume.[1] I am very glad to have them: this enables me to put them on reserve for my graduate seminar. As a matter of fact, I dealt with your and Hilary's[2] contribution in my seminar this past spring, and one of the students chose to write his paper on the subject. In all, your piece proved very stimulating and provocative and met with great interest.

I found the Minnesota conference very helpful and productive, and I think I now understand much better what you are after, though I am still perplexed by some of your ideas.[3]

It was good to see you; I hope you are setting in happily at Berkeley, and that you have found, or soon will find, an attractive place to stay.

The best wishes,

Cordially yours,
⟨UNSIGNED⟩
Carl G. Hempel

Editors' Notes

CH 16-06 (1 s., H; c)

[1] See SMART 1965, PUTNAM 1965, and FEYERABEND 1965b, respectively.

[2] Hempel refers here to Hilary Putnam.

[3] Feyerabend and Hempel participated in a conference on the "Logical Analysis of Theories in the Empirical Science" at the Minnesota Center for the Philosophy of Science (28-30 May 1966), attended by Feigl, Hanson, Maxwell, Rozeboom, Peter Achinstein, Roger Buck, William Craig, Ian G. Barbour, and Mary B. Hesse, among others. In the morning of the first day of the conference, Feigl presented a brief survey of issues, and the conference agenda, whereas Hempel delivered a paper on "The Structure of Theories"; in the afternoon, Feyerabend read a paper entitled "Critique of Nagel and Carnap on Scientific Explanation" (largely overlapping with Feyerabend 1966) and was followed by Buck's "Critique of Feyerabend's Views" and by a plenary discussion. The conference proceedings, including an edited transcription of the discussions, were published in Vol. IV of the series *Minnesota Studies in the Philosophy of Science*; however, Feyerabend's contribution to the volume was significantly different from the paper delivered at the conference. See *below*, Feyerabend to Hempel, 15 July 1966, 1966-3; and Vol. 1, Part II, 1966-1, n1; Vol. 3, Ch. 1, Feyerabend to Agassi, 2 June 1966, 1966-9; "Conference on "Logical Analysis of Theories in the Empirical Science", Tentative Program", 2 May 1966, RC 102-05-25; "Conference on the Logical Analysis of Theories in the Empirical Science, May 28, 29, 30", 10 May 1966, HF 16-02; and FEIGL 1970, HEMPEL 1970, FEYERABEND 1970, RADNER-Winokur (eds) 1970: 220n..

3. Feyerabend to Hempel, 15 July [1966]

University of California, Berkeley, Department of Philosophy

July 15 [CH]'66

Dear Peter,

thank you very much for your kind letter. I am sorry that I was not quite up to things during the Minnesota Conference, due to my illness, for I would have very much liked to talk some problems over with you in private. I am quite aware (at least in my more critical moments) that lots of the things I say and publish are far from being as clear as they ought to be and therefore I am more in need of criticism. Somehow I never manage to discuss things here in Berkeley;* my only critics here (and very good critics!) are the science graduate students. So a conference such as the one in Minnesota is a most important event for me. I hope that one day when I come East I shall have an opportunity to talk with you for a day or two. In the meantime – here is my review of Nagel, almost the same thing I read at the conference.[1] With best wishes, and again many thanks for your letter

Paul

* this might become different once Michael Scriven comes here – if he comes

Editors' Notes

CH 16-06 (1 s., H)

[1] See *above*, 1966-2, n3; and Feyerabend 1966b.

1968

1. Feyerabend to Hempel, 20 April [1968]

University of California, Berkeley, Department of Philosophy

Rec 4/24[CH]
20. April [CH]/68

Dear Peter,

enclosed a paper[1] which contains the same historical material as the paper I sent you a few days ago, but arranged differently, and made independent of the occasion for which the former paper was written (it was a reply to a paper by Karl Popper in which he said that in science the search for invariants was not everything and that one should permit irreversible, imprecise, indeterministic laws to enter the body of science. I say the very opposite, that is, I want our laws to remain precise, reversible, deterministic but want to put the imprecision and sloppiness into the method rather than into the laws. And I argue that the Copernican Revolution was made "my" way, rather than Karl's way.[X] I also argue that classical physics is far from finished and that we have not yet really exploited it to the full) ⟨sic⟩.[2]

My eulogy on Bohr which I sent you about a month ago,[3] found the agreement of various physicists who are very critical of the Copenhagen Interpretation and even of the great David Bohm himself who says that he is closer to Bohr than any-one else. I thought it was timed to unearth Bohr rather than the "Copenhagen Interpretation" which really is an insane mixture of all sorts of things, or Heisenberg, who is not half as critical as is Bohr. A look at the literature shows that hardly any of the criticisms that is today leveled against the Copenhagen Interpretation hits Bohr, and that he has thought in advance of almost every objection that has been raised against him. So the real Bohr ought to be given a chance. Popper unfortunately distorts him completely and ascribes to him the very opposite of what he says (the same is true of Margenau, in this country).

Would you like to come to Berlin (Free University) for two+½ months[XX] (salary, about 30 000 D-Mark brutto plus travelling expenses. An English-German secretary will be at your disposal and you may bring your own private assistant who will be paid by the university (or, at least, I hope to push this through). I have just accepted a chair in the History and Philosophy of Science at the Free University which I would like to be a rotating chair, with scholars from everywhere giving guest appearances. Are you inclined to come,[XXX] or is it out of the question?[4]

I shall be here in Berkeley until the end of April and shall then again go to London (c/o Dept. for the History and Philosophy of Science, University College, London W.C. 1, England)⟨.⟩[E 5]

With best wishes, also to your wife

Paul Feyerabend

[x] this is what the historical material is supposed to show
[xx] end of the April until mid July or Sep-mid-December
[xxx] if so, when at the earliest?

Editors' Notes

CH 16-06 (1 s., T+H)

[1] See *below*, Hempel to Feyerabend, 7 May 1968, 1968-2; and FEYERABEND 1970c. FEYERABEND 1970c is the result of the course "Introduction to Scientific Method", usually taught by Popper that Feyerabend delivered at the London School of Economics in the Autumn Semester 1967, in which he openly criticized Popper's conception of scientific method (see Watkins to Feyerabend, 17 May 1967; Feyerabend to Watkins, 22 May 1967; Watkins to Feyerabend, 26 May 1967; Feyerabend to Watkins, 31 May 1967; 17 December 1967; Watkins to Feyerabend, 30 December 1967; Feyerabend to Watkins, January 1968; undated, JW; and Feyerabend to Lakatos, 28 January 1968; 25 March 1968; FEYERABEND 1970c: 275n). The paper was published in Vol. 4 of the *University of Pittsburgh Series in the Philosophy of Science* which collected, among other writings, the papers delivered at a Philosophy of Science Workshop held at the University of Pittsburgh, 7-10 May 1965 (see *above*, Ch. 2, § 11, n1). The paper read by Feyerabend on this occasion – entitled "Outline of a Model for the Acquisition and Improvement of Knowledge" – was however completely different from the one that was later published in the collection (see *below*, Feyerabend to Hempel, 20 May 1968, 1968-2; COLODNY (ed.) 1970: ix). On the history of the development and publication of FEYERABEND 1970c, see *below*, Hempel to Feyerabend, 7 May 1968, 1968-2; and Vol. 3, Ch. 2, Feyerabend to Watkins, 17 April 1968, 1968-8; and Feyerabend to Lakatos, 19 April 1968; Lakatos to Feyerabend, 25 April 1968; Feyerabend to Lakatos, 30 June 1968; Lakatos to Feyerabend, 6 July 1968; Feyerabend to Lakatos, 27 July 1968; 27 July 1968; 20 August 1968; 4 September 1968; FEYERABEND 1970c: 275n).
[2] See FEYERABEND 1970a.
[3] Feyerabend refers here to the final draft of FEYERABEND 1968 and FEYERABEND 1969.
[4] In late 1966 the Free University of Berlin (*Freie Universität Berlin*) made overtures to Feyerabend concerning a newly planned Chair in the Philosophy of Science which was formally offered to him in the spring of 1967. After long consideration Feyerabend eventually accepted this offer and, with some hesitation and a delay due to his precarious health conditions, he started teaching in Berlin in the 1968 Summer Semester. However, Feyerabend kept this position only for two semesters, during which he taught jointly at the FU Berlin and at the University College London, besides co-leading a seminar at the Technical University Berlin with Hübner (see Vol. 1, Part II, 1965-3, n3 and n5) in the Winter Semester of 1968. With his resignation at the end of 1968 Feyerabend left behind himself a trail of bewilderment, tension and allegations of misconduct against him from the academic environment in Berlin. On Feyerabend and the Berlin Chair of Philosophy of Science, see *above*, n1; Ch. 2, § 11, n1; and Vol. 3, Ch. 1, Feyerabend to Agassi, 19 November 1966, 1966-22; 7 December 1966, 1966-23; 22 December 1966, 1966-25; 2 March 1967, 1967-2; 8 March 1967, 1967-3; Ch. 2, Feyerabend to Watkins, January 1968, 1968-1; 11 February 1968, 1968-5; 23 February 1968, 1968-6; 19 March 1968, 1968-7; 17 April 1968, 1968-8; Feyerabend to Lakatos, 22 May 1967; Feyerabend to Musgrave, 11 May 1967; and Feyerabend to Lakatos, 1 January 1968; Lakatos to Feyerabend, 3 January 1968; Feyerabend to Lakatos, 7 January 1968; Lakatos to Feyerabend, 20 January 1968; Feyerabend to Lakatos, 28 January 1968; 20 February 1968; 23 February 1968; 25 March 1968; 11 April 1968; 19 April 1968; 20 April 1968; Lakatos to Feyerabend, 25 April 1968; Feyerabend to Lakatos, 11 September 1968; 30 October 1968; 5 January 1969; 23 January 1969; 14 February 1969; May 1969; 2 June 1969; 5 July 1969; 28 November 1969; Feyerabend to Feigl, 11 February 1968, HF 02-133-16; FEYERABEND 1995: 127, 131-132).
[5] On Feyerabend and the University College London Chair of the History and Philosophy of Science, see Vol. 1, Part II, 1962-2, n4.

2. Hempel to Feyerabend, 7 May 1968

Department of philosophy
The Princeton University
Princeton, New Jersey 08540
May 7, 1968

Professor Paul Feyerabend
c/o Department for the History and Philosophy of Science
University College
London W.C. 1, England

Dear Paul:

Many thanks for the two long mss you sent me during the past weeks, and for your helpful letter of April 20 about your objectives in writing these pieces.[1] I have not yet been able to read all the material, but I have read a good deal of the ms, "Problems of Empiricism II"[2] and found it enormously fascinating reading. The discussion of the early history of the telescope, in particular, shows me that I had a pretty simpleminded idea of what Galileo had done, and how things looked to him and his contemporaries through the telescope. I have been so intrigued and surprised by your account that I keep talking to friends and colleagues about it, Tom Kuhn included. Where will the essay appear?

I hope your transfer to London has gone smoothly and that you are enjoying the new setting! To me, the thought of these periodic changes of ambience is dizzying – and now you have added a third stage to your total circuit, by accepting the chair in Berlin! Congratulations, and may long enjoy your zip and zest! If your envisaged rotating scheme does go, I would indeed be interested in doing a stint at the Free University; but I would not be able to get away from here for the purpose until the fall of 1970 at the earliest. I certainly hope to see you long before them and to hear something of your impressions of Berlin.

Late this summer – ca August 29 to Sept. 4 – I expect to be in Hawaii to attend, and read a paper at, the meeting of the Pacific Division of the APA.[3] I hope Diane will accompany me – we have never been to Hawaii before. – I also hope to attend the workshop conference that Herbert Feigl is planning to hold in Mpls[4] in June, this time on probability.

Best wishes to you for a pleasant stay in London, and cordial greetings from Diane and me.

Yours,
Peter

Editors' Notes

CH 16-06 (1 s., T)

[1] See *above*, Feyerabend to Hempel, 20 April 1968, 1968-1.
[2] See Feyerabend 1970c; on the history of the development and publication of Feyerabend 1970c, see *above*, 1968-1, n1.
[3] See Carl G. Hempel, "On the Limits of Scientific Understanding" (CH 54-05; CH 54-06, CH 54-07), read at the Forty-Second Annual Meeting of the Pacific Division, American Philosophical Association, held in Honolulu, under the host sponsorship of the University of Hawaii, 31 August - 4 September 1968.
[4] Abbreviation for "Minneapolis".

3. Feyerabend to Hempel, 20 May 1968

University College London, Department of the History & Philosophy of Science

20. May CH1968

Dear Peter,

thank you very much for your letter. Yes, the Galileo paper will be published, and it will appear in the Proceedings of the Pittsburgh Conference of May 1965 in which you also participated. It has little to do with the talk I gave then (and which, I am ashamed to say, was rather superficial).[1] In the meantime I became interested in the history of science and I thought I should rewrite my contribution accordingly.

I quite agree that the situation is astonishing – but it is astonishing not only for the historian. For it seems to me that Galileo, in addition to de facto progressing in a very strange manner, could not have progressed in any other way. Now, if I am right in this then we must revise not only our histories but also our methodologies, and we must admit that both Popper and Carnap (or "Popper" and "Carnap", for I am talking about two schools) would have given very bad advice to Galileo (and, generalising, that they are giving very bad advice to contemporary scientists). It is this consequence of the story which interests me most (and which I have outlined, very briefly, in sections 7, 8, and 11 of my paper). What do you think about this aspect of the affair? Am I too rash! I would be very curious to hear your opinion on this matter (but please, do not pay more attention to it than your time permits, and your own work). So far, I am sorry to say, nobody has given me a straightforward answer. The surprising character of the history is admitted, but the methodological inference I want to draw is neither denied, nor accepted. Black was the only one who agreed wholeheartedly and was said that this ideas on induction had nothing whatever to do with science but dealt with commonsense only. Mary Hesse is also prepared to agree and to say that her more recent excursions into the theory of confirmation are due to the fact that she is interested in "games" of this kind, but she does not pretend they have anything to do with science.

My suspicions did not start with Galileo, but with relativity. It is a commonplace to assume that relativity contains classical mechanics as a special case. This assumption, so it seems to me, is entirely mythical. This has nothing to do with "incommensurability" which, I agree, is a somewhat doubtful affair. It has to do with the fact that nobody has as yet succeeded to calculate say, the problem of the planets, by general relativity only. A relativistic celestial mechanics simply does not exist. Each particular prediction of planetary paths uses classical physics up to the point of the relativistic correction and then adds this correction like a small (positive, or negative) friction, assuming, without ever being able to prove this assumption, that the rest of the job but could be done relativistically also. One usually says that the assumption is supported by the fact that relativity contains classical mechanics as a special case. But where is the proof of this? True, there are derivations of the formulae of classical mechanics from the formulae of general relativity but nobody has as yet shown that the small effects one neglects in this derivation have no cumulative effect so that the stability of the planetary system is still guaranteed. Just to show the strange way in which one proceeds, consider the Schwarzschild derivation of the advance of the perihelion of Mercury. What one does is consider the path of a mass point in central symmetrical, but otherwise empty universe – one then gets the 43″. But what has a central symmetrical empty space with a hole in the middle got to do with the planetary [1|2] system? And what allows us to combine the result of this calculation with a classical result which gives us everything up to the 43″? This combined use of two inconsistent theories is a very puzzling affair indeed. (In the quantum theory the situation is even worse, as was asserted by Bohr himself. For all these matters cf. section 10 and 9 of my paper).

Thank you very much for considering Berlin. I am putting you down for Fall 1970 and shall ask the authorities to send you an official invitation. This does not commit you to anything, but keeps the door open.

Berlin is a wonderful town. Fresh air for the first time after the smog in California and the smoke in London (though London has been considerably improved in the last few years). And there are trees almost everywhere. I may settle down somewhere in a little house in Grunewald at the lakes. The Berlin people are wonderful too, especially the taxidrivers ⟨sic⟩, and among them especially the women taxidrivers ⟨sic⟩ who seem to know Karate and Judo and seem to have learned it from Emma Peel ("The Avengers", a series that appears on Berlin TV too). The unrest among the students is quite understandable if one considers how much more rigid than American Universities even this freest of all German Universities is. One observes a strange schizophrenia. Many people shout revolutionary slogans, but are servile in the face of the professors, not because they are liars, but because they do not seem to be able to behave otherwise. One of the gentlemen who during the day made a thundering peroration about "Fachidioten" and "Fachprofessoren" gave in the evening, in my seminar, a paper that was academic in the worst possible sense, having nothing to do with any option anyone ever held – a pure exercise in mental gymnastics and he did not seem to notice the difference between his ideology and his actual behavior. During my second visit to Berlin people where somewhat excited about the emergency legislation which the German parliament was discussing and many of them (though by no means the majority) were afraid this was the advent (not even the preparation to) the new kind of fascism. We had a long discussion about this in my class. In East Germany one looks somewhat wearily at these student activities, especially as they are often dictated by pure ideology, without regard to practical effect (the most noticeable difference between the radicals in Berlin and radicals in Berkeley). It is all very exciting, and one can learn a lot from it if one is not concerned too much about one's "dignity as a professor" which, however, many of my Berlin colleagues seem to be.

I shall be back in Berkeley on June 24 and shall then stay there until the end of September. Will you stop in Berkeley on your way to Hawaii? If you do, and if you have a little time, please let me know as I would like to meet you. You cannot reach me by telephone (I have a one-way telephone, I can ring everyone, no one can ring me) but please stand, at my expense, a telegram to my home address (6041A

Harwood Ave, Oakland, Calif ⟨sic⟩) and I shall come wherever you are. Best wishes also to Diana

Paul

PS: I won't continue traveling around like that for much longer. If all goes well I shall retire in about 4 years to devote myself to a pleasant live, a little study, a little writing.

Editors' Notes

CH 16-06 (1 s., T+H)

[1] See Feyerabend 1970c; on the history of the development and publication of Feyerabend 1970c, see *above*, 1968-1, n1.

1971

1. Feyerabend to Hempel, 26 December [1971]

University of California, Berkeley, Department of Philosophy

26.
26. 12. 71?[CH]

Dear Peter,

somebody sent me a draft of your paper PROBLEMS IN THE EMPIRICISTS CONSTRUAL OF THEORIES[1] which I find most interesting. I certainly agree with your conclusion in the last paragraphs of the paper. I only want to comment on your remarks and contextualism in the last but one paragraph on page 2 and then again in the last paragraph of section 2 on page 3. Contextualism, you say, can only mean "theoretical terms are and must be understood to be such as to make the corresponding theory true" and this would make the theory true apriori ⟨sic⟩. I don't think it does. It makes perfectly good sense to say "This theory is false. But if you want to understand its primitive descriptive terms, or want to explain them to others, you have to pretend that it is true." Moreover, the theory can still be refuted by experiment. For while the meaning of its terms (and of the observational terms) is determined by the postulates alone, predictions emerge only when we postulates as well as initial conditions, and so we can quite easily come up with a result that is contradicted by experiment. Example: consider a theory that has "speed" as a primitive descriptive term. We make predictions about speeds. And may predict speed 2, finds speed 1, and so reject the theory despite the fact that the meaning of "speed" is defined by its postulates. This, incidentally, seems also to solve your problem in Minnesota Studies Vol. iv, p. 236 bottom 237 top.[2] Am I right, or have I made a mistake?

Right now I have a one year sabbatical to do some research in the history of science. I may come to Princeton (but as usual my style of life is rather chaotic). If I do, I shall try to see you for a brief chat about life, things, and empiricism.

With best wishes, *also for the New Year*
Paul
Paul Feyerabend

Editors' Notes

CH 16-06 (1 s., T+H)

[1] Feyerabend refers here to a draft of HEMPEL 1973 (see *below*, Hempel to Feyerabend, 22 January 1972, 1972-1).

[2] See RADNER-WINOKUR (eds) 1970: 237.

1972

1. Hempel to Feyerabend, 22 January 1972

6 Rumbold Road
London S. W. 6
22 January 1972

Professor Paul Feyerabend
Department of Philosophy
University of California, Berkeley
Berkeley, California 94720

Dear Paul:

Thank you very much indeed for your letter about my draft of a paper on the interpretation of theoretical terms; it was roughly the text of a paper I read at the Bucharest Congress;[1] I have submitted a shortened variant of it for inclusion in the proceedings.[2]

I am not convinced by your argument. You say this in the middle of your letter: "Moreover, the theory can still be refuted by experiment. For while the meaning of its terms (and of the observational terms) is determined by the postulates alone, predictions emerge only when we use postulates as well as initial conditions…" Now what do you mean by saying that the meaning of the terms is determined by the postulates alone? I assume – and the lines in your letter preceding those just quoted seem to bear me out – that you have in mind that the extralogical ⟨sic⟩ terms occurring in the postulates are to be understood in such a way that the postulates come out to be true are under that understanding. (This does not uniquely define the meaning of each term, to be sure, but it sets limits to permissible interpretations, so to speak). But if I am right in this coastal of your assertion, then, in any application of the theory T, I can always say to myself: T is certainly true simply in view of the convention that its extralogical ⟨sic⟩ terms are to be understood in a way that makes T true. If, therefore, from T in conjunction with initial conditions A, I deduce prediction P and find it false (A•T → P, but P is false), then I must infer that A is false; T cannot be refuted; for to take T to be false would mean to ascribe to the terms of T meanings which they do not possess.

Only quite recently did I read your article "Consolations for the Specialist"[3] and was very pleased to find here, on pp. 222-225, a very full and explicit statement of [1|2] ideas just briefly touched upon in my paper. Quine and Putnam have said similar things; but as for myself, it has only slowly become clear to me what a serious presupposition underlies so much of the analytic work in logical empiricism, namely, what I call in the paper you saw the requirement of explicit linguistic interpretation. In the interest of maximal analytic clarity and neatness, it is a most attractive notion, but, strictly speaking, it was dropped as soon as the idea of strict definability of theoretical by observational terms was abandoned.

Concerning the first paragraph on p. 217 of your "Consolations...", I would think that under great stress, chaotic changes in the environment, etc., various non-scientific procedures and attitudes ma⟨y⟩[E] come to replace the scientific ones; but I would rather say that then, the scientific pursuit has been given up, so that your argument does not show the possibility of an irrational science. (Incidentally, it is not clear to me just what you mean here on p. 218, middle, by rationality: that does need to be clarified, but I'll just use the word intuitively in the next few lines) ⟨sic⟩ You may say that this reaction of mine is a dodge: I save the claim of persistent rationality of science by making it a defining trait of science. But I think it is not quite as bad as that. As a sort of minimal characteristic of science I would think that it seeks to arrive at reliable beliefs about the world, that is will therefore check its claims against the events it comes up against, and so on. This is too simpleminded a characterization, of course, but it seems to me to rule out for example such irrational procedures as simply adopting a "theory" which you invent in the manner of a fiction writer, and without any attempt ever to check it.

Perhaps one quite different consideration could be adduced in support of the expectation that rationality will not be totally abandoned: namely, on evolutionary grounds, one might expect that a society with a systematic, "rational" science is more likely to survive than a totally irrational (or a less rational) competitor.

Like you, I have a sabbatical this year, which I am spending in London with Diane; both of us enjoy it very much, as you will readily be able to imagine. I do hope your leave will prove pleasant and rewarding.

With best wishes,

Cordially yours,
Peter

Editors' Notes

CH 16-06 (1 s., T; c)

[1] Hempel refers here to the *Fourth International Congress for Logic, Methodology and Philosophy of Science*, Bucharest, 29 August - 4 September 1971.

[2] See Hempel 1973.

[3] See Feyerabend 1970a.

1977

1. Feyerabend to Hempel, [7 December 1977]

⟨7 December 1977⟩PM

Dear Peter –

every second week I explain my nonsense here at the ETH Zürich ⟨sic⟩: van der Waerden[1] occasionally comes and sits there with a fatherly smile when I argue against this and that and everything. I may get a job in Zürich ⟨sic⟩ and I may accept it – but I shall have to think hard before I leave Berkeley.[2] I have decided to do philosophy only for few more years then off to the theatre. I may see you in Pittsburgh in March.

All the best
Paul

Editors' Notes

CH 10-16 (PC, H)

[1] Feyerabend refers here to the Dutch mathematician and historian of mathematics Bartel Leendert van der Waerden (1903-1996).

[2] Feyerabend took up a part-time trial position at the *Eidgenössische Technische Hochschule Zürich* in the Fall Semester 1977, while he was visiting professor at the University of Kassel. In 1980, the ETH position was turned into a permanent professorship (see Feyerabend to Watkins, undated,; 24 July 1976; 22 December 1976; 19 November 1977, 1977-6; and FEYERABEND 1995: 154-159).

Chapter 6
Excerpts from the Feyerabend-Smart Correspondence (1963)

Paul K. Feyerabend and J. J. C. Smart

1. Feyerabend to Smart, [December 1963]

⟨UNDATED⟩

Dear Jack,

thank you very much for your letter and your most stimulating paper.[1] I very much enjoyed both and I learned quite a lot from the latter. As a matter of fact I am now planning (as a result of reading your paper) to write a paper on Science and Commonsense which will proceed from the way in which you put your problem.[2] I would have liked to write you a better letter, but being ill again I can neither use the pen nor think very well. Here, then, are a few scattered remarks for whose incoherence I must apologize.

I agree that the issue between Nagel, Sellars and "myself" ["myself" sounds funny to me – I have always thought that I was only explaining what any reasonable person would have in his very bones and what many scientists have practised. Just consider how Galileo in his Dialogues transforms commonsense with the simplest and most shrewd of arguments] may depend on empirical considerations. If it is indeed true, as Kant thought it was, that certain reaction patterns, or conceptual schemes are inborn, then it would quite obviously be impossible to replace *them* by different points of view. Still I do not believe that this possibility which as yet has not been established as a positive fact, can be used as an argument against total replacement, or at least against the initial attempt at total replacement. Two reasons for this:

[i] there seems to exist strong prima facie evidence against the existence of insuperable conceptual schemes. This evidence lies in the fact that different cultures have developed radically different "manifest image" to use a term I think is very fitting. This sociological fact should really be given more weight in philosophical arguments. To take only one example: if the Egyptian notion of mental events

M. Collodel, E. Oberheim (eds.), *Feyerabend's Formative Years. Volume 2. Feyerabend on Logical Empiricism, Bohm & Kuhn*, Vienna Circle Institute Library 10,
https://doi.org/10.1007/978-3-031-57519-8_6

(not that they were called that) should turn out to be different from the contemporary Oxford notion, then the demand that materialism be capable of explaining the latter, but not the former, completely loses its point. And so do all other demands which require synonymy, or co-extensionality with the "ordinary" notions. Now I think I can explain why this variety of manifest images is given so very little weight. It seems to be generally assumed that all these different points of view, the point of view of the Zulu, of the Hopi, of the ancient Greek etc. [1|2] are basically identical (with Strawsonism?) and that the differences belong to some theoretical superstructure, as it were. Austin, with whom I had many arguments over this matter always seemed to take it for granted that people took tables much more seriously than they took daemons and that the usages connected with table words where therefore a much more solid part of "the" common idiom than were the usages connected with the daemon words. Demons, therefore, *according to him* occupied a rather peripheral place in the manifest image. Sellars, too, did not seem to be willing to regard the Hopi point of view as a genuine alternative of our own common philosophy (in a discussion I had with him in 1957). For Kant and many other thinkers myths were a matter of fantasy and had nothing whatever to do with forms of thought. Cassirer seems to have been one of the first thinkers who adopted a different attitude here, although he, too, regarded for example mythical space "as having a position in between perceptual space and space of pure knowledge" (Philosophie der Symbolischen Formen, Vol. II p. 107).[3] Yet I think we shall have to admit, following a careful analysis of the structure of the world views which have been connected with different cultures, that they are genuine alternatives, genuinely different manifest images. I wish, ⟨sic⟩ I could present some examples at least to make this assertion more plausible. I shall certainly do this in one of my future papers. I always wanted to write a paper about the nature of myth in order to show that they are full world views. All I can do here is quote some Nietzsche. In his earlier life Nietzsche was very much interested in the philosophy of nature and he intended at some time to write a systematic treatise on this topic. He adopted for some time Mach's point of view *which he* interpreted as showing that the world in which we live (our *perceptual* world) is a mental construction, and that different kinds of such mental constructions are possible. This brings him at some places rather close to Wittgenstein. Thus in his Wahrheit und Luege im Aussermoralischen Sinn ⟨sic⟩ we find the following passage: "The waking day of a mystically excited people as are the older Greeks is indeed more similar to a dream than is the day of a sober mind; and this is due to the al- [2|3] ways present miracles as assumed by the myth. When every tree can now talk as a nymph, or when a god can rob virgins under the cover of a bull, when the goddess Athena is suddenly seen in person, driving side by side with Pisistratos ⟨sic⟩ in a splendid carriage through the market places of Athens – and this the honest Athenian did in fact believe – then like in a dream everything is possible at any moment and the whole of nature surrounds man as if it were only and masquerade of the gods." [Lynn White in an excellent paper in the Volume ⟨sic⟩ Scientific Change ed. Crombie has pointed out that the transition from this world view to the cult of Saints, who kept to themselves and left nature alone swept nature clear of all spirits and thereby made it accessible, "conceptually" as well as "manifestly" both to the technological interpretation, and to a purely mechanical interpretation. It was a change in the manifest image that made mechanism possible and thus finally led to a more modern kind of "commonsense".]

To sum up: the variety of points of view in different cultures must be taken seriously, it must be taken as an indication, that non-Strowsonian manifest images are possible, and that Strawsonism therefore cannot be inborn (the evidence referred to by you and by Chomsky[4] is much too sketchy, too weak, and too prejudiced to show that it is inborn. It is interpreted in a certain manner, and it has not been obtained under satisfactory conditions. We are not interested in how people react in a social environment that has many Strawsonian features and is therefore liable to restrict the manifold of reactions they are capable of).

There is still another point that is worth considering and this is the possible existence of *archetypes*. I'm very much in sympathy with the idea of Hughlings Jackson[5] that man contains a hierarchy of reaction patterns corresponding to a hierarchy of neurological levels. Primitive patterns are fairly universal, higher patterns may show great variation from individual to individual. Primitive patterns are usually overlayed ⟨sic⟩, and to that extent suppressed by higher patterns, but may occasionally emerge when the higher levels are either damaged, or temporarily out of action. Now it may well be that the primitive patterns in all [3|4] human beings constitute a certain rudimentary world view that is common to all. I am even convinced that this is the case. However such situation, far from making this world view the indispensable foundation of everything else would only occasionally make it raise its ugly head and would make it thereby difficult, though by no means impossible wholly to live inside a more reasonable world view. The existence if ⟨sic⟩ lower levels beneath higher levels shows that it is difficult to be *civilised* (in one's behavior, as well as in ones ⟨sic⟩ thought). The existence of higher levels, as on the other hand, as well as the possibility to modify these higher levels shows that it is possible to be civilised. Moreover it is desirable to be civilised, hence the reference to the lower levels loses all its force. However quite apart from the irrelevance, in view of the existence of a hierarchy of higher levels, to the features of some lower level, it is very doubtful to assume that these lower levels might be Strawsonian. Lower levels are those which man has in common with the higher animals. Now although animals don't speak, there are still means of exploring their manifest image (which, I gues ⟨sic⟩, would in any case possess features that are independent of speech). And it seems that this manifest image is not at all Strawsonian. (It gets worse if we go farther back on the evolutionary ladder – deep sea fish are most likely Cartesians interpreting bodies as moving modifications of space = matter). To conclude this, first, part of the argument: it seems very unlikely that God built man in Strawson's image. If he built man in any image at all, then the limits of this image have not yet been reached.

But (ii) let us now assume, contrary to what emerged as plausible, that Kant is right and that we are incapable of leaving the boundaries of a certain world view. How do we find out that this incapability exists? And how do we discover what it is we cannot shake off? Kant and many contemporary philosophers seem to take it for granted that such boundaries can be explored from the inside. They assume that the insuperability of commonsense can be discovered by procedures which move wholly inside commonsense. They therefore advise us to wait until the boundaries have been found and not to try rushing off in all directions in a vain [4|5] effort to leave behind what cannot be left behind. Now this I take to be a completely wrong approach. The attempt to find boundaries by exploration from the inside is the surest way to erect artificial boundaries, boundaries which do not correspond to the possibilities of human nature in general but only to the possibilities of a human nature that has been paralysed by a philosophical doctrine. The only way to discover whether there are indeed restrictions to our way of seeing the word is to resolutely try many different approaches and also to try applying them on all levels, on the level of commonsense as well as on the level of theoretical reason. Only if such a procedure fails again and again, only then may we try to explain these failures by the hypothesis of an unalterable inborn world view (but even here we must be careful. What has been impossible for 3000 years may become possible tomorrow. Even within the rather restricted world view of science it took about 2000 years until the atomic theory could be finally regarded as satisfactory). However what has just been said means that even *if* Kant should in fact be right we could find this out only by first adopting *my* point of view. This methodological consideration shows that there exists an interesting *asymmetry* regarding the views of Nagel, Sellars, and myself. The asymmetry consists in the methodological necessity to start with my view. Nagel and Sellars can never find out whether they are not unduly restricting research. But *I* can find out whether *I* am not unduly *expanding* it. Hence, if I am *right*, and if there exists no limit to human reason, then obviously my view should be adopted. However if I am *wrong*, that is if there do exist inborn limits to human reason, then my view and my procedure must be adopted too, for this is the only way to discover the limits. Now there either exist limits, or there don't. *Hence, my view must in any case be adopted.* * And considering the success of those who *have* adopted it in the past I have all confidence that my view is not only the necessary *first step*, *but is also right*, or at least I have all confidence that *it is not* Strawsonianism ⟨sic⟩ *which* defines the limits of human reason. [5|6]

My second point is connected with your remark that when describing macroscopic objects such as galvanometers the physicist MUST use Strawsonian terms (p 12).[6] I think that this remark is both unsupportable and false. Let me consider some possible arguments in its favour:

(1) The phy*sic*ist must prebeed ⟨sic⟩ in this way the because that is the way he is *built* – false for reasons I have just explained. (This first argument has been used both by Heisenberg and by Weizsaecker for retaining classical language at the macroscopic level and classical language is for them refined commonsense, not reformed commonsense).

(2) He must proceed in this way because he does not know how to proceed differently – both irrelevant and false. Irrelevant, because a deficiency is never an argument in favor of remaining deficient. False, because in many cases more adequate terminology, or, if the terminology is still missing, more adequate interpretations of the common words are readily available.

(3) Because ordinary language is perfectly adequate for describing macroscopic objects – both irrelevant and false. Irrelevant, because the adequacy of the ordinary

idiom should not prevent us from replacing it by a different idiom if we can in this way create greater coherence of expression. False, because many descriptive statements of commonsense are false, if taken in their commonsense interpretation. Example: material objects, such as this table in front of me are supposed by commonsense to be observer-independent in the sense that they change their shape only when interfered with physically. Sirius does not change its shape when I am sitting here and shaking my head. Relativity teaches us that the shape of all objects is observer-dependent and it therefore teaches us that "this is a table" using "table" in the usual sense of this word is always false. Now there is no need to introduce new words, such as "time slices" etc.[7] We may retain all the commonsense words – but we sure must change their meanings. Thus we shall continue to say "this milk is too damn hot", but we should now think of motion and not of substance when thinking of heat. [6|7]

(4) Because the ordinary man understands perfectly well what he means by his terms when saying *for example* "the milk is hot" and he does not mean "the average velocity etc. etc.". Irrelevant. As irrelevant as possible! What has the ordinary man's point of view got to do with the way in which the scientist is supposed to describe his galvanometer? Assume the ordinary man knows what he is talking about. Does this mean that what he is talking about does also exist? How would this argument look when applied what a Zulu says about his milk? He presumably also knows what he is talking about (spirits, I guess). Does this mean these spirits exist? Every inquisitor knew "what he was talking about" when diagnosing possession. Etc. etc. Nagel seems to use this argument. I cannot understand how he can believe for a second that it is relevant.

(5) Because without being related to the common expressions physics would not have any meaning. False. This argument assumes that theoretical terms obtain their meaning by being related to the terms of some pre-existing commonsense. However how did these terms obtain their meaning? And if it is possible to *teach* small children such rather abstract terms as "table", "chair", "god", "duty" *"mother"* ⟨sic⟩ etc. etc., then why should it not be possible to *teach* them "electron" etc. etc.?*

(6) Because different descriptions (more especially – descriptions inconsistent with the commonsense descriptions) would be false (this is the paradigm case argument). False. Reason same as in (3).

Having gone over all these reasons (and I think there is *nothing I have left out*) I must say that they are a sorry sight indeed. Is it not time to give up this illicit love affair with commonsense and to proceed to a more reasonable point of view? All these reasons also proceed from the wrong end. The question is not what some people say when talking about tables and chairs, the question is what kind of language will be *adequate* as an observation language for scientific theories. This question can be answered very simply. There are three requirements: [7|8] They are

(a) the language must be *observational*. This is pragmatic, or a psychological condition requiring that the production of certain signs of this language (in the relevant circumstances) be a reliable indicator of certain features of that surroundings.

(b) It must be *factually adequate*.

(c) It must be *relevant* for the test of the theories under review, that is its statements must be derivable (with the help of initial conditions) from the theory.

Now I challenge anyone to prove that (a)&(b)&(c) taken together entail that we must be Strawsonians. Condition (a) *taken by itself* may lead to certain exciting languages where conditioning has indeed established the needed behavioral mechanism. We may discover that Ordinary English satisfies (a) and might therefore want to retain it (after all, conditioning a person to the proper response in front of observable objects is not a simple matter and there is no reason why one should make life more difficult than it already is). But note that even if Ordinary English were to satisfy (a) this would only be an argument for retaining a *small part* of its "grammar", viz. that part that is connected with direct observation *and has nothing to do with meaning*. The remaining "theoretical" part of its grammar *that constitutes meaning* may be completely inadequate for describing what it so well indicates. Thus the shouts of a savage may signal to us the rising of the sun. However it would be incorrect to assume that what he must to convey by his shouts viz. that the sun god has just been reborn is a correct statement. Hence only *a* very *small sector* of Ordinary English is brought in by (a). Nothing has as yet been said about its meaning.

I think it is easy to see how the fact that part of ordinary English conforms to (a) could create the impression that ordinary English is allright ⟨sic⟩, or perhaps even required, as an observation language; how it could create the impression that satisfaction of (a) guarantees satisfaction of (b) and (c) also. These are the steps of unreason: first step: You can still call a table a "table" (correct); second step: the meaning of an observational term is constituted by its application in concrete situations (incorrect – this is the kind of thing Wittgenstein intended to combat); hence three: the meaning of the term actually possesses [8|9] (and this meaning is not uniquely determined by what happens in observational situations; it contains also theoretical ideas) is still adequate (incorrect result of the application of the second step to the first step); hence, four commonsense is correct, will always be correct, etc. etc. (This train of thought has also given considerable support to the paradigm case argument. Malcolm's paper on Moore and ordinary language[8] proceeds essentially in this fashion). In this connexion we can also see what is to be thought of the argument (which is similar to argument 4 above) that commonsense is essential to science *because the scientist can makes use of the readings made by his uneducated assistant.* * First, this argument might show, at the very most, that commonsense occurs in the sciences, not that it is essential to it. However even that is not established for it has not been shown that the scientist interprets *or must interpret* the *sentence* in the same manner in which it is interpreted by his uneducated assistant. As a matter of fact the interpretation used by his assistant is *completely irrelevant* as is seen from the fact that we still make use of the astronomical data which have been assembled by Babylonian priests in order to prepare them for the appearance of certain gods (their interpretation, but surely not ours!) Even better, for immediate report the voice of the laboratory assistant might be replaced by the barking of a well conditioned dog who reacts to certain smells and who, while barking, might think "what a lovely sausage".

A final remark. A view such as commonsense is connected with certain expectations, which certain attitudes, and therefore also with a certain way of perceiving the world. This way may become solidified in reaction patterns and add a new level to the more primitive and more animalistic (and animistic) ways of behavior. The attempt to replace commonsense wholly by a different view, if successful, will introduce a new pattern of attitudes, expectations. The world will be "seen" differently. That such a transformation leads to different perceptions is shown by the fact that once a new view is adopted confirming evidence turns up in great number. *Kuhn has interesting examples of this*⟨.⟩[E] Before this evidence remained un<u>noticed</u> (and not only unexplained, or un- [9|10] analysed). Such an attempt at replacement therefore is required for methodological reasons (sec [ii] on page 4, above), and it is also required for making more things "visible". All this means that the new theory will be able to provide its own manifest image. The new level of behavior created by this new manifest image is also necessary for still further levels. Hence, even physiology seems to encourage my view as opposed to weaker views. In this way we may become builders both of new kinds of human beings (more "civilized" ones ⟨sic⟩ *see page 4 top*) and builders of new kinds of commonsense instead of being built ourselves, and to this extent dominated, by the commonsense of the past. This view is also the only one that does not pretend that we know nothing when in fact we know a lot and it also takes commonsense more seriously than the other views. It admits that commonsense is a <u>*theory*</u>, that it may be <u>*knowledge*</u> and draws from this the consequence that other theories, too, might become commonsense, and should become commonsense if they have been shown to possess certain advantages.

This is all for today. I would be very grateful to have your comments on this. Please let me also know Putnam's reactions.[9] Finally, a very happy Christmas to you

⟨UNSIGNED⟩

---

[5] * *This is <u>all</u> I am prepared to accept of Reichenbach's argument.*

[7] * *Children are taught the bible ⟨sic⟩, they hear about the creation of the world out of <u>nothing</u>. If they understand <u>this</u> what prevents them from understanding the creation-operator of quantum theory?*

[9] * *<u>your</u> sins, Herbert!*

Editors' Notes

HF 02-133-11 (10 ss., T+H; c) ≈ PF 2-7-77 (10 ss., T+H; c). HF 02-133-11 and PF 2-7-77 are two different copies of Feyerabend's missing original letter to Smart. The two copies differ in Feyerabend's handwritten annotations and corrections, only some of which have been repeated in both copies, and for a few and clearly identifiable handwritten annotations by Feigl's hand in HF 02-133-1 that make explicit the identity of the author and of the recipient of the letter. It is not clear which if any of Feyerabend's handwritten annotations and corrections were also included in the original letter. This transcription is based on HF 02-133-11, with Feyerabend's handwritten annotations and corrections highlighted in italics (in all cases in which underlined phrases are italicized, only the underlining was added in Feyerabend's handwriting). A transcription of PF 2-7-77 is included in FEYERABEND 2009.

[1] Feyerabend refers here to a first draft of the paper "Explanation", to be delivered by Smart to the Boston Colloquium for the Philosophy of Science on 12 December 1963, with Hilary Putnam as official commentator (see SMART 1965).
[2] Feyerabend never wrote a paper focused on the topic of science and commonsense. However, the discussion with Smart led eventually to the publication of SMART 1965, PUTNAM 1965, SELLARS 1965 and FEYERABEND 1965b (the latter, completed in late December 1964; see FEYERABEND 1965b: 231, 235, 247, 259n28) in Vol. 2 of the *Boston Studies in the Philosophy of Science, Proceedings of the Boston Colloquium for the Philosophy of Science, 1962-1964*.
[3] See CASSIRER 1925: 177 [83].
[4] See SMART 1965: 168.
[5] Feyerabend refers here to the British neurologist John Hughlings Jackson (1835-1911).
[6] See SMART 1965: 164-165.
[7] See SMART 1965: 164.
[8] See MALCOLM 1942.
[9] See *above*, n1.

2. Smart to Feyerabend, 11 December 1963

Department of Philosophy
Emerson Hall
Cambridge 38, Mass.

December 11, 1963

Dear Paul

Thanks very much for your long and interesting letter. It was very good of you to take so much trouble. I certainly hope that you soon write and publish the paper you plan on "Science and Commonsense".[10]

I haven't heard from Sellars yet. I'll put an extra footnote into my paper pointing out that you would argue against my view that Strowsonianism may be inborn with evidence from sociology (the beliefs and myths of different cultures) and physiology.[11]

On p. 4 you say that Kant [1|2] and many contemporary philosophers say that the boundaries of common sense have to be explored from the inside. But surely Sellars need not say this. He could use physics, physiology, etc. – he could explain in the scientific image why men grow up with a certain manifest image.

For this reason I do not see why Sellars should be restricting research. Nor Nagel either. (I had some talk with Nagel and have slightly modified my essay. But only trivially. Playing down his pragmatic and instrumentalist concession which he now wishes to repudiate. [2|3] He seems to defend himself against you as I predicted he might try to do, by using a weaker criterion for "deducibility" of one theory from another than that of formal deducibility.)

I don't think Sellars or Nagel want to restrict enquiry – even if they (in different ways) retain common sense they don't want to inhibit theory. Sellars' view is a bit different from Heisemberg's, I think.

I enjoyed your considerations from methodology. I do hope you soon write your article. (That will keep you off aesthetics,[12]→ too!!!) I'll show your letter to Hilary.[13] I'll let you know how [3|4] the meeting goes tomorrow.

I was sorry to hear that you have been ill again. You do have bad luck. I've been pretty well, but am limping a bit this week after rupturing a calf muscle playing squash – not quite enough to prevent me getting about though. I went down to Columbia to read a paper on Time and had some talk with Nagel about the paper on Explanation.

I have got a house at Short Beach near New Haven for next term at Yale when the family will be here. The back yard goes on to the beach which will thrill the children.

Very many thanks once more and I'll write again after the meeting[14]

Jack

---

[3] → Not being an aesthetician I'd like to see you keep to metaphysics and phil. of science!

Editors' Notes

PF 2-7-49 (2 ss., H)

[1] See *above*, Feyerabend to Smart, undated, 1963-1, n2 in particular.
[2] No such footnote is present in SMART 1965.
[3] Most likely, Smart is here commenting on a handwritten note of Feyerabend that is not present in the copy of the Feyerabend's letter to Smart that was sent to Feigl (see *above*, 1963-1, introductory note).
[4] Smart refers here to Hilary Putnam (see *above*, Feyerabend to Smart, undated, 1963-1, n1 and n2 in particular; *below*, Smart to Feyerabend, 16 December 1963, 1963-4).
[5] See *above*, 1963-1, n1.

3. Feyerabend to Smart, [December 1963]

⟨UNDATED⟩

Dear Jack,

as always, also the last letter from you makes me jump (from excitement, not from pain). I have not studied Sellars (I ought to get his book)[1] and I do not know the details of his point of view. I therefore really cannot say much about him. However I can say something about Nagel's attempt to get out of trouble by using a more "liberal" notion of derivability. This is really quite incredible. Why is it that commonsense must be changed? Or that the phenomenological theories must be given up? Because (1) some of the laws used here are incompatible with better theories; and (2) because some of the grammatical rules defining basic terms are incompatible with better theories. Both the laws and the grammatical rules should therefore be given up as refuted. After all, we refute theories, laws etc. not only by a direct comparison with the facts, but also by indirect comparison, involving other theories. Now what am I to think of a method that revises the notion of derivability in order to escape such refutation? I thought that devising ad hoc hypotheses was the epitomy of bad procedure. But I forgot of course the even worse crime of ad hoc revisions of logic (compare on that my review of Reichenbach's book on quantum theory[2] which I am sending you with the same mail). Are we now allowed to say that a white raven refutes "all ravens are black" only for those poor, naive, and unsophisticated people who still use the classical notion of derivability? Quite clearly the envisaged revision of logic is incompatible with sound scientific method which demands that testability be increased and not decreased (by the way, how far are we to go in this? Has it not perhaps been foolish to give up the Aristotelian physics which is surely compatible with Newtonian physics if we only get rid of that stupid old relation of derivability). Only very strong arguments could remove this d⟨e⟩[E]mand for maximum testability and the corollary demand to stick to old derivability, very strong arguments, that is, in favor of either commonsense, or of some older theories. However what has the defender of the past to show on this point? All he has (and refute me, please, if I am wrong) are the six "arguments" I mentioned in my last letter. Now this sorny ⟨sic⟩ assembly of philosophical corpses surely is not strong enough to eliminate the well founded demand for maximum testability and the corollary demand that classical logic be retained (this last remark, by the way, is my answer to Henkin's criticism of my paper on Reichenbach*, in Journal of Symbolic Logic[3]). Moreover, if deducibility is really relaxed, then an interpretation of more recent theories via the new and now considerably weaker correspondence rules will not be rich enough to give us what the physicists have in mind. Such an interpretation will then have to be established in an independent fashion – and this brings us back in any case to the procedure I recommend – forget the past, treat the present as if it were as good as the past (especially as you have arguments showing that it is better) and prepare for the future.

An exasperated
Paul

* Phil. Studies[4]

Editors' Notes

HF 02-133-10 (1 s., T+H). HF 02-133-10 is the copy of Feyerabend's letter to Smart which Feyerabend sent to Feigl.

[1] See Sellars 1963.
[2] See Feyerabend 1958b.
[3] See Henkin 1960.
[4] See Feyerabend 1958b.

4. Smart to Feyerabend, 16 December 1963

Dept. of Philosophy
Emerson Hall
Harvard University
Cambridge 38, Mass

December 16, 1963

Dear Paul

Sorry to have exasperated you! But I am unrepentant if this helps to speed up your article on Science and Commonsense! I did enjoy your last letter. No one is so able to stick up for Feyerabend so well as Feyerabend himself!

Putnam is suspicious (in the light of recent phil. of language) of your use of the notion of "meaning". At least as a first approximation he applauds Quine's desire to banish it from philosophy. I [1|2] think, however, that your essential argument could be restated without use of the notion of "meaning". That's my hunch anyway. Putnam would want to say not that the meaning of "table" has changed but that now we have different beliefs about tables. (Tables are what you eat off.)[1] Don't quote me here, however – I'm speaking from memory only of chat he said.

I can guess some of your rejoinders: I look forward to some exciting philosophical controversy when you've written your new article.[2]

Thank again for writing so fully. (I haven't heard from Sellars yet.)

Yours
Jack

Editors' Notes

PF 2-7-46 (1 s., H)

[1] See Putnam 1965.
[2] See *above*, Feyerabend to Smart, undated, 1963-1, n2 in particular.

Part II
Feyerabend and Bohm

Chapter 7
Autobiographical Remarks (1993)

Paul K. Feyerabend

1. Peat to Feyerabend, 2 July 1993

Prof. Paul Feyerabend | CH 8706 Meilen
90 Fentiman Ave. | Ontatio
In Der Au 30 | Switzerland
Ottawa | K1S OT8
613-730-1998
e-mail ad454@freenet.carelton.ca

2 July 1993

Dear Prof Feyerabend,

You may have heard that David Bohm[1] died of a heart attack in October of the last year. I had known Dave since 1971 when I spent sabbatical at Birbeck college and since that time we maintained regular contact.

Before his death Dave had asked me if I would help him write his autobiography, or if I would take on the task of writing a biography. I am now engaged upon the research phase of a biography and am contacting colleagues and friends who knew Dave.[2] A few years ago Dave and I were at a conference with Herbert and Francon Frochlich and there was some talk about Dave's time at Bristol and how he had enjoyed talking with you. I would be particularly interested to learn about this period in Dave's life.

M. Collodel, E. Oberheim (eds.), *Feyerabend's Formative Years. Volume 2. Feyerabend on Logical Empiricism, Bohm & Kuhn*, Vienna Circle Institute Library 10,
https://doi.org/10.1007/978-3-031-57519-8_7

Let me pose some specific questions: –

I gather that he was not particularly happy with conditions at Bristol, in particular I was given the impression that few people were interested in fundamental questions. Does this accord with your own impressions?

Dave was subject to periods of depression and discouragement during his life, did he strike you as being particularly depressed at Bristol?

Dave had been forced to leave the US because of his political associations in Berkeley. While in Brazil he still seemed sympathetic to Marxist philosophy yet later in his life, when I knew him, he seemed to be disenchanted. Did he touch on his political beliefs in Bristol?

In his later years Dave seemed to be thinking a great deal about Hegel. Did he have that same passion in Bristol? Indeed, how much philosophy had Dave read at that time and what level was his understanding of philosophy? Those who knew Dave in his later years paint the picture of a philosophical innocent who came upon insights in a totally original way – without any knowledge of the history of philosophy. However, from Dave's own reminiscences it would [1|2] appear that he was always interested in philosophy. How would you have assessed his knowledge during his period in Bristol?

His early work on the Hidden Variables/Causal Interpretation was done at Princeton and in Brazil. When I knew him in 1971 he was thinking about pre-space, cohomology etc and appeared to have set the Causal Interpretation to one side. Only later did he begin revive interest and write the book with Basil Hiley that will be published later this year.[3] What sort of problems in physics was Bohm discussion in Bristol?

Did Dave speak of his interest in Ouspensky, Gurdjieff or Krishnamurti while at Bristol?

I would also be interested in anything else you could add about Dave or Saral Bohm[4] at that time.

Yours sincerely,
David Peat

Editors' Notes

PF 2-10-5 (2 ss. T)

[1] See *above*, Vol. 1, Part I, Ch. 1, n40.
[2] See PEAT 1996.
[3] See BOHM-HILEY 1993.
[4] Sarah "Saral" Woolfson was Bohm's wife.

2. Feyerabend to Peat, 9 July 1993

07/09/93

Dear Dr. Peat,

I have no records, I have only my memory which has large lacunae. This is what I remember:

Interest in fundamental questions at Bristol: in 1955/56 I gave a lecture course (not a very good one) on philosophical issues of quantum mechanics. Except for Cecil Powell[1] all the main physicists were there, Pryce[2] and Lang[3] included, also lots of chemists, mathematicians – about 80 people. Pryce who was not exactly an easygoing guy and who constantly mocked people interested in more general questions was himself very interested and tried hard to understand what Bohr was all about. He had the feeling that Bohr was a Good Guy, even when he started talking in general terms, but he did not quite know what his goodness consisted in. Dave came, I think, a year later. At any rate, he was there at the Colston Symposium in Bristol, in 1957, giving a talk and battling with Rosenfeld and others. The symposium was organised by the head of the philosophy department, Stefan Koerner together with Pryce and had philosophers and physicists in about the relation 1:1.[4] You can find the text of Bohm's talk and a record of the debates in Observation and Interpretation in the Philosophy of Physics With Special Reference to Quantum Mechanics, S. Koerner and M.H.L. Pryce eds. Colston Research Society, Bristol 1957.[5]

Dave occasionally visited me at home – but he could be very distracted. He would come in, hang up his coat, come to the main room, start talking – and it took him some time to hear (from my wife) and to see that I was not there. At that time he either read Hegel's logic, or had just read it and like Lenin interpreted it materialistically. Some of his articles (e.g. his contribution to the symposium Scientific Change which was later edited by A. Crombie)[6] are rather Hegelian. Dave and I also visited Popper. I remember a debate where Popper, saying "I am a logician" criticized Bohr's reply to Einstein's closed-box-with-the-escaping-photon argument for irrelevantly introducing the general theory of relativity while Dave replied that physicists are more tolerant in that respect and, after all, "that's how things are". After a talk which Bohm gave at the philosophical association Dave, Roy Edlgey ⟨sic⟩, a colleague of mine,[7] Joske Agassi, who was then Popper's assistant and I sat in a car, in the middle of a thick fog and discussed Gilbert Ryle, Wittgenstein and Popper.[8] Dave who knew about all of them remarked that Popper relied on logic which was OK, but too rigid to aid scientific research and that 'ordinary language philosophy' was not really using ordinary language but an artificial lingo which was also too rigid. He agreed that there could be 'category mistakes' in the sense that predicates could be applied to inappropriate subjects (such as calling a the ⟨sic⟩ number two blue) but that regarding category mistakes as boundaries to talk, as Ryle suggested, was going too far: scientific research often went right through such 'mistakes'. Dave's knowledge of Ryle and Wittgenstein was not very detailed but he had an amazing ability of getting the whole picture from a few hints. After that he criticized the whole picture, not just a few embroideries here and there as is the custom of many philosophers.

I did not notice any depressive features in Bristol, but I saw him only on and off, on special occasions. We did not at all talk about politics, not in Bristol.

I saw him again, twice, once in Berkeley, and once in Zuerich (1986) where he gave a talk at the Institute for Physical Chemistry with Hans Primas in the chair, a more general talk about fragmentation and where he participated in a general discussion on his talk. The general talk and the discussion were part of a public seminar called "Living with the eight deadly sins of civilised Humanity ⟨sic⟩". This was after a book by Konrad Lorenz, the ethologist, with the title The Eight Deadly Sin ⟨sic⟩ of Civilised Humanity (Die acht Todsuenden der zivilisierten Menschheit – 1972).[9] Bohm's talk and part of his debate with the audience have been published, in German, on pages 331ff of the Book ⟨sic⟩ Leben mit den "Acht Todsuenden der Zivilisierten ⟨sic⟩ Menschheit"? VDF Zuerich 1987.[10] At Primas ⟨sic⟩ he repeated his old hidden variable model with the addition that the wave function, without the particle, was like information that would gradually get more and more confused and thus, in the end, would lose its informative content. In his general talk he made the point that everywhere, in the sciences, in the arts, in public life things are being separated that are parts of a uniform Whole. Relativity and the quantum theory, properly interpreted, conflict with this tendency – but that is not realised by many physicists. He discussed fragmentation and some of its causes in private life, groups, public life. Role of concepts, etc. afterwards we went to have dinner and talked about the consequences in economics, next day in the discussion I mentioned Jesse Jackson' Rainbow Coalition and some of the opposition it encountered and the sly ways in which the Reagan Conservatives extended their radius of appeal. Wholeness through rhetoric? It was no real wholeness, Dave said, for hidden behind the greater unification was an antagonistic position while hidden behind the "Rainbow" was the desire for a greater unity – so there was a fragmentation between rhetoric and ideology. Being asked about Buddhism he described its unifying capacities. On that occasion I noticed that Dave was rather depressed, especially because of his state of halth ⟨sic⟩.

Let me add, though that was not among your questions, that I came to know about Dave by way of his textbook on quantum mechanics which I read as soon as it appeared, in 1951, I think.[11] For the first time all this business about measurement made some kind of sense. I did not yet know about his hidden variable paper and I heard about that from Bohr[12] whom I asked, after seminar, in Denish ⟨sic⟩, which I had not quite understood ⟨sic⟩ to repeat some of the arguments and whose first reaction was "Have you read Bohm?". It seemed that, for him, the sky was falling down.[13]

Is Saral still alive? Are you in touch with her? If yes to both – please send her my love. It's a long time since we saw each other but I remember always with pleasure and, now, with sadness, our meetings.

Best wishes

Paul Feyerabend

Editors' Notes

DB A.21 (2 ss. T+H) ≈ PF 2-10-6 (2 ss., T; c)

[1] See *above*, Vol. 1, Part II, 1955-29, n1.
[2] See *above*, Vol. 1, Part II, 1955-29, n1.
[3] Feyerabend refers here to the British physicist and crystallographer Andrew R. Lang (1924-2008).
[4] On the 1957 Symposium of the Colston Research Society, see *above*, Vol. 1, Part II, 1956-14, n4.
[5] See BOHM 1957b.
[6] See BOHM 1963a.
[7] See FEYERABEND 1995: 106; EDGLEY 1996.
[8] Edgley recollects the episode in these terms: «Sitting in my car in a fog recently I recalled that occasion, I suppose during the Philosophy and Physics ⟨sic⟩ conference at Bristol, when I drove you and Bohm and Agassi home one evening to your several lodgings and the four of us sat in my car in the fog outside your flat in Tyndall's Park Road and argued about Wittgenstein and linguistic analysis. In those days I was still an unreconstructed analytical philosopher, so I was on the receiving end of a lot of flak from the three of you. The argument went on for three hours, till 2 a.m., and would no doubt have gone on till dawn if the fog hadn't mercifully cleared» (Edgley to Feyerabend, 15 November 1988, PF 3-9-1).
[9] See LORENZ 1973.
[10] See BOHM 1987b.
[11] See BOHM 1951; FEYERABEND 1995: 88.
[12] See *above*, Vol 1, Part II, 1953-6, n5.
[13] See BOHM 1952a; BOHM 1952b; FEYERABEND 1995: 78.

3. Peat to Feyerabend, 4 August 1993

Prof. Paul Feyerabend

90 Fentiman Ave.
ETH Zurich
Ottawa
Ontatio
K1S OT8
613-730-1998
e-mail ad454@freenet.carelton.ca

4 August 1993

Dear Prof Feyerabend,

Thank you for your very interesting and informative letter about David Bohm. By coincidence his colleague Basil Hiley[1] was visiting from England at the time and was fascinated to read your letter.

In particular we were excited by what you said in the last paragraph concerning your visit to Bohr. You say in your letter "Bohr … whose first reaction was "Have you read Bohm?" It seem that, for him, the sky was falling in."

I would be very interested to know more, for this implies that Bohr was in some way disturbed by Bohm's paper, rather than simply being dismissive. This is a particularly important point.

The official Copenhagen Interpretation, as expressed by Pauli, Heisenberg and Rosenfeld[2] seems to have been along the lines that Bohm's work was not to be taken seriously and should be dismissed. Received opinion is that the great Ivory Tower of orthodoxy had been in no way shaken. Following a seminar at Princeton (when Bohm was already in Brazil) Oppenheimer said that it would be best if they all agreed to ignore Bohm's work! Max Dresden recalls that criticism was no ⟨sic⟩ so much directed to scientific ideas but to Bohm's political beliefs!

However, your impression of Bohr's reaction was that Bohm's papers had caused him some unease – do you think that Bohr was actually shaken? Did he feel that there was something to answer in Bohm's work? And why, I wonder, did Bohr never reply or offer any public criticism of Bohm's work, but left it to Pauli, Heisenberg and Rosenfeld? As one looks through the literature on Quantum Theory one finds a number of "interpretations" such as Wigner's, Wheeler-Everett, etc, but hardly ever any reference to Bohm's. It is as if <u>it does not</u> exist as a serious interpretation of the <u>Quantum Theory</u>.[3] The remark in your letter certainly therefore opens up a number of [1|2] fascinating possibilities. I would be very glad to hear more on this subject.

Yes, Saral is still alive and lives in Edgware. She is very active, making visits to her relatives in Israel and has just spent a month or so in the US. I am sire that she would like to hear from you, her address is 20 Gibbs Green, Edgware, Middlesex, HA8 9RH.

Warm regards

David Peat
David Peat

Editors' Notes

PF 2-15-61 (2 ss. T+H)

[1] Peat refers to the British quantum physicist Basil J. Hiley (b. 1935).
[2] See *above*, Vol. 1, Part II, 1956-14, n4; 1956-9, n3; and 1948-2, n3, respectively.
[3] Most likely, the underlining in this sentence is by Feyerabend's hand.

4. Feyerabend to Peat, 30 August 1993

08/30/93

Dear David Peat –

sorry for replying so late – I just returned from Rome.

Bohr was neither dismissive, nor shaken. He was amazed. He wanted to explain why but he was called away to an Important Social Function ⟨sic⟩ and so I never heard the end of it.

Incidentally, it was interesting to listen to the discussion after Bohr's talk (he gave a talk and was present for two hours at a ⟨s⟩eminar – I attended both events). Bohr had to leave, so he did not participate in the discussion after his talk. There were quite a few people who thought that his mostly qualitative remarks were not conclusive and found holes into them. The Bohrians for a while tried to close the holes in the Bohrian fashion. When they did not succeed they said: [1|2] "But von Neumann has proved …" – and that ended the discussion. Nobody said another word though I had the impression that none of the Bohrians would have been able to explain what it was that von Neumann had proved and how he had achieved it. It seems that Bohr himself never used von Neumann as a crutch.

If I understand Dave correctly then his 1952 paper was never supposed to be the outline of a physical theory – for him a physical theory would have to be much more complex, and much richer. All he wanted to do was to contest von Neumann's claim to have eliminated not only physical hidden variables, but even unphysical and abstract definitions of such variables.* [2|3]

Having removed this apparently insuperable obstacle he could then start with the physical task of constructing a suitable theory.* The fact, however, that Bohm's model was pushed aside while all sorts of weird ideas flourished is very interesting and I hope that some fine day a historian or sociologist of science takes a closer look at the matter. There exists an older study by Paul Forman (Historical Studies in the Physical Sciences, Vol. 3, 1971, 1-115)[1] that discusses the situation between 1918-1927 and explains to what extent subjectivistic tendencies were at that time [3|4] supported ⟨sic⟩ by the general intellectual and political situation, but there is nothing like that for the Fifties and later, neither for inside science, nor for the political + intellectual ambience.* Pauli, of course, was interested in astrology and hermeticism, he was analysed by a pupil of C. G. Jung and discussed his dreams, astrology, alchemy etc. with Jung in a series of letters that have just been published (ed. C.A. Meier, Springer 1992).[2]

All the best!

Paul Feyerabend
IN DER AU 38
CH-8706 MEILEN, SWITZERLAND

---

[2] * von Neumann himself thought he had done that[3]
[3] * Einstein (his contribution to a volume in honor of Max Born) was not very enthusiastic about Bohm's model, regarded as a physical theory.[4]
[4] * an exception is J. T. Cushing "Causal quantum theory: Why a non-starter?" in F. Selleri (ed.) The Wave-Particle Duality Plenum, London 1992. He has also a book forthcoming: Quantum Theory, Historical Contingency and the 'Copenhagen' Hegemony. See also his contribution to the Conference on Bell's Theorem, World Scientific Publishing 1992, A. van der Merwe, F ⟨sic⟩ Selleri eds.[5]

Editors' Notes

DB A.21 (4 ss., H)

[1] See Forman 1971.
[2] See Meier (ed.) 1992.
[3] On Neumann's "Impossibility Proof", see *above*, Vol. 1, Part II, 1953-1, n3.
[4] See Einstein 1953.
[5] See Cushing 1992; Cushing 1994; Cushing 1993, respectively.

Chapter 8
On Certain Epistemological and Cosmological Problems Raised in Dr. Feyerabend's Review of Causality and Chance in Modern Physics (1959)

David Bohm

On Certain Epistemological and Cosmological Problems Raised in Dr. Feyeraben's Review of Causality and Chance in Modern Physics.

by David Bohm
H. H. Wills Physics Laboratory, University of Bristol

In a very thorough and interesting review of Causality and Chance in Modern Physics(1), Dr. Feyerabend(2) has made a number of criticisms of the main notions that have been developed in that book. These criticisms are worth further discussion in some detail, not only because they ought to be answered, but also because they raise several basic epistemological and cosmological problems that are implicit in the theory that is presented in the book. In the process of answering these criticisms, the author has been led to develop and elucidate a number of further interesting consequences of the notions of the qualitative infinity of nature, especially in the treatment of the relationship of human thinking to the cosmos, and of the question of what is the meaning of truth, in the context of theories that do not have a completely universal domain of validity.

Since the notion of the qualitative infinity of nature is discussed at length in the book (especially in Chapter V), as well as in Dr. Feyerabend's review, it will not be necessary to describe it in detail here. Rather, we shall list only a few of the salient aspects that are essential for discussing the questions raised by Dr. Feyerabend. [1|2]

(1) For reasons that are developed in the book, one is led to consider the possibility that in the description of nature (the cosmos as a whole) there may be required a limitless number of potentially or actually significant qualities.

(2) These qualities have a certain kind of systematic organisation into domains, levels, regions, species, classes, etc. (hereafter to be referred to collectively as "segments") each having a certain relative autonomy, stability and indifference to the others. It is this characteristic that permits each segment to be treated adequately

M. Collodel, E. Oberheim (eds.), *Feyerabend's Formative Years. Volume 2. Feyerabend on Logical Empiricism, Bohm & Kuhn*, Vienna Circle Institute Library 10, https://doi.org/10.1007/978-3-031-57519-8_8

within certain limits and to a certain degree of approximation with the aid of finite conceptions of it, in spite of the infinite background in which it exists.

(3) This autonomy, stability and indifference is incomplete, so that even under normal conditions, a very accurate and detailed treatment of a given segment may require us to take into account the way in which its behaviour reflects the other segments that have previously been ignored. Moreover, under new conditions, these corrections can become so important that the basic qualities of the segment can change fundamentally (and indeed, the segment may even eventually cease to exist, to be replaced by something else).

(4) From the above characteristics, it follows that every law has a limited domain of validity, these limits being determined objectively by the effects of domains that this law necessarily leaves out. When a theory is first proposed, these limits are not known, but eventually their existence is disclosed, when contradictions between theory and [2|3] further experiments are discovered. Later, these limits are defined more precisely in terms of newer theories, which explain the new experimental facts (e.g. as relativity defines the limits of validity of Newtonian mechanics as the domain of velocities small compared with that of light).

With the aid of the above four characteristics, we can begin the discussion of Dr. Feyerabend's review. Further details concerning the notion of the qualitative infinity of nature will be given throughout this article, as the occasion arises.

1. On the Problem of Residual Intrinsic and Objective Error that Belongs to a Law.

The first of Dr. Feyerabend's criticisms centres on the conclusion that if nature is infinite in the way suggested in the book, then it follows (as we have seen in the previous section) that the laws of finite systems will have intrinsic and objective errors. These errors will reflect the effects of the infinity of factors that have necessarily been left out of the laws in question and are therefore not merely subjective mistakes in our ideas about nature.

Dr. Feyerabend objects to the above conclusions, on the basis of the assumption that the author wished to regard all errors as having this objective and intrinsic nature.

As a matter of fact, the author never did make the proposal described above, nor would he even have considered such an idea seriously. It is true that the idea in the form suggested by Dr. Feyerabend was not explicitly denied in the text. Nevertheless, the author's intention [3|4] (which he perhaps did not bring out with sufficient clarity) had always been to consider the correction of errors as carried out by individuals who were seriously and honestly trying to arrive as close as possible to the truth, using all the methods at their disposal (including those suggested by Professor Popper, to which Dr. Feyerabend refers in his article). It was never his intention to imply that progress toward greater truth in our knowledge is in any sense automatic, or that it can be achieved without the conscious and critical effort of the people who are involved in it. It was always implied that all kinds of subjective factors might interfere with the process described above (emotions, mistaken ideas hold too rigidly, and in some cases, as Dr. Feyerabend suggests, a plain desire to avoid the truth).

Generally speaking, the procedures of the scientific method have developed in such a way as to minimize the effects of subjective factors of the kind described above. Naturally, these procedures will not be useful for this purpose unless the individuals who apply them really want to arrive at the truth. As to why they should want to arrive at the truth, this is a very difficult and complex question which it is not necessary to discuss in great detail here. No doubt, as Dr. Feyerabend suggests, the survival value of this behaviour for the race as a whole has something to do with the reason for the evolution of such a characteristic possibility in the behaviour of human beings. But (as Dr. Feyerabend seems also to imply) there is almost certainly more to it than the mere question of survival. For the sake of the present discussion, [4|5] I propose that we accept as a fact that for one reason or another, most scientists are actually trying, in their researches, to come closer to the truth about something. They may not always be fully effective in their efforts, partly because of various factors, emotional, philosophical, volitional, etc. but it cannot be denied that this is what most such individuals are generally trying to do.

Accepting the above presuppositions, the author then intended to show that even if, with the aid of appropriate scientific methods, we are able gradually to eliminate all subjective sources of error, there will still remain, associated with any given theory, an objective residual error, representing the effects of the reflection, within the domain covered properly by the theory, of an infinity of features that are necessarily left out of account in any finite conception of a given segment of the universe. In other words, even if we operate with the best possible intentions and use the best possible methods, we still cannot obtain any item of knowledge that is completely free of error. Rather, we can only go on to the next domains, and in the theory of these, we can begin the task of defining and delimiting the errors of the domain in question.

2. On the Problem of the Relation of Knowledge to its Object.

At this point, it is worthwhile to consider in some details the questions, implicit in Dr. Feyerabend's discussion, of broadening the theory of errors, so as to take into account subjective as well as objective sources of error. Of course, it was not the author's intention to treat such questions in the book, as they lead us into the problem of what is meant by the term "knowledge" and as to what is the relationship between knowledge [5|6] and the object that is known. These questions involve some of the deepest and most difficult unsolved problems which now confront us in philosophy, and which have indeed been the central problems ever since philosophical thinking began. Nevertheless, it may perhaps be of interest to sketch in a very preliminary way a method by which such questions can be considered on the basis of the qualitative infinity of nature.

In order to carry out such a discussion effectively, we must begin by indicating what is the role of the mind and the processes of thought, within the framework of the general cosmological point of view that we are discussing here. Now a great deal of philosophical thinking on this subject, especially in the West is based (either implicitly or explicitly) on the notion of the dualism of mind and matter, initiated mainly by Descartes, but developed much more systematically by Kant. In the point of view of Descartes, mind and matter are conceived as two kinds of substances, each completely separate from and alien to the other. In Kant's point of view, one similarly begins with the notion of an unbridgeable gulf between mind and matter. With both of these notions, there arises immediately then the problem of how the inner world of our thoughts is related to the external world of matter in such a way that we are somehow able, in many cases, to have essentially correct knowledge about the latter within the former.

Once the problem is put in the way described above it is not capable of a rational solution, within the framework that has been given. For if mind and matter are completely alien to each other, separated by an unbridgeable gulf, then any relationship between them must be impossible [6|7] in terms of mind and matter, so that at best, it could perhaps be explained as due to something outside of those. Indeed, Descartes saw the point clearly and proposed that the relation was due to God, who, being external to both mind and matter, was able to put essentially correct modes of thinking into the minds of men, modes that were able, if followed faithfully, to reflect the behaviour of matter more or less correctly. In Kant's point of view, no solution for the problem was even proposed. Rather, he restricted himself to the conclusion that all that could be in our minds were the phenomena, whose basic forms and modalities (space, time, causality, etc.) where characteristic of the mind only, while matter (the "things in themselves") was completely unknowable. The unsolved problem was then, of course, why certain aspects of thinking (such as space, time and causality), while representing nothing but modes of the human mind, could serve so systematically as correct guides in our efforts to deal with the supposedly unknowable external world.

The other side of this problem is to understand why our thinking (in these modes, or among other modes that we might develop) is not always correct, i.e. why it frequently contains errors. The consideration of the two questions together then gives rise to the problem of what is the relation between truth and error. And this is, of course, just the problem on which Dr. Feyerabend's review centres, as indeed does also Professor Popper's well known methodology of falsification(3) on the basis of which many of Dr. Feyerabend's arguments are founded.

It is clear that if we are even to consider the question of obtaining a rational account of the relationship of knowledge to what is known [7|8] (or what is a closely related question, that of the subject and object), we cannot begin with the notion of an unbridgeable gulf between the two. Rather, we must begin with the notion that most fundamentally, they are united, in the sense of being parts or sides of the broader world, containing both "mind" and "matter". Now, in terms of physical theories based on the assumption of some finite set of basic qualities, it has never been possible to do this convincingly. Such attempts must necessarily involve, in one way or another, the mechanistic hypotheses that the qualities of the human mind, which have variously been labelled "consciousness", "thought", "spirit", "ego", "personality", etc., could, in principle, be reduced, completely, perfectly, unconditionally, in every detail, and without any approximation, to some sort of over-all or collective effect of the operation of a mass of organised and interacting basic elements (such as atoms, molecules, etc.). This notion is so implausible and so full of inconsistencies (which is not the author's purpose to discuss in detail here) that it has been rejected, either explicitly or implicitly, by a great many philosophers who have thought deeply on the subject.

3. A Possible Relation of Human Thinking to the Cosmos, According to the Qualitative Infinity of Nature.

If we start from the qualitative infinity of nature, we do not have to accept the mechanistic hypothesis that the laws of any given segment or level are, in principle reducible completely to those of a final set of laws of assumed basic elements out of which the universe is supposed to be constituted. Indeed, if the qualities are infinite, then each [8|9] level is seen to make its own contribution to the totality, which cannot be fully reduced to that of any finite number of others*. Thus, we have, as Dr. Feyerabend pointed out, a way to explain the "emergence" of new qualities, within the framework of a rational over-all system of natural law. Such new qualities can come into being, not only in the physical and chemical levels, but also in the biological level (associated with a great degree of certain kinds of integration ⟨sic⟩ and even more, in the psychological level of conscious mental processes. Thus, we do not have to suppose that the conscious, thinking, human being can be reduced fully to his constituent atoms. Rather, since all laws holding in the atomic (and other) levels are necessarily incomplete, it is quite possible that the individual atoms constituting a living being each behave in a way a slightly different from that of an atom in a relatively poorly integrated system. This difference might be so slight that it would be extremely difficult (but not in principle impossible) to detect when one observed individual atoms, even in living beings. Nevertheless, the cumulative (collective) effect of the differences would be just such as to make it possible for a new quality (e.g. life) to come into being in the system as a whole. Similarly for thinking consciousness, etc. And in order to understand all aspects of the behaviour of a living being, one would have to take into account those new qualities, as factors having some degree of relative autonomy.

It is instructive to think of the higher level qualities as being, in a certain sense, embedded in a matrix, furnished by a dynamic, ever [9|10] changing organization of matter in lower levels. The higher level qualities cannot exist without this matrix; and yet, they are not reducible to it. Because of the very close relationship implied by embedding, the form of the matrix will reveal a great deal about that which is embedded in it, without, however, necessarily constituting all the essential features of the latter.

With this background, we can now see that whatever the human mind may be (and it must be admitted that this is for the most part unknown or else very poorly known) there will, in the point of view described above, never be any problem of how mind and matter can become related. For mind is now seen as one relatively autonomous and stable level of qualities, immersed in the same infinite totality of the cosmos, in which is immersed what we have become accustomed to call matter. The notion of mind and matter as absolutely alien to each other, separated from each other, by an unbridgeable gulf, therefore simply does not arise. Far from being separated by such a gulf, they interpenetrate each other so thoroughly that the main problem now before us is to explain the extent to which they are actually different and separate from each other. In its most essential features, however, this is the same problem that arises in every level, and with every relatively stable, autonomous and indifferent system (such as an electron). We begin with the infinity and discuss the processes by which what is outside the system under discussion produce, in the system, a [10|11] relative balance in the resulting changes. For many kinds of systems (e.g. the liquid as composed of moving atoms) we can already carry out such an analysis, at least in principle.* At present, it is certainly premature to attempt to do a similar analysis for the mind, because we know too little about it. But for the sake of argument, let us suppose that the analysis could actually be carried far enough so that we could see how it is possible for at least the main qualities of a mind to emerge as relatively indifferent (within limits) to the details of what is happening in the structure of its matrix (which is the body, with its system of nerves, muscles, glands, bloodstream, etc.). We would then understand how the mind could begin to function according to its own laws while outside disturbances would then be reflected as further alterations within these laws.

The question now arises "What are the laws of the mind?" First of all, it is clear that the mind is able, somehow, to make various kinds of <u>images</u> of the objects with which the individual is confronted. But these images (or reflections) have certain peculiar properties, which do not appear in any other known kind of reflection.

In a very broad sense, every object, can <u>in some way</u> reflect anything else. For it is (potentially or actually) the converging point of areas waves, such as those of heat and light, as well as [11|12] of particles, some of which may come from far away. This reflection is, however, confused, in a sense that can be clearly defined. Thus, while it is true that heat, light, etc., impinging on an object will set its molecules in random motion, this motion with evidently be a very bad image of the objects from which the heat and light came. It is a bad image, because it does not, for example, reflect adequately the <u>topological</u> and <u>metrical</u> relationships of these objects. To illustrate what is meant here in more detail let us consider instead, an image formed by a good lens. Thus if (with two semi-transparent objects, A and B), we have A inside B in the original space, then the image of A will be inside the image of B. Similarly, the relative sizes of the images of A and B will be the same as those of the objects, etc., etc.

Now it is evidently true that the human mind can form images which more or less faithfully reflect the immediate relationships of things in space, images which resemble those of a lens. But the mind can do much more than this. For it can also image what we may call <u>the general laws</u> of the objects. In other words, when we correctly understand an object, we do not only know where it is, how it is related to other things, etc., but also how it must move, change and develop under a wide variety of conditions not only if left to itself, but also if it is altered (either by means of an experiment, or as a result of the effects of other objects that it may meet in the natural course of its motions). [12|13]

Let us now consider in more detail how the mind comes to reflect the general laws of the objects. If we begin by considering our sensations and perceptions, we note that they produce in the mind a set of ever-changing images, which we may call "immediate appearances". These appearances cannot be accepted at face value, because they alter, transform (and respond to our own actions)* in a way that contradicts the notion that each of them is the separate, fixed, self-maintaining and substantial thing that it seems to be in our most immediate perceptions. We then try to form a new kind of mental image, in which each of these things and the phases of development of each thing (e.g. egg-chicken) are represented as different sides or aspects of a more permanent more general kind of thing.

But now something new has come into the image. For the various sides, parts and phases of development, are understood as related to each other in a logically necessary way, by means of a general law. Thus, we see the various aspects of the surface of a solid figure as different sides of a single three-dimensional object, the faces of which are related by the laws (perhaps sensed only intuitively) of a suitable geometry.** As a result, we can often predict [13|14] with a considerable degree of success, what the object will look like if we view it from another angle. Even more, we eventually come to understand its movement (inertia, momentum, pattern of growth, etc.) as an essential quality, i.e. as an aspect of the certain general all. Hence, we can predict how it will alter its position, size, shape, etc., how it will respond to disturbances produced by us, as by those that it will meet in the natural course of its motion and development.*

To sum up, then, thinking can tell us something new, something that is not completely contained in our immediate perceptions and sense impressions. This new feature is introduced by the possibility of reflecting the <u>general laws</u> of the object within our mental processes.

Where does this capacity of thought to reflect general laws come from? From the point of view of the qualitative infinity of nature, the existence of such a capacity is not surprising. Indeed, it is clear that if each kind of thing is embedded in the infinite totality of the cosmos, and obtains its relative autonomy and indifference only by the balance of the changes reflected into it from the rest of the world, then not only its immediate and particular constitution, but also its general laws cannot do anything else but reflect, <u>in some ways</u>, <u>at least</u>, the laws of the cosmos in which it is thus embedded. In this most general and essential aspect, thought is the same as anything else. But as in the case of images of objects in space, some systems can [14|15] more effectively reflect general relationships than others can. What distinguishes the thinking process is its exceptionally great potentialities for reflecting faithfully, i.e. with a certain degree of homeomorphism* both the immediate particular forms of its objects and their general laws. These potentialities must doubtless, at least in part, owe their origin to the extremely complex and delicately organized structure of the brain and nervous system. However, for the purpose of further discussion here, it is not necessary to enquire into the precise reasons why this potential capacity exists. All that we need to do is to note that factual experience alone indicates that the human mind has shown itself capable of reflecting essentially correctly the laws of a tremendous range of kinds of things, and that there is no reason known to suspect that the reason is any ultimate limit to this capacity.

In spite of its enormous potentialities, the human mind does not, however, immediately reflect (e.g., by instinct) the laws of the objects with which it is confronted. This will happen only if a certain procedure is followed. First, the individual concerned must be prepared by education. After this, he must be ready to make suitable investigations, researches, etc., utilizing methods (e.g., scientific) which have been developed over centuries in society. Indeed, the human mind, left to itself, generally follows [15|16] its own laws, producing reveries, dreams, fantasies, etc., much more readily than it produces a rational understanding of the objects that confront it. Thus, we are not at all justified in supposing that human thinking is nothing else but a reflection of the laws of its objects. Such a point of view would, in the last analysis, deny any degree of autonomous character to the mind. Rather, it would be more accurate to say that the mind has its own general laws (just as everything else does). Among these laws, one of the most important and characteristic is this: If an individual chooses, he can direct his thinking processes in such a way that its laws develop a certain kind of homeomorphism with the laws of its object. As is wellknown ⟨sic⟩, the result can in general be arrived at successfully, only if the individual concerned pays careful attention to his perceptual images of the object, and tests his proposed laws by applying them and by doing suitable experiments. (Of course, more generally, the individual can also take into account the perceptions of other people and he can allow his ideas to be tested by other people, too).

It must now be emphasized once again that in the point of view under discussion here, ideas, emotions, desires, etc., are to be regarded as parts of reality, which belong to the psychological level of human beings. Thus, an idea held by a certain individual is just as much a real and essential part or side of the cosmos, as is, for example, a molecule. Both can be [16|17] causes of future developments, both are to a certain extent the result of past causes, both have a certain limited degree of autonomy, etc. In other words, both are real, and in certain ways, irreducible links in the chains of cosmic development. The main further special characteristic of an idea is its capacity to reflect (with a certain kind of homeomorphism) the particular forms and general laws of objects other than itself (and in introspective thinking, it may even be able to make some kind of reflection of itself, i.e. to have itself as object).

It is in the context of the question of the extent, degree, and general character of the homeomorphism described above that one can discuss the problem of the truth of an idea. Of course, no idea is a perfect reflection of the particular form and general laws of its object. In other words, there is always some error in any ideas. As we have already indicated in earlier sections, these errors divide roughly into two types, viz., subjective and objective.

Now when an individual begins to think of any specified object presented to him in his perceptions, (this object may even be an internal mental or emotional state which he perceives) his special subjective starting point inevitably enters into any idea that he may form. Thus, in addition to reflecting the object in some ways, his thoughts will also reflect his peculiar conditions, relationships, emotions, desires, etc. As a result, [17|18] there is no reason why his ideas should at this stage provide a very faithful reflection of the laws of his object. They will therefore, in general, contain a great deal of error, mostly subjective in nature. If he chooses, however, the individual concerned can follow appropriate procedures of thinking, action, research, etc., which will step by step lead him to correct his errors. If he is studying a well defined segment of nature, he can eventually eliminate most (if not all) of the subjective error. There will remain, however, as we indicated in Section 1, a residual objective component of the error of any particular law, coming from factors that it necessarily leaves out of account (because nature is infinite). This objective part of the error can be corrected when new laws are developed, which treat some of the factors that have been left out of account. In this connection, it should moreover be noted that another individual starting from a different point may well go through a different series in his development of the laws, having different kinds of subjectively based errors. In the long run, however, anyone studying a given segment of nature and following appropriate scientific procedures should arrive at laws having essentially the same content (within the limits of the objectively based errors in the law).

Finally, we remakr ⟨sic⟩ that in the point of view under discussion here, subjective error (as well as objective error) always has a reason. This reason can be found in principle by understanding [18|19] the peculiar conditions, relationships, etc., appertaining to the starting point of the individual subject in question. (He may often be able to examine them himself if he so chooses). Thus, even illusions and delusions have a rationally understandable basis, either in some nervous defect of the subject, as in some idea, emotion or desire which he happens to have for these reasons. Such an idea, emotion, or desire, may even lead to an intentional effort to avoid the truth (of the kind mentioned by Feyerabend).

Of course it is generally admitted that in the long run, it is not satisfactory to live on the principle of trying to avoid the truth, either through deception of other people or through self-deception (which, as is well known, are in any case very difficult to keep separated from each other). It is not appropriate here, however, to go more deeply into the interesting question raised by Dr. Feyerabend, as to why truth is preferable to falsity and error. We shall allow the problem to rest here with the statement that even the answer to this question can be shown to have a rational basis, as can indeed also the answer to the question of why we do not always seek to approach the truth even though it would doubtless be best for us in the long run if we did.

4. Dr. Feyerabend's Second Criticism – On the Justification of the Notion of the Qualitative Infinity of Nature.

We now come to Dr. Feyerabend's second basic criticism of the notion of the qualitative infinity of nature. In essence, this [19|20] criticism is directed against the explanation with the aid of the above notion of the necessity of a certain characteristic of the scientific method; viz., the need to test every feature of every theory, no matter how fundamental and well-founded that feature may seem to be.

Briefly the question is simply this: It is true that if nature were known for certain to be infinite in the way assumed here (i.e. organized into a set of relatively autonomous and indifferent domains, levels, etc., of various kinds) then one would have no doubt of the need, in scientific research, always to test every feature of every theory in the manner described above. For one would know that every domain of law has its limits, and yet one would never know precisely where the limits of a given domain lie. One would therefore have no choice but to be prepared, at any point, to discover errors in the theory in question. On the other hand, as Dr. Feyerabend points out, there is no way to prove that nature really is infinite in the way described above. Thus, the mere fact that the hypothesis of the qualitative infinity of nature is in accord with the actual historical development of science is no proof that it is really correct (even though the latter has shown a long series of revolutionary changes in which theories previously thought to be fundamental where demonstrated to be special cases of still more fundamental theories).

Indeed, by applying to the notion of the qualitative infinity itself the same criterion that this notion suggests for other theories (i.e., the need for caution and avoidance of the unlimited [20|21] extrapolation of previous historical trends), one is led to consider the possibility that eventually, one may have, so to speak, to complete the circle and return to a finite point of view. Thus, Dr. Feyerabend argues that the notion of the qualitative infinity of nature cannot satisfactorily be regarded as a justification of the characteristic of the scientific method under discussion. He proposes instead that for this purpose we accept Professor Popper's[3] methodology of falsification, viz., that "we are prepared to collect refuting instances from all domains[2]".

In answer to this criticism, the author wishes first to make it clear that he never intended the notion of the qualitative infinity to be regarded as a final doctrine. Indeed it was expressly stated* that this notion would probably eventually fit into some more general point of view, "which retains its positive achievements and which goes beyond them". All that was intended by the author was to suggest that the idea of the qualitative infinity of nature has a number of advantages over any other known to him, for the purpose of solving certain cosmological, epistemological, and methodological problems, relating to a theory of scientific research. These advantages are explained in some detail in the book (especially in Chapter V). In the next section of this article, the author will, however, elaborate on this question and will compare the notion of the qualitative infinity of nature and Professor Popper's methodology of falsification, with [21|22] regard to their use as fundamental starting points in a theory of scientific research. Thus, the question of an absolute proof of the correctness of the notion of the qualitative infinity of nature is not at stake here, any more than (as we shall see in the next section) is the question of an absolute proof of the correctness of Professor Popper's methodology of falsification. As in the case of the comparison of any two scientific theories, the only relevant point at issue is to determine which of these theories gives a better treatment of the problem under consideration.

5. A Critical Consideration of Professor Popper's Methodology of Falsification.

Of course, that can be no doubt that Professor Popper's methodology of falsification must be an important part of any systematic account of scientific research. But even if this is admitted, it does not follow that this methodology, by itself is able to give a full treatment of all aspects of the scientific method. Indeed, a very preliminary consideration of the problem leads us to ask whether the method is really universally valid, in the sense that (a) its sole use could never lead one to do scientific research in a wrong or inadequate way, and (b) there could be no complementary aspects of the method involving other fundamental and necessary procedures not taken into account in Professor Popper's formulation of this methodology. Unless these questions can be answered in the negative, then the possibility must [22|23] remain open that the methodology of falsification (like all of the theories) has its limits of applicability, and that it is perhaps a special case of a proper method. If this were so, then under suitable conditions, one might even be led to adopt a methodology that in some ways or in certain instances contradict that of falsification (as relativity, containing Newtonian mechanics as a limiting case, leads to conclusions contradicting some aspects of the latter).

The problem then immediately arises as to how we can prove the universal applicability of the methodology of falsification. As in the case of the infinity of nature, the mere agreement of this method with a broad domain of historical experience in scientific research could not provide such proof. Nor does the method itself indicate any way to prove its correctness, or indeed, to indicate that it is false (not universally valid) if, in fact, this latter should really be the case. Thus, any person applying this method to a particular theory will in each instance either confirm or refute that theory in any given experiment. If he confirms it, he continues to apply the theory to new cases. If he refutes it, he seeks a newer and better theory. But suppose that the methodology of falsification were to be wrong or inadequate, in some domains, or in certain cases. It is clear that nothing in the method could suggest a procedure by which one would be able to "collect refuting instances" that apply to the principle of falsification itself. Hence, the application of this principle in [23|24] experiment cannot serve as the ground of the principle (i.e., its logical justification).

Of course, as in the case of the qualitative infinity of nature, the above conclusion does not constitute any kind of argument against the methodology of falsification, provided that one admits that this methodology is to be regarded as a proposed theory, claiming to reflect adequately certain aspects of the general laws of scientific research. The only question that remains open is then that of just how adequately this methodology actually does reflect scientific research, as we have known it to date. And this is the question that we shall discuss in the remaining section of the article.

6. On the problem of truth of general laws.

One of the principal logical difficulties within the methodology of falsification is that it provides no way to define what is meant by the notion of truth of a general law. As is well known, in this methodology, one is directed to treat any given hypothesis as if it were universally valid, until one obtains instances in which it is falsified. After this, the hypothesis in question is discarded. One then seeks a better hypothesis, fitting the new facts; and thereafter, one treats this hypothesis in turn as if it were universally valid, until it too is eventually falsified. At no stage does one ever assert of any theory or hypotheses that it is true, or even that it contains some elements of truth. At most [24|25] one can say of a theory that it has or has not yet been falsified.

It takes only a brief analysis to show that in the point of view described above, falsification must be considered to be just a kind of experience, and can in no sense be regarded as referring to some objective characteristic of the relationship between the hypothesis in question and the things to which the hypothesis refers. For any attempt to give falsification a further meaning will lead us into the problem of defining the sense in which general statements or laws can be true, a problem that Professor Popper does not regard as capable of being solved.

To illustrate the difficulties involved, let us suppose, for example, that a given individual, A, has on a certain occasion (or set of occasions) done specified experiments leading to the conclusion that the hypotheses, H, is false. When we say that the hypothesis H has been falsified do we mean only that the individual, A, has had the experiences described on the specific instances mentioned? Or do we mean also that anyone doing a similar experiment would arrive at the conclusion that the hypothesis, H, is false? If the latter is our meaning, then we have made a general statement. Such a statement in essence amounts to a restricted law, which certainly does not follow necessarily from the experiences mentioned. That this is so can be seen by considering our reaction to the further statement that the individual, A, had never been trained in the subject under investigation, and had simply jumped into the experiment without [25|26] any preparation whatsoever. We would then say that this was not really or truly a falsification. But similar difficulties have frequently arisen even in experiments carried out by competent and well trained scientists. It has often happened that preliminary experiments disagreed with a given theory, which was later confirmed by better experiments. In other words, as experience in scientific research itself has shown, any falsification may itself later be falsified.

It is clear then that it is not enough to say that a hypothesis has on certain occasion been falsified. Rather, before this falsification carries any weight in scientific research, one must be able to say that the hypothesis has been, in some sense, truly falsified. And by this, one refers not to a particular (or existential) statement as to the fact that several falsifications have actually occurred in the experiences of various individuals. Rather, one refers to some kind of general statement as to its falsity. For example, it would be a consequence of the "true falsity" of a hypothesis that any competent person doing a certain kind of experiment would discover this falsity. But if one accepts this definition of falsity, one can equally well put even the positive content of a theory into the same form. Thus, if the law of gravity is true, it follows that any competent person releasing an object in a vacuum will discover that it falls. It is not then rather arbitrary to adopt the idea that the notion of truth may be [26|27] accepted without further question when it is applied to the falsification of hypotheses, but not when it is applied to the hypotheses themselves? Indeed, once one accepts the idea that general statements concerning falsifications can be true in some sense, then there is no logical reasons to stop here; one may just as well go on to consider the possible truth of the hypotheses themselves.

Thus, if with Professor Popper, we avoid the problem of attempting to say what could be meant by the truth of hypotheses, i.e., of trying to find a sense in which this notion could have an empirically adequate and logically consistent definition, then there is, it would seem, no choice but to regard each instance of falsification as a particular experience, which a given individual has had, on a certain occasion. On the other hand, as Dr. Feyerabend quite rightly points out, general laws should not be regarded as mere "relationships between the manifold aspects of our experience". He admits that "experience as well as theories" must be criticized. Of course, experience can be criticized on the basis of "well-confirmed" hypotheses concerning general laws. But unless "well-confirmed" means also "in some sense objectively true", then is it not arbitrary to reject or criticize any item of experience whatsoever, simply because it does not conform to certain hypotheses in connection with which various individuals on specified occasions happen to have had a rather large number of other experiences of confirmation? Indeed, the possible danger of rejecting or criticizing experiences just because they do not conform to [27|28] hypotheses is clearly recognized by almost everyone who has tried to think about the methodology of scientific research. Nevertheless, as Dr. Feyerabend and Professor Popper both agree, this is just what we must do on certain occasions. But there does not seem to be anything in the methodology of falsification which would justify such a procedure, in a rational way, without an appeal to some kind of implicit notion of truth (e.g., at least the truth of falsifications), a notion which stands in contradiction with the explicit statements that define the essential aspects of this methodology.

As important a step forward as the development of the methodology of falsification has been, there is evidently something fundamental that is still lacking in it. What is missing is some way of explicitly and openly facing the question that an adequate account of the scientific method requires a suitable definition of what is the significance of the term "objective truth" in the context of a situation in which any given law may (and probably eventually will) be shown to have instances in which it is falsified. That is to say, it must consider the problem of what can be meant by the term "truth" as applied to a law that is not completely universally valid.

In terms of the qualitative infinity of nature, this problem is dealt with in a natural way. We begin by supposing that a general law can be true in some ways, within certain limits, and to a certain degree of approximation without having to be completely [28|29] universal in its domain of validity. Of course, we do not immediately know the definition of its domain of validity. This, however, cannot prevent the existence of such a domain. (To assert otherwise would be to fall back into the positivist philosophy, which both Professor Popper and Dr. Feyerabend reject). This domain can later be defined in its essentials, when we progress to further theories. In this way, we are led to develop the notion of a series of in principle knowable objective truths, each having objective limits of validity, which are determined by other terms in the same series (and which are therefore also in principle eventually knowable).

In the above definition, the idea of truth is that developed in the previous section. In other words, an idea is true to the extent that it constitutes a reflection having some kind of homeomorphism with the immediate form and with the general law of its object. Thus, an idea which is true in a certain domain or within certain limits is one that reflects with such a homeomorphism only within the domain and limits in question.

In connection with the above problem it is instructive to introduce an idea frequently met with in mathematics; viz., that of a relationship which holds if elements outside a given class be left out of account. For example, in the theory of numbers, one may seek relations which holds modulo the nearest integer. In topology, one may discuss relations which hold, modulo the cells that arise in a certain mode of "covering" of a space by means of finite number of "open sets". Such a notion is easily [29|30] generalized to apply to any kind of statement. Thus, we may then propose that a certain theory holds modulo everything that is outside a given level, domain, class, etc. In this way the notion of statements having a relative and limited kind of validity could be formalized and given a quite definite meaning*.

Having defined the notion of relative and yet objective truth in the manner described above, we can now see immediately why the falsification of "well-confirmed" theories is such an exceptionally good way to be led to correct new hypotheses that are likely to be fruitful. For as we approach the limits of a given theory we will not only discover that it breaks down, but much more important, <u>the precise way in which it breaks down will reflect many features of the new domains that are mainly responsible for this breakdown.</u> (For example, classical mechanics broke down in such a way as to indicate discrete orbits, quantized energy and momentum, and a wave-like as well as the particle-like character for all matter and field systems). Here it is important to note that such a reflection is possible only if we have, to begin with, a theory possessing a certain domain of validity (i.e., in Professor Popper's terminology "well-confirmed"). Evidently, it will generally be of little use to falsify an arbitrary not previously well-confirmed hypothesis. And the notion of objective truth with limits to its domain of validity explains why this is so. For as we have seen, only a theory that is systematically valid in some domain is able, by its precise mode of breakdown, to reflect the [30|31] features of a new domain well enough, so that we obtain in this way some clue as to the nature of the next domain. On the other hand, in a methodology of falsification, there is no reason, <u>intrinsic to the methodology itself</u>, (and not just arbitrarily injected on the basis of further general experience) which will show the necessity for falsifying theories in just this way. Indeed, if, as is implied in Professor Popper's point of view, science were concerned, most essentially only with the falsification of hypotheses, and not also with their verification, then why should it matter whether or not the hypotheses to be falsified are first well confirmed? Why would not <u>any</u> hypotheses serve for this purpose?

It is clear from the above discussion that some additional essential factor enters into scientific research, not fully dealt with by taking into account only the falsification of hypotheses. It is true that falsification plays, at a certain stage, a crucial role: and it is the great merit of Professor Popper's methodology of falsification to have called attention to this previously neglected role, and to have treated it in a systematic way. Nevertheless, the question of verification still retains a fundamental kind of significance, even if only because, before one can usefully raise the question of falsifying a theory, one must, in general, first confirm it, at least to a certain degree of approximation, within certain limits, and in a certain domain of phenomena. But the role of confirmation is much more than that of a [31|32] mere preparation for future falsification. For after all, it cannot be denied that what scientists are actually seeking in their researches is most fundamentally, not to know in what ways their ideas are false, but rather, to try to arrive closer, to what is, in some sense, a set of truths concerning the objects of their investigations. Even more, the demonstration of the falsity of certain aspects of previously accepted ideas generally has value only to the extent that it brings one closer to truth. Thus, if for example, it had turned out to be the general rule that experiments would simply disclose that certain ideas were false, without helping to suggest new ideas that were more nearly true, then scientific research would not have the interest that it actually has for most of those who are engaged in it. For after all, the mere knowledge of the falsity of a certain theory has, in itself, no value except in so far as it tells us that the truth is to be sought elsewhere. But the essential question is then "Where?". And here, we have seen that what is significant is not just the falsification of a given theory, but, rather, the relationship between the theory in its domain of validity and the precise way in which it shows itself to be false outside its domain. It is this relationship which makes the method of falsification something more than a mere negation of the universal validity of the theory in question, but rather, as we have seen, a very good clue as to what are the main features of the next [32|33] domain.

To sum up, then, one can criticize the methodology of falsification, because it presents an unbalanced account of scientific research as a whole. It is unbalanced in the sense that the role of falsification is overemphasized, while the at least equally fundamental role of confirmation is almost entirely excluded from the basic features of the theory, with the result that a number of important questions concerned with the relationship between falsification and confirmation cannot adequately be considered.

The relationship between falsification and confirmation can be brought out still further by asking ourselves whether a scientist always discards a theory and seeks a broader one when he is confronted with a well-established falsification of this theory. No doubt, this is frequently what he does. But quite often, what is done instead is to keep the theory, and instead, to restrict the domain in which it is assumed to be valid. This restriction is often of crucial importance, especially in the early period of development of a given science, when the main problem is to obtain a body of organized knowledge with which to start. Thus, as an illustration, it is well-known that primitive peoples tend to regard all objects as alive, and possessed of individuality, wishes, desires, etc. In a sense, one may say that they apply the laws of living beings (in so far as they know these [33|34] laws with the aid of a certain amount of experience) to all kinds of things. The eventual realization that, for example, stones are not alive could be regarded as a refutation of the universal validity of the laws of living beings. One possible procedure would thus have been to seek a set of still more nearly universal laws, which would have applied both to stones and to living beings. But another equally significant step closer to the truth was the simple realization that the laws of living beings still apply, but to a restricted domain. In this way, there gradually developed the knowledge that the world can, with a certain degree of correctness, be studied in its various domains separately. Within each domain, subdomains were fund in a similar way. Thus, for example, one arrived at the notion of a natural mode of classifying things according to genus, species, etc. In this whole procedure, the essential goal was not so much to try to show the falsity of previously accepted ideas, but rather, to try to define and delimit in more and more detail a set of domains in which certain kinds of laws are true, certain kinds and properties actually do exist, etc.

Now there is no doubt that eventually this whole procedure of classification was shown to be inadequate and one-sided (e.g., in biology, by the theory of evolution of the species). In other words, it was, in a certain sense, falsified. But this kind of falsification was both possible and fruitful only because it had already been demonstrated that there is some degree of validity to the previously accepted schemes of classification. [34|35]

With regard to the above described problem of the restriction of hypotheses and the search for a domain of confirmation it is true that in Professor Popper's point of view one might in principle regard such a restriction as a new or modified hypothesis, which is made in response to the falsification of the assumption that a given law is universally valid. Such an interpretation is not, however, adequate to the problem. It is inadequate because there are no grounds, intrinsic to the methodology of falsification, which could ever lead us to the notion that what is most significant at certain stages of the development of science is not the extension of theories, but rather the use of certain laws for the purpose of finding the natural division of nature into domains, levels, and the natural development of various systems of classification into genus, species, etc. Indeed the very idea of a domain of validity is a notion that is intrinsically foreign to the methodology of falsification, which, as we have seen, does not admit the application of the concept of truth to scientific theories. At best, one would have to deal with the division of nature into autonomous levels, segments, etc., as the result of arbitrary presuppositions which were inserted a priority into hypotheses. On the other hand, in terms of the notion of truths with objectively limited domains of validity, such a division is seen to arise naturally and inevitably, in a well defined ⟨sic⟩ way, from the character of the laws themselves. [35|36]

We have seen that the confirmation and falsification play a complementary set of dual rules. Sometimes one of these procedures and sometimes the other comes to the fore; but in any case, wherever the one is, the other is in the background, and is ready to come into major importance later.

Now it follows from the above that the first step towards really systematic knowledge does not necessarily lead immediately in the direction of more and more nearly universal laws. Rather, the first step is often in the opposite direction of restricting the domain of validity of a law previously regarded as universal, or in other words, of particularizing the way in which we understand things. Indeed, in certain periods (e.g., that of scholasticism in the Middle Ages) this tendency had been carried so far as almost to constitute a caricature of the search for truth. Partly as a result of this, there developed later a tendency to move again in the direction of a search for greater and greater universality. Here, the demonstration of the limited validity (the falsity) of a division of nature according to certain classifications played a fundamental role. In our reaction against the scholastic tendency toward rigid and arbitrary systems of classification, however, we must be careful not to go all the way to the opposite extreme, and to ignore the importance of these classifications as an essential side of our knowledge, (especially when we must start in a field in which well confirmed ⟨sic⟩ theories do not as yet exist). And in the [36|37] development of such a system of classification it is in general more important to emphasize the search for a domain of confirmation than it is to emphasize the effort to falsify hypotheses.

To sum up, then, it is necessary to have a theory of scientific research which pays adequate attention to confirmation as well as to falsification, because both are equally significant sides of the process by which scientific knowledge progresses toward greater truth. And even more important, it is necessary to formulate clearly an empirically adequate and logically consistent notion of what is meant by truth, which is after all an essential implicit goal of all scientific research in the first place.

In the opinion of the author, the notion of the qualitative infinity of nature provides a better treatment of the above problems than does the methodology of falsification. It does so in the following ways:

(1) It begins by making possible logically consistent definition of truth as applied to a law that does not have a completely universal domain of validity. This definition, which takes into account both subjective and objective sources of error leads to an adequate treatment of the problems that arise in the consideration of scientific research as it has actually been carried out to date.

(2) It brings out in a clear way the importance of having a body of theory, verified (or confirmed) within some limits, before the question of falsification can in general usefully be raised. [37|38]

(3) As a result, it leads immediately to an explanation of how and why a study of the precise manner of falsification of a theory previously demonstrated to be valid within certain limits is such a good method for obtaining definite clues as to the nature of the next as yet unknown domain. In this way, it provides a logical justification of the methodology of falsification, but only for those cases in which this methodology is actually likely to be appropriate, and not for arbitrary cases (e.g., those of poorly confirmed or unconfirmed hypotheses).

(4) It brings out the importance of a certain aspect of the scientific method which is not adequately treated in the methodology of falsification, viz., the restriction or particularization of the domain of validity of a theory, and of the positive search for such a domain. In this way, it leads explicitly to a recognition of the importance of verification (or confirmation), as opposed to a one-sided emphasis on falsification alone. Moreover, it shows that scientific research is not confined only to the search for theories having a greater and greater degree of universality, because the movement toward particularization (the intentional restriction of the domain of laws and

the resulting tendency to study only the details of very narrow spheres) also plays, in the long run, a very essential role.

---

[1] (1) D. Bohm, Causality and Chance in Modern Physics, Routledge and Kegan Paul, London, 1957.[1]
[4] (2) P. Feyerabend. ⟨sic⟩[2]
[7] (3) K. Popper, The Logic of Scientific Discovery, Hutchinson, London 1959.[3]
[9] * See reference (1), Page 145.[4]
[11] * See, for example, reference (1), p.50.[5]
[13] * E.g., the apparent shape of an object changes as one changes his position or he turns his head.
[13] ** It is wellknown ⟨sic⟩, for example, that people born blind who obtained their sight later were not immediately able to interpret their visual images in this way. Thus, a table was seen as a confused and changing two-dimensional figure, and only later was it seen as a fixed three-dimensional object.
[14] * E.g., when we see an object teetering on the edge of a cliff, the image in our minds moves in such a way as to anticipate its subsequent fall.
[15] * Homeomorphism signifies a kind of mapping, not necessarily one-to-one, in which certain group relations of the elements are preserved.
[21] * See Ref. (1), P136.[6]
[30] * The author is in fact engaged in developing an example of such a formalization in more detail. It is hoped that this will be published soon.

Editors' Notes

KP 278.2 (38 ss., T+H)

[1] See Bohm 1957a.
[2] See Feyerabend 1960b.
[3] See Popper 1959a.
[4] See Bohm 1957a: 145.
[5] See Bohm 1957a: 50.
[6] See Popper 1959a: 136.

Part III
Feyerabend and Kuhn

Chapter 9
The Feyerabend-Kuhn Correspondence (1960-1988)

Paul K. Feyerabend

1960

1. Feyerabend to Kuhn, [1960]

University of California, Berkeley, Department of Philosophy

⟨UNDATED⟩

Dear Tom,

(1) enclosed my comments on Hanson's "Logic of Discovery". The comments are only partly about the relevant passages in his book – in the main they are about an article he published in the Journal for Philosophy + about a very similar paper he read in Chicago. Please return my comments as soon as you have read them – they are the only copy I possess.[1]
(2) Could I please get a copy of the item you refer to in footnote 3 of your paper on the "Essential Tension"?[2]
(3) As regards this latter paper, I have one remark and one question.
The remark: On page 6 you deplore the absence of collections of readings in the natural sciences.[3] This does not seem to me to be entirely correct. Collections which come to my mind are (a) Beyer's collection of papers on Nuclear Physics (English, Italian, French, German); (b) Sommerfeld's collection of papers on Relativity (Lorentz-Einstein-Minkowski); (c) Wax's Collection of Papers on Stochastic Processes; (d) Schwinger's collection of papers on Quantum Electrodynamics, with its very interesting and highly critical introduction.[4] Again, when asserting the "convergent" character of the textbooks you have not mentioned that some [1|2] Encyclopedias, such as the famous "Blue" and "Green" Handbuch der Physik,[5] as well as the Encyclopaedie der mathematischen Wissenschaften ⟨sic⟩ (in which the

M. Collodel, E. Oberheim (eds.), *Feyerabend's Formative Years. Volume 2. Feyerabend on Logical Empiricism, Bohm & Kuhn*, Vienna Circle Institute Library 10,
https://doi.org/10.1007/978-3-031-57519-8_9

Ehrenfest's article on statistical mechanics originally appeared)[6] are written in a sometimes very different spirit. The same applies to the review-articles in Reviews of Modern Physics where all difficulties are explained.
(4) The thesis of your paper, if I understand it correctly, is this: revolutionary advances in the Greek Style do not occur until some kind of convergence has been established. I wonder whether it is not the case that here cause and effect have been interchanged. Could it not be the case that the first time a universal and testable (not metaphysical) theory is put forth which is detailed enough to account with precision for a lot of things, that in the very same moment many reasonable people will become interested, will try to work out consequences, formally improve the theory, to test it and that thereby a kind of convergence is established which of course will in its turn soon lead to the discovery of weaknesses of the theory and thereby to further revolutions.
If that is correct then the main emphasis should not be upon convergence, but [2|3] upon theories that are likely to bring about convergence, i.e. upon theories which are universal, testable, detailed and therefore, because promising to many, liable to create convergence, i.e. concentrated effort which of course will then lead quickly to new discoveries. Have I made myself clear? And am I right?

Kind regards
Paul

Editors' Notes

TK MC 240-21/28 (3 ss., H)

[1] See HANSON 1958a; HANSON 1958b; and HANSON 1961, respectively. The latter was read at 1959 Annual Meeting of the *American Association for the Advancement of Science*, Section L (see Vol. 1, Part II, 1960-1, n4), and was officially commented by Feyerabend (see FEYERABEND 1961).
[2] See RONCHI 1956; KUHN 1959: [231n3].
[3] See KUHN 1959: [228-229].
[4] See BEYER (ed.) 1949, SOMMERFELD (ed.) 1923, WAX (ed.) 1954; SCHWINGER (ed.) 1958.
[5] See GEIGER-SCHEEL (eds) 1926-33 and FLÜGGE (ed.) 1954-84, respectively.
[6] See KLEIN-MEYER (eds) 1904-35; HASSE-HECKE-DEURING-SPERNER (eds) 1939-67.

2. Feyerabend to Kuhn, [1960]

⟨UNDATED⟩

Dear Tom,

I have read your paper on thought experiments[1] but I shall not yet return it as I want to reread it most carefully. Let me only make some very general remarks. I found the paper extremely illuminating and I think you should publish it, for the philosopher's benefit, in either Philosophical Review, or Review of Metaphysics, or in Philosophy of Science. However my elation left me completely when I turned to the last chapter where you seem to suggest that the procedure adopted by those who discuss thought experiments is similar to the procedure adopted by the Ordinary Language Philosophers.[2] For I believe that the attitude is completely different! Let me start with a secondary remark, secondary, that is, to the adopted procedure. Hardly anybody within the Ordinary Language Circle is acquainted with the work that has been done with respect to the "puzzles" he is dealing with. I got the shock of my life to hear Austin seriously discuss the following "puzzle":[3] a dog D circles around a cow AB (A: the head of the cow).

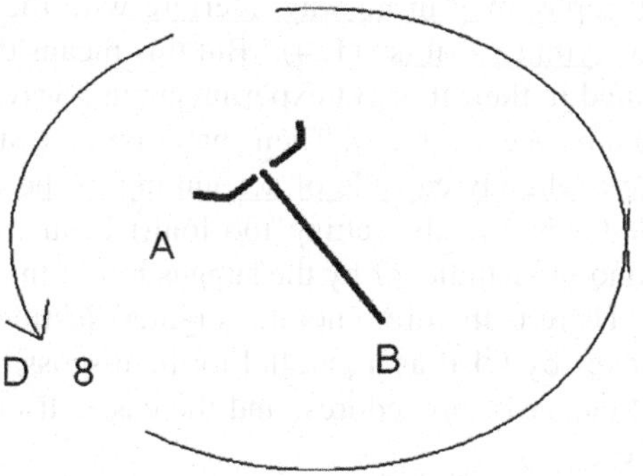

The cow, afraid of letting the dog get out of sight, turns round with him: does the dog, or does he not run "around" the cow? This was not meant to be an example of a thought experiment [1|2] carried out two thousand years ago, but an example of a "puzzle" to which a solution has not yet been found. Has this man, who is supposed to be one of the greatest philosophers of our time (how on earth could it happen that a charlatan became so famous?) never heard of the relativity of motion? Also, in his "Dilemmas" Ryle discusses at length Zeno's paradoxes with not the slightest reference to modern ways of resolving them (quite possible that he would not be able to understand any of these solutions!).[4] I think that it is not too far from the truth if we say that as regards knowledge of suggested solutions, knowledge of the history of a problem, knowledge of the particular circumstances under which it arose, it is not too far from the truth to say that with respect to all these things the Ordinary Language Philosophers are a bunch of Ignoramuses. (again ⟨sic⟩ the problem arises why they could become so famous). However this remark is only secondary. The most important difference between the OLPs and those scientists who frame consideration of thought experiments proceeded to a recasting of our ways of thinking about nature is the following. The scientist is aware that our ordinary means of expression may be faulty and may need revision. Once this need of revision [2|3] is recognized, and presented, most clearly, in an imaginary experiment, he will resolutely start out to tear down large pieces of common idiom and to replace it by what at first sight will seem a fairly artificial idiom, but what as a matter of fact will be much more adequate. For him a thought experiment is a starting point of the revision of language. Most OLP's ⟨sic⟩ I have read, or met, proceed in a completely different manner, and this by the way is also Wittgenstein's manner. They will consider paradoxes; quite correct. But first of all their paradoxes very rarely can be regarded as Centers ⟨sic⟩ of decisive innovations. And secondly the purpose is not revision of language, but realization that, properly understood, the language one is talking can already take care of the situation. Take Wittgenstein. What he thinks ought to be done (within what he calls "philosophy") is to "assemble reminders for a particular purpose" (Investigations, 127).[5] What is the purpose? The purpose is "complete clarity" (133). However how is this clarity to be achieved? Not [3|4] by retiring and completing the system of rules for the use of our words (133).[6] As a matter of fact "Philosophy may in no way interfere with the actual usage of language … it leaves everything as it is" (124).[7] But this means that the OLPs remove the paradoxes exhibited in their thought experiments in a way completely different from the way adopted by the scientists. Their intention is to show that without any change our language is already capable of accounting of the case. Without further argument (for my letter is already getting too long) I must say that this is done always either by a kind of swindle, or by the suggestion of most primitive and unilluminating theories. There is therefore not the slightest justification to consider the discussion of paradoxes by OLP as a parallel to the discussion of thought experiment by scientists. Standards, procedures, and the cases discussed are completely different in both cases.

Forgive, please, my strong words, but the fact that uninformed cranks are taken seriously has always aroused, first my curiosity, then my anger, then again my curiosity. However quite apart [4|5] from this fatal last section I think your paper a masterpiece and I am again curious about the fact that all you do with it is hide it in you ⟨sic⟩ filing cabinet and occasionally give it to your friends.

Paul

P.S. As a matter of fact, your paper could be a most important contribution to exhibiting the complete barrenness of OLP. This could be done by comparing discussions of the example of the inclined plane

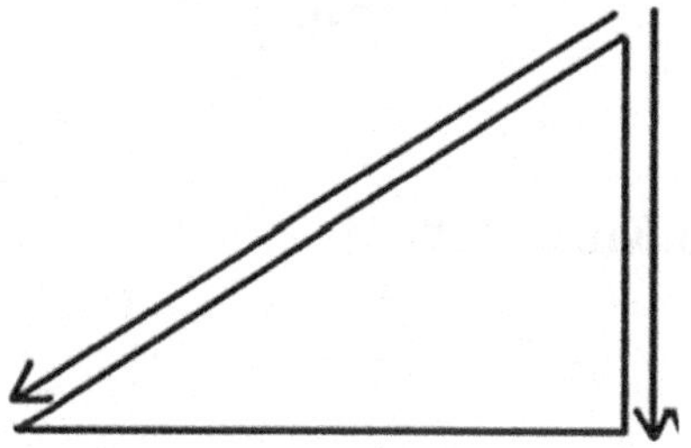

with a similar discussion of a concrete "thought experiment" by an OLPer. Such comparison would, I think, most convincingly show the tremendous advantage of the "Galilean method" of revision as opposed to the other method of talking around and around the problem until one has lost track of it (one has "dissolved" it).

Editors' Notes

TK MC 240-21/28 (3 ss., H)

[1] See KUHN 1964; and *below*, n2.
[2] Feyerabend is referring to a draft of KUHN 1964 since the section he comments upon and criticizes is completely missing from the published version of the paper. Most likely, the draft to which Feyerabend is referring is from 1960 or 1961, as KUHN 1964 refers to papers from the early sixties but only once and in a very superficial way to KUHN 1962 (see KUHN 1964: [262n31]).
[3] The British philosopher John L. Austin (1911-1960) was visiting professor at the University of California, Berkeley in 1958-59. On Feyerabend and Austin, see *above*, Vol. 1, Part II, 1958-12, n10.
[4] See RYLE 1954: Ch. III.
[5] WITTGENSTEIN 1953: § 127.
[6] WITTGENSTEIN 1953: § 133.
[7] WITTGENSTEIN 1953: § 124.

1961

1. Feyerabend to Kuhn, [1961]

⟨UNDATED⟩

Dear Tom,

my affairs are not yet settled, and will not be settled for another six weeks at least (unless a miracle happens). This means that I am still surrounded by an ocean of noise^{X} and not very capable of thinking.[1] However I cannot much longer keep back the comments I want to make on your essay. The longer I delay their deliverance, the more rebellious they become, and the more violent. Also there is a Kuhnian fifth column forming in my class (consisting, fortunately enough, of two members only) and the best thing to stop them is, I think, to stop their source. This is why despite my present predicament I still have decided to invade the lion's den with THREE COMMENTS, two critical, one not so critical.

(1) RELATION BETWEEN METHODOLOGY AND HISTORY. My main assertion is here: reference to history plays no role whatever in the arguments used for establishing and defending (or attacking) a certain set of methodological rules. Or, to express it in different words: history cannot help us a little bit in questions concerning the validity of such methodological rules.

Before going on I should perhaps make clear what I do not deny. I do not deny that history may make us aware of methodological rules of which we have not thought so far. Nor do I deny that history is capable of showing us which people have adopted which rules and who may therefore be regarded as a friend, or as an enemy. However I do deny that history can make a contribution to considerations of validity; that is, I deny that discovery of friends or enemies has got anything to do with the validity of the procedure shared by the friends, or disputed by the enemies. The reason why I deny this is very simple. It is due to my belief that the validity of any rule of procedure does not at all depend on the number of people who act in accordance with it, nor on the extent to which powerful or efficient institutions practice applying these rules (murder does not cease to be murder when practiced by the inquisition and rationalised by saying that souls are being saved that way). Imagine what would happen if the validity of certain demands (such as the validity of ethical rules, of methodological principles, of the rules of logic) were indeed dependent upon their incorporation into practice, or upon their realization in history. Anybody defending a different procedure could then argue for its validity only after he had converted a sufficient number of people, sufficient, that is, for making his school much more powerful and impressive than the school of his opponents. This being the case he could start collecting followers only by deception. But it is reasonable to assume that the Copernicans deceived, and deceived, and deceived, until they were finally strong enough not to need to resort to deception any more? And if this is not a reasonable assumption, does it then not follow that the arguments for, or against the validity of a certain procedure must be completely free from any reference to the extent to which these practices have been, or are being realised by some school, or community or other?

By the way, it is interesting to note that the idea of the methodological relevance of historical studies has a parallel in certain ideas that may be found in Wittgenstein: acc. to W. ⟨sic⟩[2] rules, whether methodological or otherwise, are such things as can be obtained by an analysis of existing practices or 'forms of life', as he also expresses himself [by the way –you should really at some time read Wittgenstein in order entirely to eliminate his influence]. What has no foundation in such a practice is a 'castle in the air' that should be abandoned in order not to disturb the firm ground of language, or the practices, upon which it is built. However, if such a 'castle in the air' is built by enough people, then this constitutes sufficient grounds for adopting the rules implicit in it. Again, the transition from a stage where the 'castle' was not adopted by many to its almost universal acceptance cannot be brought about by arguments, it can be brought about only by a kind of deception for which there are found various names [1|2] in Wittgenstein. This point of view of course makes it completely impossible to give a rational account of the development of our knowledge. More especially, it makes it impossible to write a rational history of any subject concerned with the advancement of knowledge [it does, of course, not exclude the possibility of a Hegelian kind of history, where ideas develop miraculously out of other ideas and where the arguments of individuals and thereby the rational side of the subject is more or less irrelevant]. This is the way in which too strong an emphasis on the role of historical research may lead to the deterioration of historical research itself.

Now let me substantiate what I have said in a more concrete manner. In my lectures on the history of epistemology (144) I usually proceed as follows: I start by giving an outline of two traditions which I call the hypothetical tradition and the dogmatic tradition respectively. In the dogmatic tradition a doctrine stands in the center. The aim is to apply this doctrine to all parts of life and to show how well it represents the way things are. The doctrine is never questioned. What is questioned is either man's proper understanding of it, or his capability adequately to apply it to practical matters. Changes of doctrine, if they occur, are due either to change, forgetfulness, misreading of earlier texts; or they are the result of the victory of a tribe subscribing to a different doctrine. A rational, i.e. argued development or improvement of doctrine does not, and cannot exist. In the hypothetical tradition a completely different attitude is adopted with respect to theories. It is realised that theories, being the result of human considerations will be as fallible as the human beings who created them. Hence, the main thing is to find out their limitations, to improve them in cases where they have been found wanting and in this way to come to a better and better understanding of the world. In order to facilitate improvement theories must be formulated in such a manner that inadequacies, should they exist, can be easily recognised, i.e. they should be formulated in a clear language that allows for easy testing of the theory, and so on, and so on. I also point out that the first people to introduce, and to stick to the hypothetical tradition were the Ionian philosophers of nature (in this respect I agree with both Popper's article 'Back to the Presocratics' – Presidential Address of the Arist. Soc., published in the Proceedings 1959; and I also agree with the very similar point of view which Wally Matson has developed in his review of Cornford's Principium Sapientiae[3] and in other papers of his; only, he makes the Ionians inventors of the method connected with cosmology, and not with the method characteristic for all parts of the sciences). That is, these philosophers not only invented cosmological theories, they also invented the method proper for the discussion of such theories, i.e. the method of introducing, and constantly improving (by criticism) hypothetical conjectures. When I say that they introduced this methodology I do not mean to say that they explicitly stated its rules. This was not the case. They introduced this methodology of test and improvement implicitly by creating a tradition, which was very different from the traditions then prevalent, where the doctrine did not form the center of understanding and was supposed to remain untouched, but where it was the starting point of criticism which latter ⟨sic⟩ was regarded as essential. It seems to me that it is the introduction of this method rather than the particular theories used in accordance with this method which must be regarded as the most essential contribution of the Greeks. When we say that we owe our knowledge to the Greeks, we should think of this feature above all. [2|3]

Now having introduced these two traditions, and having discussed early instantiations of either of them, I then trace their development to more recent times. More especially, I point out that science belongs to the hypothetical tradition (specifications will be made later on)⟨*⟩[E]. In this way I try to show that the two traditions with their so very different approach to almost all problems still prevail and that the question as to which tradition we shall support is a very real question. It is also the most fundamental question of all epistemology (philosophy of science, of course, included). It is, to repeat, the question as to which tradition we hall support; or, to formulate it in a way that makes very evident its independence of the existence of either of the two traditions: which methodology shall we adopt, and defend – the methodology of proof in a dogmatic system, or the methodology of the improvement of hypotheses. Still another formulation of the question is this: shall we choose absolute truth and a method leading up to the final establishment of the one, true system; or shall we choose hypothetical truth and a method which enables us to find out the shortcomings of our hypotheses.

Before this question is answered we must become clear about the status of absolute knowledge and of hypothetical knowledge. If we assume, as is done by many philosophers, that absolute truth is there in the world and is adequately represented by absolute knowledge, if this is our assumption, then we shall have to take this into account, and we shall have to admit that absolute knowledge is required in order adequately to describe part of the world. On the basis of the same assumption the hypothetical tradition may be then said to represent only part of what is there; it is incomplete and, if it is supposed to be the only tradition worth considering, prejudiced. Now I argue that certainty, or hypothetical character is a formal characteristic of our theories rather than a material characteristic of a part of the world, a characteristic, moreover, which cannot be defended by reference to any features in the world. Of course, once we have adopted the dogmatic procedure, in the very same moment we shall discover a host of 'absolute truths'. But the important thing is that these 'truths' will be discovered only after one has decided to proceed in the dogmatic manner. Reference to their external existence therefore cannot be regarded as a motive for adopting this procedure in the first place. Or, to express it in a more intuitive manner: certainty, or hypothetical character of a system of knowledge are characteristics of this system that are entirely manmade. This being the case the decision to be made between absolute knowledge and hypothetical knowledge is a decision that cannot at all be supported by reference to characteristics of the world we live in. It is a decision which we have to make prior to any factual discovery and which, if followed through, uniquely determines the kind of factual knowledge (or of knowledge) we are going to obtain. As will turn out a little later this decision influences our attitude to human beings. It is therefore an ethical decision.

To repeat: prior to any factual inquiry a decision has to be made concerning the method to be followed. As this decision makes factual inquiry possible, it cannot be derived from factual arguments. But how shall we now proceed?

The procedure is not too difficult. One step consists in pointing out that absolute knowledge cannot be factually relevant. Other steps consist in pointing out the consequences for man, and for society, of the adoption of the one or the other procedure. It will be pointed out that the absolutist tradition, if it is to be successful, must not allow democratic institutions. It will be pointed out, that the hypothetical tradition increases freedom, power of imagination, and so on. On this basis the decision is made (by myself) in favor of the hypothetical tradition [footnote: all this has far-reaching consequences: for example I shall show in my lectures that having chosen the hypothetical procedure one will have to defend Brecht's discursive and non-aristotelian drama as opposed to the usual dramatic form; hence the decision is not a narrow, and specialised decision; it is bound to affect our whole life, our relation to others, and so on. But this only by the way]. We have therefore decided in favor of the hypothetical procedure. [3|4]

Now please note that historical considerations did not at all enter the considerations leading up to this decision. Of course, I started with the discussion of the two traditions which is a historical matter. But I did so only in order to make it very clear that a very fundamental decision is demanded and not in order already to decide one way or another. The existence of the two traditions presents us with the problem: whom shall we support? The one, the other, both, or neither? And in the discussion of this problem the existence of these traditions, the number of their followers and all these things are entirely irrelevant. Of course, once we have made the decision, in this very same moment we shall look at history, present or past in order to find out who our friends are, and who our enemies are. If there are no friends, then we shall try to persuade people to our point of view and in this way create a tradition just as the Ionians created the hypothetical tradition with its particular ways of argument, with its particular relation between master and pupil (the master encouraged the pupil to criticise him; he does not demand from him absolute following as seemed to be the case in the Aegyptian priesthood). If there are friends, we shall join them. If there are friends who, however, are not quite sure in their ways and who show inclinations to become dogmatic, then we shall criticise them severely. In any case, what happens in history will not at all influence the decision we already made. It will only influence our behavior *on the basis of this decision*. Hence, historical matters are completely irrelevant in matters concerning the validity of this decision.

This brings me to a very important point in your essay. More than once you emphasize (I do not remember your exact words) that *existence of a discipline*, such as physics, *does not presuppose the existence of a set of rules*. More specifically you say that the continuity of the development of physics is guaranteed already by the existence of institutions, of a tradition in the sense of membership in certain organizations, publication in certain journals, and so on, and that it is not based upon the existence of a set of methodological rules to which all participants of the tradition conform. This latter point may also be made by pointing out that continuity of development does not presuppose preservation of *any* rules throughout the course of the development. Now this point I am ready to concede (although I think there will be difficulties if one tries to formulate it in a more precise manner, but I shall neglect these). That is, I am ready to concede, for example, that the feeling of belonging to a certain tradition may exist in a group of people whose beliefs have nothing whatever to do with the beliefs of those who according to common opinion originated the tradition. Or, to express it in different words, I believe that preservation of doctrine, or of method is not a necessary presupposition for certain people's feeling that they belong to a very powerful tradition. Examples can be immediately given; the one is Christianity (the development from primitive Christianity to, say, the Roman Catholic Church); another example is Marxism; still another example is Greek philosophy (a continuity is perceived from the Ionians through Plato to Aristotle although methodologically the latter have very little in common with the former); also Descartes is by some felt to be the forefather of existentialism; and so on, and so on. But all we can say here is that sometimes people *think* they have something in common with some of their predecessors when they have nothing at all in common. Of course, the fact that people think they have something in common with some earlier thinkers will be expressed in their actions; they will share certain superficial similarities with these predecessors (such as similarities of terminology, or similarities of institutions) – but even this is not essential as the example of the Roman Catholic Church shows. The only connecting link between the early forms of Christianity and the latter is the fact that both put emphasis on the bible – but they interpret it differently, give it a different place; that is, the difference could not be greater. Now I do not think that it is a sound principle for a historian to make the mistakes of the people he investigates the basis for the selection of his material. That is, he should not make the results of mistaken assumptions of identity and continuity *his* criterion for identity or continuity. To be more specific: assume that physics starts becoming dogmatic and that it develops some doctrine which by many physicists is held to be fundamental for research and which they are not prepared to give up, *come what may*. The mechanicism of the 19th century provides an example of what I mean (think of the much quoted passage from the *Erhaltung der Kraft*).[4] Physicists may still assume, mistakenly, that they belong to the same tradition as belonged Galileo and even Faraday. It will *therefore* still be possible to expound the new dogmatism in the usual physical journals (as it is possible, for Bohr and others, to [4|5] expound their dogmatism on the pages of Phys. Rev.)[5] Note, that the possibility of doing this and the continuity of practice (of publication, of getting a salary as a physicist rather than as a dogmatic ontologist) following from it is the result of the mistaken belief that what Bohr does (I am now thinking especially of his answer to Einstein in Phys. Rev. Vol. 48)[6] and what Pauli does (some passages in his Handbuch article),[7] and what Heisenberg does (his contribution to Niels Bohr and the Development of Physics)[8] is not very different from what Faraday did, what Einstein did, and what all the earlier physicists did. Now look what happens when a historian enters the scene and proceeds as I think you do proceed. He will notice that Bohr's "rules" are not exactly identical with Einstein's "rules" and that the latter are again very different from those to which Helmholtz and Kelvin adhered. As a matter of fact he will notice that there is a very great difference between any two of these thinkers. He will also notice that both Bohr and Einstein publish in Phys. Rev. (that is, it is not the case that Bohr publishes in Phil. Rev. and Einstein in Phys. Rev. or the other way round) and that both of them occasionally make reference to observational results, and that both of them are paid as physicists and listened to at physical congresses (where philosophers are usually ridiculed). His first discovery (difference of rules) apparently clashes with his second discovery (unity of tradition) unless the assumption is given up that a tradition presupposes the existence of stable rules – and this is how I think you have arrived at your result. But this result is not correct as your second discovery is not correct. What criterion would you apply when establishing unity of tradition? The criterion which you apply is identical with the criterion of the physicists who believe that Bohr and Einstein are doing the same thing and who are mistaken. Now their mistake has very interesting consequences. It has the consequence of creating an outward and apparent continuity (such as usage of instruments, journals used for publication, communities established). This continuity, which is the result of a mistake of judgment on the part of physicists you take as your measure. Small wonder that you arrive at this astounding conclusion that one does not need to follow any rules if one wants to be a physicist. Of course, as a physicist who makes mistakes of judgment one does not need to follow any rules. But assume that the mistakes are corrected; and assume that the tremendous difference is recognised, and emphasised, between Bohr and Einstein (the former belongs to the dogmatic tradition, the latter belongs to the hypothetical tradition). Then, of course, the continuity will not come into existence as people will refuse to do what Bohr is doing. – All this brings me back to my initial point: you want to give history a prominent position in considerations methodological; as a matter of fact you want to found methodology upon considerations historical (history, you say, is the material for methodology). History, for you, comes first. But the result of this attitude is now that you are not any longer able to give a reasonable account of historical developments. Instead of analysing the reasons for continuity you take continuity, which, as I have tried to show, is the historical result of an error of judgment in those creating the continuity, as an unanalysable thing and you are therefore prevented from giving a rational account of the history of the sciences, i.e. an account which includes all motives, mistakes included. That is, your over-evaluation of history leads to something like historical irrationalism, where things develop, where this development is the measure of truth and unity, but where we cannot understand the reasons which brought about the development. This is just as in Hegel: everything is history, and history is a mess.

Assume, however, that methodology is given prevalence. That is that we emphasise different possibilities of procedure and the importance of the difference between these possibilities. We shall then be able to put a well defined yardstick to history. We shall be able to see how certain ideals are given [5|6] ⟨up⟩^{E} partly because of opportunism, partly because of sloppiness and we shall also be able to judge the mistakes of those who assert that there is an unbroken line from Galileo to Bohr. We shall be able to discover when theology is preached in the trappings of physics and we shall also be able to raise a warning voice when the first signs of such deviations become apparent. We shall not any more be impressed by existing continuity and choose contemporary science as an example just because it has somehow developed out of the ideas of Galileo (or of More?). We shall be able to criticise, and thereby to advance the cause of science because our yardstick is not part of the very indefinite mass of historical facts. What this difference amounts to, i.e. the difference between you and myself is of course the difference between immanent, and therefore relative values on the one side; and transcendent, and therefore absolute values on the other. I myself prefer the absolute point of view not only because of its absolutism, but because of the clarity it leads to in history, science, and everywhere else. I do hope I have made myself clear on this point. And if I have – do you agree?

(2) FALSIFIABILITY. It is on this point that I got reactions from my students. As you may know I have abandoned questionnaire-midterms. This is too uninteresting a kind of examination. For the important thing is not to be able to give answers to questions; the important thing is to be able to select problems. Hence I ask my students to write about a topic of their own choice which to them seems to be important and somehow related to the topic of the course. Two papers which I got seem to be inspired by your teaching and they made it very clear to me what the issue really is (by the way, the papers were quite good, though not very critical). I shall split discussion of this issue into three parts viz⟨.⟩^{E} (a) relation between method and actual practice; (b) proper interpretation of what actually happens in the sciences; (c) possibility of doing science on the basis of the methodology of falsifiability.

Regarding the first point I have already made clear my position: if it should indeed happen that scientists do not take falsifications seriously and are content with ad hoc hypotheses, if this should indeed be the case, then I would say "the worse for the scientists". Such procedure would be an indication for me that scientists have become dogmatic and this would finish them for me. (I am not interested in dogmatic systems, even if such systems are taught in physics departments, and even if they get the Nobel-prize which would then correspond to the donation of a bishoprice, ⟨sic⟩ or some other remunerative thing in the Roman Catholicism of the Middle Ages). By the way, it should be clear what I understand by an ad hoc hypothesis. Copernicus' assumption to the effect that the fixed stars are very far away is not an ad hoc hypothesis as it admits of independent test. An ad hoc hypothesis concerning the theory T is a hypothesis whose only criterion of truth is that it would be incorrect if the hypothesis were not true. Such hypotheses are characteristic of dogmatic systems. They abound in primitive myth where the myth is indeed the measure of truth. They are not frequently found in the sciences. A good instance is the later form of Lorentz' theory which includes not only the Lorentz-contraction, but also time dilatation and all the other effects to be expected if an aether exists. In this later form Lorentz' theory is indeed nothing but a hypothesis which preserves the aether assumption without adding new tests to the theory. Hypotheses of this kind must be excluded (I think that the impetus theory is of a similar kind). But, as I repeat, they are very rare within the sciences. But if they should occur my attitude would be as indicated: too bad for the sciences. And I would start becoming one of the most violent critics of scientific procedure thus degenerated. And I think that you, too, would join me in my criticism. For the cases which you do discuss are not at all of the kind envisaged here. However they cannot be regarded as an objection against the methodology of falsifiability either. As a matter of fact they are perfectly compatible with this methodology. This brings me to the second one of the three points separated above. [6|7]

In your essay you point out more than once that falsification is too simple and naive an account of what happens in the sciences. I do not remember your exact words, but this, at least, is their spirit: when we investigate what happens to theories we do not find that they are first introduced, and then given up immediately after the first contrary experimental result. They are retained a considerable time, people try to adapt the experimental results to the theory, or they simply wait, hoping that at some future time a satisfactory way will be found of incorporating the resisting experimental results. Nor is the theory given up if it is found to be in real trouble, unless a new theory is available which can take over. This being the case, you infer, the idea of falsifiability gives much too superficial and naive an account of actual scientific practice. For this idea, so you seem to assume (and it will turn out that it is this assumption, which is erroneous, which leads you to criticising the theory of falsification) demands that a theory be given up immediately after the first falsifying instance is found. Not only is this not the case, you say, but it is also desirable, that it should not be the case (item c above): for if we were indeed to proceed in such a manner that we give up a theory as soon as a difficult case arises, if we were indeed to abandon a theory as soon as the first falsifying instance turns up, then theories would come and go, they would never be worked out in detail, they would never obtain that imposing shape we know from classical mechanics, from classical electrodynamics, nor would they be ever developed in such detail as to give us a universal picture of the world. The idea of falsifiability, you conclude, is therefore not only naive as regards the actual procedure of science; adoption of it would also be detrimental to the state of the sciences; and it would be detrimental to the state of science as falsifying instances always turn up almost immediately and as no theory could therefore live very long and be developed in sufficient detail. This, so it seems to me, is your position. My reply is that this argument of yours is due to too naive an interpretation of Popper's methodology. Hesitation to give up a conceptual scheme, retainment of it even after decisive falsifications have occurred and all the other instances you refer to are perfectly compatible with the idea of falsification and perhaps even demanded by it. The only thing that is not compatible with the theory is conscious use of ad hoc hypotheses in the sense explained above and refusal to worry in the face of admittedly refuting instances. It is true that these demands are very modest, and almost trivial and at some time in our discussion you have said to me that you find them trivial. However it must not be overlooked that despite their triviality these demands are consciously violated by all members of the dogmatic tradition where the theory is the only measure of truth and where contradicting statements, even if they should happen to be empirical, are excluded just because they contradict the adopted theory. But let me now analyse a little more in detail the historical cases which you have adduced with the purpose of showing the limitations of the idea of falsifiability and let me also discuss some of the points which arose in our oral arguments. To repeat, my contention is that all your findings are perfectly compatible with the idea of falsifiability and that even the demand not immediately to give up a theory may be made compatible with it.

I shall proceed in small steps in order to make things as clear as possible. Let me therefore start with the most extreme example. Let me suppose that a theory T is found to be in trouble and that it is admitted by everybody that it has been refuted by the facts. This was the situation in classical mechanics before Bohr had started his investigations. You assert that even in this case T will be retained unless a new theory has been found which takes its place. My reply is (α) that it will be retained even after such a new theory has been found; and (β) that such a procedure need not be incompatible with the idea of falsifiability. Concerning (α) I need [7|8] only point out that after all we still use the classical mechanics for the calculation of the behavior of the upper planets. But (β) such use is perfectly compatible with falsifiability. For this idea does not deny that a false theory may be a good instrument of prediction, and it therefore does not prohibit its continued use as such. It only asserts that a refuted theory must be regarded as false. What it forbids therefore is drawing realistic consequences from a theory that admittedly has been falsified. As soon as a theory has been falsified we must stop making it the basis of our interpretation of the world. On this point we must completely suspend judgment until we are able to understand what are the reasons for the failure of the theory. The theory is therefore finished in one sense: it is finished as a picture of the world. Which, as I said above, does not preclude its use as a predictive instrument or as a heuristic guide in the search for a new theory.

However the cases which you discuss are not cases where a theory has been admittedly refuted. They are cases where certain difficulties exist but where it is not yet admitted that these difficulties amount to the refutation of the theory. Such a situation is not excluded by the idea of falsifiability. Quite on the contrary: the idea of falsifiability demands that criticism be applied not only to the theories we possess, but also to the falsifying instances which we believe we possess. And a prolonged hesitation with respect to the latter simply means that we are not yet convinced that their falsifying character has been definitely established. Such prolonged hesitation and the attempts to make the instances compatible with the theory, all this amounts to vigorously testing the apparently falsifying instances themselves. If all such attempts misfire, then the falsifying instances have been confirmed to a very high degree and therefore very decisively argue against the theory. This being the case hesitation, and the attempt to show that what prima facie seems to run against the theory as a matter of fact is perfectly compatible with it ⟨sic⟩ must be regarded as completely in accordance with the idea of falsifiability. For this idea says that a theory should be given up (that is, that the realistic interpretation of a theory should be given up) as soon as a falsifying instance has been discovered. It does not say that the theory should be given up as soon as the first inconsistency with facts arises. Quite on the contrary, it demands that a falsifying instance should not be trusted more than a theory to be tested, unless it has been subjected to at least as vigorous tests as the theory itself. In other words, it demands that it should not be accepted too lightly. It should not be accepted before it has been tested in all possible ways. And testing a falsifying instance in all possible ways is just the same as trying out all possibilities of making it compatible with the proposed theory. This is why the attempt to try out a theory as much as possible is compatible with the idea of falsifiability. You seem to assume that accepting this idea means giving up a theory on the slightest pretext. But doing this would mean accepting some other lower-level hypothesis on the slightest pretext, viz. the refuting instance. And this we shall not do once we have decided vigorously to test all elements of our knowledge. Also Popper's theory of falsifiability makes not special assumptions concerning the way in which falsifying instances are produced. We both agree that sometimes a falsifying instance can be produced only by confronting the theory under investigation with an alternative theory; and as a matter of fact I (and perhaps also you) would go even further and say this is what always happens (although the alternative theory very often is not explicitly formulated, but is part of our 'background knowledge'). The general theory of falsification has nothing to do with this specific ways ⟨sic⟩ of producing falsifying instances. It makes this trivial, but still important statement that once a falsifying instance has been produced the theory falsified will have to go. Taking all this into consideration I cannot [8|9] agree with you when you say that adopting the theory of falsification would mean giving up theories on the slightest pretext, that theories are not given up on the slightest pretext, and that therefore falsification is too narrow an epistemology. Such a criticism applies to any theory of basic statements: if we assume that there exist absolutely certain statements of observation, and that the theory is about them, then 'a disagreement with facts' indeed would be the end of the theory. But if we drop the idea that statements of observation can be established with certainty, if we admit that any scientific statement is but a hypothesis, in the very same moment we may try to retain the theory in the attempt to test the test statement itself. One thing, of course, must not happen: it must not happen that we take the truth of the theory as a criterion (and perhaps the only criterion) of the truth of the test statement. This would mean accepting dogmatism.

A final remark about what one of your students wrote in his paper. It is essential for Popper's theory that we must not attempt to show that a theory is correct. What we must do is rather to test it; and as every test is an attempt at falsification we may therefore also say that what a scientist tries is to overthrow the theories he possesses. To this your student objected when a man who deals with a theory and wants to use it for giving an account of what happens in the universe must first of all believe in it. He emphasises that it is psychologically impossible for a person to be concerned about a theory and at the same time to try to overthrow it and that for this reason Popper's methodology cannot work. I also got the same argument in class with reference to a passage in your 'Copernican Revolution' where you say (I quote) that "a scientist must believe his system before he will trust it as a guide to fruitful investigations of the unknown". Now this is of course very correct, and almost trivial. A scientist will consider a theory only if he is interested in it (by the way – he does not at all need to believe it, he may try out a theory he does not believe in in order to check his non-belief and discover to his surprise that it works). But at the same time he knows that his interest, and his belief is not sufficient to guarantee truth. He must therefore test the theory and he will test it exactly because he believes in it and because he is interested in it [we may perhaps test the virtue of a woman we want to marry; but we shall not test the virtue of a woman in whom we are not interested; in the latter case her faithfulness is of no interest to us; as a matter of fact we could not care less]. Now what does it mean to test the theory? A test is an experiment one of whose outcomes would refute the theory. Hence a test is an attempt to overthrow the theory. This attempt is of course undertaken with the belief, and the hope that the theory will come out alright. But a possible overthrow it is nevertheless. For if it would not be a possible overthrow there would have been no need to carry out the experiment. We could have foreseen its result in advance and we could have outlined in advance the structure of the world as seen by the theory. However apart from being possible refutations the tests, if successful, i.e. if confirming the theory, have also another function. They are successful explorations of the world according to a definite map, i.e. according to the theory. They therefore shape a certain picture of the world; this picture will be more definite and clear the more decisive the test and the more confidence we may possess in the correctness of its outcome. I.e. this picture will be the more definite, the more critical we are with respect to the outcome of the test itself. This is how belief in a theory leads to attempts to overthrow it and how attempts to overthrow a theory lead to the construction of a picture of the world which becomes more and more definite until finally we as it were see the world from the point of view of the theory in a very similar manner as Kant thought we always see the world from the point of view of the Newtonian framework. [9|10]

I do not know whether I have made my points clear. If I have succeeded in doing so you should have realised that the description of scientific development which you give is not only not incompatible with the idea of falsifiability, but that such development is even demanded by the idea of falsifiability.

There are various other things I would have like ⟨sic⟩ to say, but I leave all this for later. I shall send this letter by a student. At the present moment I am confined home with a swollen foot (infection), an affair which may easily last another three days. I am curious as to your reactions.

Be well!

Paul

---

[1] X since I gave notice the noise seems to have disappeared! I am not quite clear as to what are the causes of this phenomenon.
[3] ⟨*⟩E alas, the letter grew too long, so the specifications could not be made. But I did make them in class.

Editors' Notes

TK MC 240-21/28 (10 ss. T+H)

1 Feyerabend is probably referring to concerns that eventually led him transfer from Benvenue Avenue, Berkeley to Harwood Avenue, Oakland. In his autobiography Feyerabend reports that from his arrival at Berkeley and up till the late 1960s, he lived in four different places: first "a smelly room in a flophouse on Telegraph Avenue"; then "a small apartment atop a set-back garage on Hillegass" that eventually "the landlady took over"; finally, "[a]fter a brief stay on Benvenue, [he] rented a huge wooden box at 6041A Harwood Avenue in Oakland" (FEYERABEND 1995: p. 113). Feyerabend's correspondences and diaries allow to identify the time of his relocations rather precisely. In particular, he moved to Benvenue Avenue in the summer of 1960 (see Vol. 3, Ch. 1, Feyerabend to Watkins, 6 September 1960; Ch. 3, Feyerabend to Lakatos, 11 October 1960) and to Harwood Avenue on 15 June 1961. The reason behind the latter quick relocation was the noise that Feyerabend experienced on Benvenue: this was so unbearable for him that he took the decision to leave as early as October 1960 (see Feyerabend's Notebook 1960-1968, PF 3-5-1). This would imply that the letter was certainly written before mid-June 1961, but not too many weeks earlier. Indeed, if Feyerabend's expectations expressed there were approximately correct, Feyerabend (2006b) dates to late April-early May, namely, at least a couple of weeks after Proto-Structure was mimeographed (cf. above, p. 6 and n17). On the other hand, Feyerabend (1995b) was undoubtedly written before November 1961 (cf. Feyerabend, 1995b, p. 372).
2 Abbreviation for "according to Wittgenstein".
3 See POPPER 1959b; MATSON 1955; and FEYERABEND 1962b: 73-74 [*removed*].
4 It is not clear whether Feyerabend is referring to HELMMHOLTZ 1847 or to HELMMHOLTZ 1871.
5 Standard abbreviation for the title of the scholarly journal *Physical Review*.
6 See BOHR 1935.
7 It is not clear whether Feyerabend is referring to Pauli 1926 or to Pauli 1933.
8 See HEISENBERG 1955.

2. Feyerabend to Kuhn, [1961]

⟨UNDATED⟩

Dear Tom,

I shall start my comments on your essay with some general remarks which will in part repeat what I have told you before. I more than ever think that your essay is quite unique as regards the contribution it makes both to the history and to the philosophy of science. On the other hand my impression of danger, and my misgivings have been very much increased. What you are writing is not just history. It is ideology covered up as history. Now please, do not misunderstand me. I do not say that we should live without ideology – quite on the contrary, I think that living without an ideology (the correct one, of course) would be both impossible and inadvisable. Nor do I say that history should not be written without a point of view in mind – quite on the contrary, I think that such a history, if it would be possible, would be the most drab and uninteresting affair imaginable. Nor, thirdly, do I pretend that in history a nice distinction can be drawn between what is regarded as a factual report, and what is regarded as interpretation according to some point of view. But points of view can be made explicit, and it is possible to write history in such a manner that the reader is always aware of one's ideology or point of view as well as of the possibility of an alternative interpretation of the historical facts. That is, history can be written in such a manner that what is factual and what is reasonable appear as two clearly distinct affairs. In some cases it is very easy to keep this distinction clear. Nobody will think that the history of crime justifies crime, or shows that crime possesses an inherent 'reason' or an inherent morality of its own. In the case of the sciences or of other disciplines of which we have respect the situation is much more difficult and the distinction cannot be drawn with equal ease. But in these cases it is of paramount importance to make the reader realise that it still exists. You have not done so. Quite on the contrary, you use a kind of double-talk where every assertion may be read in two ways, as the report of a historical fact, and as a methodological rule. You thereby take your readers in. You present your material in such a manner that (at least for the periods following the introduction of paradigms) history seems to satisfy the principle that ALLES WIRKLICHE IST VERNUENFTIG[1] so that evaluations can then be directly derived from historical study. It is this way of presentation which I find objectionable. I do not object to your belief that once a paradigm has been found a scientist should not waste his time looking for alternatives but try working it out. That is I do object to this belief, and I shall have very soon to say a little more about it. What I do object do ⟨sic⟩ most emphatically is the way in which you present this belief of yours; you present it not as a demand, but as something that is an obvious consequence of historical facts. Or rather, you do not even talk about this belief, you let it as it were emerge from history as if history could tell you anything about the way you should run science (is does not imply ought!)⟨.⟩[E] It is this bewitching way of representation to which I object most, the fact that you take your readers in rather than trying to persuade them. This manner of presentation you share with Hegel and with Wittgenstein; and with all those who say, when engaging in a political enterprise, that "history will be our judge". Do you really believe this? (You differ from them only insofar as you refer to the past whereas they refer to the history of the future.) Or have I perhaps completely misunderstood you?

Just a further remark before I turn to the more specific points. You have expressed to me your belief, and you express it again in your essay that it is only by concentrating on a single paradigm, by trying to fit nature into it despite all apparent difficulties, that scientific progress is achieved. You seem to think it psychologically impossible for a scientist and, indeed, for any human being to be able to entertain various [1|2] alternative hypotheses and to discuss them impassionately. I think you are a little too pessimistic. Faraday did so, if I am correctly informed, and so did the Presocratics, so did Einstein (partly classical, partly non-classical considerations). Also I think I have shown in my own essay[2] that considering a set of mutually inconsistent but factually adequate theories increases the empirical content of any element of the set and this for the simple reasons that many tests presuppose the existence of an alternative! (they are crucial tests). If this is the case then we must make the decision what do we prefer, increased empirical content of the theories we possess, or that unanimity of research and the close fitting produced by it in the periods which you call the normal periods. History cannot help us in this decision. Many scientists seem to prefer the latter alternative – but quite obviously this decision of theirs is not binding to anybody – unless he wants to be a member of one community rather than of another. But advance of knowledge, so I would have thought, has nothing to do with membership in communities (Wittgenstein notwithstanding).

I now turn to the details. page 1, line 4: "Image of science"[3] – this very difficult affair for science has always been at least two things⟨,⟩[E] viz. an intention to find out about the universe, and a practice which is just the result of the attempt to realise the intention. A scientist wants knowledge and he thinks that this is something good and valuable. And if he knows a little, he wants to improve what little he knows, he wants to know more. He wants to conquer fear and superstition. All this is part of the "image of science". But in the attempt to carry out these intentions a practice develops. Institutions are founded with laws of their own. And it may well be that the practice deteriorates and completely gets removed from the original aim⟨,⟩[E] viz. of carrying out the said intentions. The ideal which the scientist has in mind may be something extremely valuable⟨,⟩[E] something which every human being should try to achieve. The practice, on the other side, may be corrupt, degenerate, and not fit to realise the ideal. Therefore we must speak of at least two things when speaking of "the image of science". History can give us only one of them and this is why every historical account of the sciences is bound to be incomplete. "The monograph" you say "is a sketch of the quite different conception of science that can emerge from the historical record of the research activity itself".[4] This conception, this is my main contention, may be completely misleading as it may tell us the results of the degeneration of a valiant effort to achieve something very valuable. Clearly the attempt to know the universe may degenerate gradually and it is quite conceivable that the way

in which integrals and other mathematical symbols will be used at some future time will be very similar to the way numbers were used by the number mystics (this is already partly the case). Shall we, then, regard this as the most "modern" "image" of "science"? One could of course make this assertion but it would be very clear that the words would then have lost any meaning. I repeat, therefore, that an investigation of the whole research activity is not enough and does not tell us all about the sciences. It is much more than what can ever emerge from an investigation of "finished research achievements"– but it is still far from being enough.
page 3, line 7:[5] we must take into account also the methodological difference. That is much of Aristotelianism does not only offer a description of nature which is different of ⟨sic⟩ the description accepted today, or in the 17th century. This Aristotelian description is also handled in a completely different manner, according to a different methodology. I am very much against calling a discipline "unscientific" or "mythological" on account of the fact that what it says about nature is different from what we are inclined to say today. But if a discipline is built up in such a manner that any possible occurrence is bound to confirm it, i.e. if it is built up in such a manner that one can (and does) say in advance of [2|3] the carrying out of tests that it will be, and must be successful, then I perceive a great difference indeed between this discipline and science today. Superstitions are that way. You cannot dispute them away and contrary evidence will be handled in a manner which finally turns it into confirming evidence. It is for this reason (and not because it makes reference to ghosts etc.) that I would like to keep it apart from today's science. Hence your assertion to the effect that "Aristotelian dynamics .... [was] neither less scientific nor more the product of human idiosyncrasy than our own"[6] does not seem to me to be completely correct. Of course, it is always possible to take a theory⟨,⟩[E] however dogmatic, to turn it around and to use it in such a manner that it is capable of refutation by experiment. This is what Galileo did with the Aristotelian theory of motion. But the important thing is that many Aristotelians used this theory in a completely different manner, not as something whose validity should be found out by repeated test, but as a basic theory that could not possibly be false (I always thought that the impetus theory, despite its ingenuity, was nothing but an ad hoc manouvre ⟨sic⟩ designed to save the basic axiom that any motion needs a mover).
page 4, line 3 of second paragraph: "insufficiency of methodological directives etc."[7] This is completely correct. But it should perhaps be pointed out that methodological directives are necessary in order to guarantee that what has been introduced in a very arbitrary manner "compounded of personal and historical accident" (page 5, line 12)[8] is not just a fairytale⟨,⟩[E] but a description of properties of nature. The fact that scientific method is not sufficient for arriving at a certain belief, theory, piece of knowledge is reflected by the existence of a variety of schools and the existence of "incommensurable ways of seeing the world and of doing science in it" (page 5).[9] But this variety cannot be unlimited, and the limits are set by the methodological rules which exclude some of the "ways of seeing the world" on account of the fact that they are not about the world at all, but interesting fairy tales (their "logic" is different from the "logic" of scientific theories). Also whom are you attacking on pages 4 and 5?[10] Who, apart from some very radical Baconians,

Cartesians and Platonists has ever believed that the correct method will lead to the one and only correct theory?
page 6, line 7 for me the passage "so long as those commitments retain an element of the arbitrary"[11] is most important. Only on that condition can I swallow the "suppresses fundamental novelties".
page 7[12] I would like to add here that a scientific revolution leads to an increase of empirical content of the theory that is being replaced. The practice of repressing novelties, the attempt to let the theory penetrate every part of nature (and of thinking) usually leads to a situation where people are psychologically unable to imagine alternatives, where procedures such as transcendental deduction will be very successful and where the impression arises that "we are always chasing our own tails in a formal system" (as Eddington has expressed it). This practice therefore finally leads to the establishment of a language which can be used for the description of every and any situation and a situation which does not fit the theory cannot even be imagined. This means, however that this method of excluding novelties, this attempt to press nature into the boxes of the theory will gradually decrease the empirical content of the theory until it is finally almost zero. A scientific revolution which shows the limitation of the theory and which points out very plainly where it is wrong therefore gives back empirical content to the theory, i.e. makes it again a theory of nature. Conclusion: better live in permanent revolution than in the state of normalcy. Revolution in Permanence should be the battle cry of every empiricist. The 'close fitting' between theory and nature [3|4] which normal science tries to achieve decreases the empirical content of the theory and thereby also decreases its ontological significance. 'Normal Science' therefore tries to achieve something which is very undesirable and its undesirable aftereffects ⟨sic⟩ are eliminated only by a revolution⟨.⟩[E] The closer we try to get to nature (by the method of trying to fit nature into boxes designed by us) the further we get away from it.
page 9, line 4 from bottom:[13] good example of what I called "double talk" in my general introduction. Now you are not content with simply stating the historical fact as to how science actually proceeded – now you add a MUST, as if history could ever force us to repeat what others have done before us.
page 12, line 6:[14] perhaps it would be better to leave out Planck who had nothing to do with the photon theory.
page 13, line 10: "Those men were scientists"[15] – I am not quite clear what you want to assert here and on what grounds you want to make this assertion. Is the Timaios scientific? I do not know. But it seems to me that you make "being scientific" a fetish and a fetish, moreover, which you apply rather arbitrarily. I would like to see a little more explanation here. Surely the fact that they talked about "light" does not make them scientific. So what are your grounds for calling them thus? I have a great suspicion that it is in such phrases that you are trying to smuggle in your own ideal of what does and does not constitute a science, i.e. your own methodology and you do this not explicitly, but in an indirect way, by taking your readers in. Also the trouble of these earlier schools does not seem to me to lie in the fact that there were many of them and that people did not concentrate upon the elaboration of a single paradigm. The trouble of these earlier schools seems to me to lie in the fact that their

assertions were incapable of test, that crucial experiments could therefore not be staged. Had these earlier theories been developed in greater detail such that crucial experiments were possible, then many of them could have been eliminated by tests (just as the corpuscle theory of light was eliminated by test, and just as the monism of the Ionians was eliminated by test: change does exist – to be replaced by the pluralism of the atomists, of Aristotle, of Anaxagoras etc. etc.) Not the absence of a paradigm makes these earlier researches seem too chaotic, but the absence of clear methods of test and elimination. [4|5]

page 15, 16:[16] electricity, the morass of Baconism: excellent!

page 17:[17] this (that is a morass of partly irrelevant facts with a minimum of theoretical coherence) is not the only possible initial stage of science⟨.⟩[E] Another initial stage is just the opposite: a wealth of theories, or points of view (myths) with a minimum of fact connected with it. This second initial stage was the stage of the Ionian science (of the Ionian theory of matter of cosmology, that is). However this second initial stage seems to be much more promising and fruitful than the first. When there is a morass of fact and no theory – what can you discuss? What arguments can you produce? On the other hand when there are many theories, then one can discuss their plausibility, their ability to order facts, their comprehensiveness. Hence a wealth of theories seems to be a much more promising starting point of science (which is, after all, an intellectual enterprise) than a heap of facts.

p 19: a "new, and more rigid definition of the field"[18] – yes! But remember page 6 where you said that the additional commitments leading to this more rigid definition "retain an element of the arbitrary". The only non-arbitrary elements are the methodological ones, i.e. the stipulations which demand such relatively trivial things as that the theories be testable, that ad hoc hypotheses (i.e. hypotheses for the saving of appearances which do not admit of any independent test) be avoided etc. etc.

page 20, second paragraph, line 2:[19] the scientist who writes popular stuff finds his reputation impaired: this does not seem to me to be quite correct especially as nowadays there cannot be drawn a sharp line between fundamental inquiry and popularisation: Bohr's articles (for example, Licht und Leben) are certainly popular – and yet they contribute to the understanding of the theories. The same is true of Atomic Theory and the Description of Nature: in these works fundamental inquiry turns out to be so far removed from technicality that it can be understood by the layman. This is a quite curious phenomenon: Bohr's investigations have always advanced the subject; but at the same time they were understandable by people who were not masters of the formalism. But even popularisation in the proper sense – i.e. the attempt to explain what is technical in non technical terms is not frowned upon today – if it is done properly. What is frowned upon are the Jeans-Eddington excursions into theology and mysticism; but their books are bad popularisation. [5|6]

page 21, second paragraph: 'history and prehistory'.[20] Here it is very obvious that your 'history' is ideologically infected double talk. What is the criterion according to which you distinguish between 'history' and 'prehistory⟨'?⟩[E] It cannot be succession in time, for a discipline may disintegrate, i.e. become worse, whereas the word 'history' suggests that it has improved since its 'prehistory'. The amount of activity, the amount of publications, the number of scientists in the society cannot be decisive either. For it may well be that science becomes influential, recognised only after it has become degenerate. Hence, when differentiating between the 'history' and the 'prehistory' you must, at least implicitly, refer to some ideal of scientific procedure by which you measure progress (or regress). If I understand you correctly the ideal is 'normal science' or pattern-guided science (science guided by a single pattern which everybody accepts with the sole exception of some people you would perhaps be inclined to call cranks). But you never state clearly that this is your ideal. Quite on the contrary – you insinuate that this is what historical research teaches you. Now quite obviously you mean here historical research which concentrates upon the 'history' (as conceived by you) rather than on the 'prehistory' which makes your whole procedure circular and also present the reader with a conceiled ⟨sic⟩ ideology rather than with clear history. But doing this you do a disservice both to history and to philosophy. You do a disservice to history because you misuse it to wrap up your own predilections in it without saying so. You falsify history just as Hegel falsified it in order to finally arrive at the Prussian state. And you make philosophy irrational by presenting the philosopher not with a doctrine, an ideal concerning knowledge which he would then be able rationally to discuss; you present him rather with a socalled ⟨sic⟩ fact (which is not really a fact at all but a misleading report – misleading, I believe, even to you – of your own predilections) which he simply must swallow for facts cannot be different from what they are (your own defence in a discussion with me: this is how things are, you said. But the important thing here is that designating a certain period as 'history' means adding value judgement to facts!) As far as I am concerned (following here Popper and Matson) history starts with the Presocratics and this for the reason that critical discussion which we regard as the essence of knowledge, science or whatever you call it, starts with the Presocratics. [In the above paragraph I may have been a little violent – but please take this in the proper spirit. My violence is really the result of an effort to say things as clearly as possible. And also you ought to know that despite all this I find your essay very important, very stimulating]. [6|7]

page 23:[21] your definition of 'scientific' is much too wide; as a matter of fact it is so wide that it coincides with 'what can be said in some language'. Any language includes 'implicit laws' and hence what is excluded by your definition is something that is not a language at all. This becomes even clearer by your reference to paradigms. Any language contains implicitly a paradigm (this has been made very clear by Wittgenstein). Now it is my opinion that what adherence to scientific method amounts to is making a distinction within a given language between theories (in a very wide sense) about the world that possess certain likeable qualities and other theories which do not possess such qualities. This is what the empiricists tried to do; they were not content with just having a theory, but they wanted this theory to be related to experience in a manner that allowed for tests. Now I would be the last to argue about the meanings of words (such as, for example, the word 'scientific'). Still it seems to me that what scientists regarded as valuable were testable, and refutable theories and that they were not content with any old scheme that brought order into phenomena. The theory of witchcraft brought order into natural phenomena but in such a manner that no conceivable counter argument existed. All this

shows, of course, the importance of methodological arguments in the attempt to delineate the domain of the sciences. To repeat – your explanation is much too wide, as it includes all sorts of dogmas, non empirical statements, ad hoc hypotheses, circular hypotheses and so on.
page 24:[22] completely agree with first paragraph. Second paragraph is true as a description. But I feel uneasy about the influence this description may have upon the reader, or the student of physics. Will he not be inclined to feel that he should not upset the applecart and use his strength mainly for mopping-up-operations? Historical presentations have a curious influence. They tell what is the case. But sometimes they make people feel that this is what ought to be done. And they make people feel that way especially when the writer of the history has this belief himself. Do not overlook, by the way, how very well what you call the 'normal science' fits into the pattern of the methodology of falsification. For this activity may just as well be regarded as a continued test procedure and of course the test procedure will be the stronger, the more decisive, the harder one tries to establish a fit between the theory and nature and the narrower the boxes into which one tries to fit nature. From the point of view of the methodology of falsification all this can be justified as it leads to very decisive, and very sharp tests. Exactly the same applies to the fact that initial failure of fitting is not immediately taken as a sign that the theory must be abandoned. This means that also the refuting instances are severely tested, that they are not adopted at once, but again and again investigated and are regarded as decisive failures of the theory only after many attempts to accommodate them have failed. This is a methodological justification of normal science and only after such a justification has been given, only then can we be allowed to be impressed by the phenomenon of normal science, and only then can we defend it. Otherwise our defence would be of the form: do this, because this is what everybody is doing.
page 25: 'defects which have accompanying virtues'[23] – here some explanation is missing: why should detailed investigation of nature be a virtue? Or, to put it in different terms: in what respect is the quantum theory (which is very detailed) a better theory than the theory of relativity (which is not so detailed)? Some people would prefer the quantum theory because it produces more facts. Some people would prefer the theory of relativity because it produces a few very simple principles for the understanding of the whole universe (and not only of some small parts of it). So what is the [7|8] advantage of being detailed? I know, some people take it for granted that it is better to be detailed than to possess a Grand General Theory – but I just cannot swallow their assertions without hearing some reason for it. I can give a reason: being more detailed means providing stronger tests. And as a theory is the more valuable the more testable it is, detailed research is to be applauded. But this again shows that in order to be able to give reasons for ones predilections one has to refer to methodological considerations. You take much too much for granted!
page 25: "built in mechanism"[24] – you describe science as if it were a big organism which falls asleep when it drinks too much alcohol. I.e. you describe the fact that theories are abandoned when things get complicated as a natural phenomenon. You are here even more naive than Hesiod who was already able to distinguish between laws of nature which one cannot change, and laws of society which one can change.

The laws of scientific development are a result of human decisions. Human decisions are being made in accordance with certain ideals. One can persuade people to give up certain ideals and to accept others. Thus the 'built in mechanism' is the result of the fact that people do not want a dogma which they would defend come what may, but a body of knowledge which has a chance of being refuted. However once this has been made clear, in the very same moment we have a possibility to discuss this 'built in mechanism' and perhaps even a possibility to replace it by a different mechanism. You create the impression, by your use of the phrase 'built in mechanism' that there are historical forces at work which, without any help from thinking beings bring about the downfall of a theory once too many counter instances have accumulated. Or that there is a law of human behaviour according to which faithful acceptance of a theory can be shaken by accumulation of counter instances. I do not believe in the existence of any such law. Quite on the contrary I believe that there may exist points of view where accumulation of counter instances is regarded proof to the fallibility of human nature and of its incapability to understand the one, unchangeable and absolutely true theory of which the cases discussed are apparently counter instances.

p 26:[25] As regards E. O. Lawrence I have heard a rumour to the effect that theoreticians did not believe in the possibility of a machine as Lawrence set out to build. It would be interesting to know whether this rumour is correct. [What follows is excellent and I completely agree.]

p 35: "Deserting the science it defines"[26] – this reminds me of Mach's reply to Planck on the occasion of the latter's having said 'no scientist would do what Mach suggests'. Mach said 'Scientists have now become a church and I do not regard it as an honour to be a member of this or of any church.'

p 41:[27] concerning rules: I think the matter is very simple. Consider the case of chess. No set of rules is here sufficient to define the many special problems which a chess player may encounter, and try to solve, in the course of a game. These special problems arise from the special way in which certain players start out and they are therefore due, to a large degree, to the individual predilections of the participants of the game. However despite this failure of any set of rules to define the 'puzzles' that chessplayers are confronted with there do exist rules which must be followed if one wants to play chess. Only these rules are not detailed enough to explain everything that might arise in the course of the game. They leave considerable leeway. They are mainly negative rules, excluding certain moves, but leaving many other possibilities open. [8|9]

Now in science, I think, the situation is exactly the same. No number of rules ever 'seems sufficient to define the puzzles that scientists normally undertake, or to restrict scientific attention to their pursuit' (page 41).[28] Just as in the case of chess it does not follow at all from this that science is not played according to certain rules. Although there is no set of rules capable of explaining every move that is being made, there are rules which definitely forbid that certain moves are made. For example they forbid the use of ad hoc hypotheses, i.e. of hypotheses used for the adaption of a troublesome case to the paradigm which are framed in such a manner that independent tests are not possible (independent, that is, of the fact that they turn the troublesome case into a case that is not so troublesome). They also forbid dogmatic moves, i.e. moves which decrease the testability of a given hypothesis; and so on and so on. In short, the rules according to which the game of science is played are simply the rules of scientific method. To sum up: although there are no rules which explain and define each and every part of the scientific activity it does not follow that science is not played according to rules.

p 44 bottom:[29] very good (the various traditions which can exist inside the quantum theory).

p 45:[30] it seems to me to be very important to point out, as you do, that there does exist accumulation, viz. during what you call 'normal science'. It also exists in what you call the prescientific stage, if this prescientific stage is guided by an empiristic methodology.

p 46: "recognition that nature has somehow violated …"[31] – this recognition can take place only if one has first adopted the attitude that paradigms should be testable, and that they are therefore fallible. Churchdogma, for example, will not be treated that way. Any difficulty that arises is in this case regarded as an indication of insufficient understanding rather than of failure of the theory. Even Ptolemaic astronomy could have been saved in the face of any possible difficulty relying, as it did, upon the method of Fourier analysis (which will enable the astronomer to analyse any motion, even non periodic motions, in terms of motions around a fictitious center with constant angular velocity).

p 49: the reference to the "questionable" account of seeing.[32] I agree that there is a complete parallel: when seeing some object one may perceive the outlines and not the object – and even the outlines may appear as a rather scattered affair without much coherence. Next step: the outlines are perceived as a whole, but the object is not yet perceived. Next step: the object is perceived, but not as something that is known. Next step: the object is perceived as something known. What seems to be important even here is that the switch from one attitude to the next may be influenced by one's belief that one is dealing, in the world, only with objects of a certain kind, and with no other objects (dogmatism). In this case certain appearances will look as if they were not of real things, but illusions. Hence even the 'perception of anomaly' (page 51, bottom)[33] will be method-dependent. [I believe that people in the 15th century saw ghosts with the same intensity as they saw real things – the phenomenal character was the same. This may have changed today. Users of Mescaline report that they are fully aware of the unreal character of what they perceive, not only mentally, but even in the perception itself. I bet that somebody of the 15th century who believed in the devil would have seen the phenomena occurring on the occasion of swallowing mescaline in a very different manner, namely as real occurrences provided he did not know of the effect of drugs.] [9|10]

page 53:[34] I think one must be careful not to interpret any reluctance as indicative of the existence of either a paradigm, or of a part of a paradigm that is being defended. One must take into account that sometimes people are simply unreasonable.

page 58:[35] the 'awareness of anomaly' does not seem to me to precede either the special, or the general theory of relativity. The constancy of the velocity of light in all coordinate systems was considered by Einstein when he was 16 and knew neither

the investigations of Lorentz, nor the Michelson-Morley experiment. Taking this together with the principle of relativity we obtain at once special relativity and, including the elevator, general relativity (principle of equivalence). Of course, Einstein was aware of the breakdown of the classical point of view in the field of radiation. But what he was looking for was not a new theory to replace the old theory of radiation but a point of view that was general enough to survive the collapse of more specific theories (he refers to thermodynamics in this connection).
page 59: "Without the special apparatus that is constructed only for anticipated functions the results that lead ultimately to novelty could not have occurred"[36] – I think this is simply false and it is false for a reason to which you, too, would agree. The reason is that sometimes, and in very interesting cases at that, a test of a theory T can be carried out only with the help of an alternative theory T′ which agrees with T where it has been successful, but is incompatible with it outside this domain. A very good example is provided by the classical thermodynamics. We know now that the second law is not strictly valid; there are perpetua mobilia of the second kind, and the Brownian particle is one of them. However could this fact have been discovered in a direct way, i.e. by simply sticking to the phenomenological theory and trying to make it fit reality more and more closely? I. e. could anomalies have been brought about in this way? Definitely not. Consider the physical situation: in order to discover in a direct way that the Brownian particle moves by absorbing heat from the surrounding medium one would have to be able to measure the heat disappearing from the medium, the changes of kinetic energy of the particle, the resistance overcome in the fluid (viscosity) and one would have to show that the heat lost by the medium balanced the gain of kinetic energy of the particle. It is quite obvious that the measurements involved in this procedure cannot be carried out as the thermometer would be itself involved in the fluctuations of temperature to be measured. Elaboration of the paradigm of phenomenological thermodynamics therefore will never be able to discover the anomaly presented by the existence of the Brownian particle. But this anomaly was discovered. It was discovered by the elaboration of an alternative account, viz. of the kinetic theory which then produced predictions that could be tested by experiment. Now I think that this is not just an isolated case. I think that very often anomalies can be discovered only in this way; i.e. not by further and further elaborating a given paradigm, but by elaborating an alternative paradigm and producing with its help testable predictions which, if they are confirmed, show that the first paradigm is in trouble. Your insistence upon faithfulness to one and only one paradigm is bound to result in the elimination of otherwise very important tests and it is bound in this way to reduce the empirical content of the paradigm you want to be accepted. It may well be – and Bohm and Vigier are definitely of this opinion – that the situation is the same in the present quantum theory. The [10|11] 'orthodox' refuse considering alternatives and their argument is that the present point of view has not yet encountered anomalies which would necessitate reconsideration of it in its entirety. Bohm points out that the limitations of the present point of view will become evident only if one has first introduced an alternative and shown that it is preferable. Hence if the absence of limitations is taken as a reason for not considering alternatives, then trouble will never be discovered,

simply because it could be discovered only with the help of alternatives. This, then, would make the present quantum theory a wonderful metaphysics.
page 70: "historical setting" of the theory of relativity[37] –cf. what I said above in the remark about page 58.
page 72: "Invention of alternatives is just what scientists do not, and probably ought not undertake"[38] – ought not? See above. You say 'probably' – what are your reasons? Also the situation is not quite so simple. In some sense Maxwell's electrodynamics is really an alternative to the Newtonian point of view (contact action as opposed to action at a distance). Should it perhaps not have been developed? Should the wave theory of light not have been developed which is also a theory working with contact action. Should thermodynamics not have been developed which is irreversible whereas Newton's theory is reversible? All these theories: the wave theory of light; the electrodynamics of Maxwell; the phenomenological thermodynamics are alternatives to Newton's theory which were developed even before that theory had got into trouble. They were developed by articulation of a very different paradigm (Faraday!) and only after they had been developed, only then did it turn out that they meant trouble for Newton: reversibility objection; recurrence objection; the different transformation properties of Newton's mechanics (Galilei transformation) and of Maxwell's theory (Lorentz transformation). This shows what I have said above: that the shortcomings of a paradigm very often can be demonstrated only by comparing it with other paradigm cases; and that this also takes place in history. The whole development of relativity is the attempt to resolve a clash between two already existing paradigms rather than the attempt to make sense out of the isolated breakdown of one paradigm. This is true both of the special theory (the two paradigms are here Newton's theory and Maxwell's electrodynamics) and of general relativity (the paradigms are here either special relativity and Euclidian geometry [rotating disk]; or special relativity and Newton's theory of gravitation [action at a distance forbidden by special relativity)). From all this it follows not only that alternatives should be considered as they are necessary to accelerate the breakdown of whatever paradigm is in the center of attention; it also follows that they have been considered in the history of thought.

This result is, of course, very decisive for the evaluation of the last chapter. What is the result: alternatives are both used, and needed; and they are needed as it is only with their help that it is possible to find anomalies in whatever theory is being held at a special moment.
page 73: "what scientists never do"[39] – of course, they continue playing around with the theory they have. This is quite correct. For without this theory they would have nothing to do. However one thing must be changed and this quite irrespective of whether the scientists as a matter of fact carry out, or do not carry out the change: the theory cannot any more be regarded as a description of reality. Or at least it must now be regarded as questionable whether it is a correct representation of reality. It can now only be used as an instrument for prediction – but this is quite alright, for even false theories may be good instruments of prediction in certain domains. [11|12] However as I have pointed out above the situation is not always as bad as that. Very often the downfall of a theory is brought about by comparing it and its results with

an alternative and if the comparison is against the theory then the alternative can take over at once.

Now as regards the "methodological stereotype of falsification"[40] I have to say two things. First, that history is irrelevant to methodology. And secondly, that the history of the sciences supports, rather that is opposed to, that "stereotype".

History is an irrelevant judge: How do we arrive at the "stereotype"? We arrive at it in somewhat the following manner: we compare testable theories with non-testable theories (dogmas) i.e. with theories which from their very nature can never get into trouble when being compared with experience. This comparison is, of course, motivated by the existence of dogmas and by the existence of theories that are capable of empirical test. However what is important is that the <u>result</u> of the comparison and the <u>arguments</u> leading up to that result have nothing to do with history. What are the arguments? The arguments are very simple, almost trivial. We want our theories to be about the world, we want them to have factual relevance. A dogmatic theory does not have factual relevance. Hence we cannot use non-falsifiable theories. The thing is, of course, not quite as simple as that (which is shown by the discussion of the synthetic apriori). However you will understand what I mean – as a matter of fact I remember that you were prepared, at some time, to accept the argument in the simple form in which I have presented it above. Now look at the argument. Do historical considerations occur in it? Not at all. And would it be relevant to object to the result by pointing out that science does contain dogmatic elements? It would not. Quite on the contrary – the brief argument given above would now give us grounds for <u>criticising</u> whatever dogmatic elements exist in the sciences. And the criticism would simply consist in the remark that a theory is the less about the world, the smaller its degree of testability. On the other hand the demand for factual reference leads at once to the demand to use theories that possess a high degree of testability <u>and this quite irrespective of what theories are actually being used</u>. Note that the very same demand implies <u>that alternatives should be considered</u>. The reason which I have outlined above is that the consideration of alternatives provides <u>additional tests</u> and thereby increases the factual relevance and the factual content of the theory. I regard this as a very important result. <u>I can give reasons</u> why alternatives should be considered. <u>You apparently cannot</u> give reasons for the opposite position held by yourself. For you either simply report what has happened, or, when you turn to methodological considerations, you use quite frequently words such as "probably", "in all likelihood" which show, at least to me, that you really do not know what to say. But the situation is even worse. Your hidden predilection for monism (for <u>one</u> paradigm) leads you to a false report of historical events. You regard as <u>one</u> paradigm (classical physics, for example) which is in fact a bundle of alternatives (contact action: Maxwell vs action at a distance: Newton; reversibility: Newton vs irreversibility: Clausius; Galilei transformation: Newton vs Lorentz invariance: Maxwell). Which only confirms what I have said on the first page, viz. that you do not write history plain and simple, but that you present an ideology, and a very questionable monolithic ideology at that, in the covers of history. In this respect you are really very similar to those who point to history in order to justify their crimes. You are a mystic. an irrationalist. And by this I mean that you not only hold certain beliefs (conservative character [12|13] of normal science) but that you are not prepared to let these beliefs speak for themselves; you rather present them in a manner which suggests that they are facts and thereby force people to swallow them without criticising them. What are you afraid of? Are you afraid that people will oppose at once when your beliefs are presented to them in their proper form, viz. as demands as to how science ought to be run? When discussing such demands you are very careful and give the appearance of a critical person ('probably' etc. etc.). But this is just a trick (of which you yourself may not be aware). For you present the very same demands a little later as facts ('what scientists never do' etc. etc.) and then with the assurance of the historian who knows. Again, it is this kind of double-talk to which I object most. You really are like a witch doctor. That is, I do not object to your findings, but to the manner in which you represent them – as if they were (α) undisputable [which they are not – see above about the many alternatives contained in the classical point of view] and (β) unescapable; neither of which is the case.

However history is not only irrelevant, you have also interpreted your findings in an incorrect manner when suggesting that they refute the "stereotype of falsification". What does the stereotype say? It does not say that a theory that has been refuted cannot be used as an instrument of prediction. Quite on the contrary, the stereotype admits that false theories may be very valuable instruments for prediction. It only insists that it be realised that they are false, i.e. that they are not more than just an instrument. Nor does the stereotype assert that any theory that is in trouble must be regarded as refuted. Quite on the contrary it demands that the falsifying instance (which is described by a lower-level hypothesis) be treated just as critically as is the theory. I.e. it demands that the possibility of a fault of experimentation or calculation be not dismissed too readily. After all, the stereotype applies the demand for falsifiability and thereby the demand for the application of critical tests to all statements alike, and not only to theories. Only when it is agreed that the counterinstances cannot be dealt with in any other way, only when it becomes clear that we have found a genuine counterinstance, only when the statement expressing the counterinstance has been tested to such a degree that it may be regarded as a highly confirmed hypothesis, only then falsification sets in and this means that the theory must now be regarded as false and that it must no longer be used as a means of representing the world. It must no longer be interpreted realistically. Which, as I said above, does not exclude its use in the building of instruments, for prediction etc. etc. To sum up: the fact that theories are not given up the moment some difficulty arises does not at all show that scientific practice does not conform to the stereotype of falsification. Quite on the contrary, immediate abandonment of a theory as soon as the first difficulty is perceived would mean that an uncritical attitude is adopted with respect to the test statement itself which a falsificationist would never allow. I conclude, then, that your statement to the effect that "no process yet disclosed by the historical study of scientific development at all resembles the methodological stereotype of falsification by direct comparison with nature"[41] is both false (second point above), and irrelevant (first point above). You will really have to change your attitude on this or else I am looking forward to writing a blistering review.

This immediately applies to your page 74[42] where you say that as long as a new paradigm is not available scientists will try to patch up the old paradigm by ad hoc procedures etc. etc. If that were really the case science would be in a bad spot. If that were really the case you should at once abandon science or try to change it rather than make propaganda for it. [13|14] After all, you would not really like to accept something like witchcraft (although sometimes I am not quite sure about you in this respect; yes sometimes I am afraid that in 1500 you would have been one of the foremost defenders, on historical grounds, of Sprenger and Co. You are too much impressed by the existence of powerful tradition. Where there is such a tradition you at once suspect there is reason too and you object to anybody who tries to start completely afresh). But what difference does it make that the modem witchcraft is being called 'science'? After all, a name does not make the difference. Nor does the existence of institutions, of journals, of instruments, of technicalities. What distinguishes reason from folly is the kind of procedure adopted: are difficulties taken seriously? Is it admitted that the theory may, after all, be false? Or are they only taken as an incentive to invent ingenious ad hoc hypotheses? Hence, if your description were really correct, if it were really correct that scientists devise ad hoc hypothesis as long as a successful, or a better alternative is not available [a presumption which, as I have tried to show above is hardly ever fulfilled – alternatives do always exist and are considered!], if that were really the case, then to hell with science. But I do not believe that your historical description is correct. What is an ad hoc hypothesis? It is a hypothesis whose only confirming instances is what otherwise would have to be regarded as a failure of the theory. Such hypotheses are never framed by scientists. Consider Copernicus's argument, as against those who attacked his theory on account of the absence of a parallax: the stars are too far away for a parallax to be noticeable. If this hypothesis were ad hoc in the true sense then the absence of the parallax would have to be taken as the only measure for the distance of the stars. But in cases such as the one mentioned one immediately tries to find independent confirmation, independent, that is, of the occasion for which the hypothesis has been invented.

page 75:[43] I quite agree: there is never anything like research without a paradigm. This is the reason why one should always cultivate alternatives in order to be able both to drop a falsified theory, and to continue realistic research into the properties of the universe.

page 76: "either no scientific theory ever confronts a counterinstance, or all such theories confront counterinstances at all time"[44] – this is patently false. What do you mean by a counterinstance? A difficulty? An inconsistency between a theory and an experimental statement? In this case the second part of your statement would be correct: there exist always experimental statements which seem to be inconsistent with some theory. However such a situation is just the starting point. For we want to know who fares worse in this inconsistency, the theory, or the experimental result. And we also want to know whether the appearance of inconsistency will remain on closer examination. This means, of course, that we must test the apparent

counterinstances. Now if by a counterinstance we mean a falsifying fact whose accurateness has been definitely established by many careful tests, then it may well be that at a given time no counterinstance in this sense exists for a given theory although this does not exclude that a counterinstance in this strict sense may be discovered tomorrow. Now it seems to me sensible to mean by a counterinstance not just anything that seems to run counter the theory (just as Berkeley thought that the fact that plants grow upwards refutes Newton's theory of gravitation) but a low level hypothesis that has been carefully tested and highly confirmed. But if we mean this by counterinstance, then your surprising dictum is quite obviously false.
page 77: "It is a poor carpenter who blames his tools"[45] – sounds great, especially in view of the slightly sentimental and disciplinary note it carries. But I would like to add that it is even a poorer carpenter who works with crooked nails and blames himself for not achieving a good job. And it is also a poor carpenter [14|15] who uses only one type of nails pointing out (wrongly) that this is what every carpenter does and adding that one cannot do a good job if one shifts back and forth from one type of nail to a different type of nail. "Failure to achieve the solution discredits only the scientist, and not the theory"[46] – who says so? You? The community of scientists? Assume the latter does. You refer to those who tried to change Newton's law of gravitation and you point out that their efforts were completely neglected. And quite unjustly so we must say today. For Newton's law needs revision (general relativity). Now non Euclidian geometries could well have been invented much earlier. As a matter of fact they were (Saccheri) only people did not realise it – they regarded the implausible and counterintuitive character of the theorems they had derived from their non Euclidian assumptions as a refutation of non Euclidian geometries [Euclid's ab omni naevo vindicatus]. A nominalist would not have drawn such a conclusion. A nominalist could have invented non Euclidian geometries, and perhaps even applied them to problems of physics. This would have led to a better theory than was Newton's theory. So the people who tried to change Newton's law were on the right track – a change was necessary – only they had the wrong ideas about what kind of change should be brought about. You seem to take a completely different point of view. You seem to say that what I say is completely unrealistic. Now this I readily admit: the real, the actual development went along different lines. Newton's theory was only abandoned after a lot of trouble had arrived [by the way your insistence that the trouble must be big enough seems to be circular; there is always great trouble and maybe that in the beginning there was greater trouble than there was at any time later on; the case of the moon equation could have been hardly worse; by "great trouble" you seem to mean "trouble that leads to the search for a new paradigm" – now this is purely psychological; people may look for a new paradigm on the slightest pretext, and they may on some other occasion stick to a theory much longer than would seem to be reasonable; this all depends upon the climate of the time, upon the character of those carrying the main burden of the research; there is no fixable amount of trouble after which a theory should be abandoned and before which it should not be abandoned. The fact is that it very often is abandoned when some trouble arises and only in this sense the trouble can be called big enough; recall the story of the lethal illness: it is an illness which is grave enough to end in

death]. But the truth is that it should, and could have been abandoned much earlier. The same is true of Aristarch's theory versus Ptolemy's. It is a historical accident rather than a fact giving us insight into the nature of the scientific enterprise that it was the geocentric theory that was elaborated in sufficient detail to become of practical value and that it was this theory which therefore gained great popularity. At some place you seem to insinuate (I have forgotten the passage) that this was a much more natural way to start. That is you not only say that this is how things went, but that it was better that they went that way. I do not see any reason in this. Had the heliocentric theory been elaborated in Greece we would perhaps be farther now than we actually are. It was not. Why? Because of Aristotle's triumph over Democritos. And why did Aristotle triumph? Well, this is again one of the historical accidents. Perhaps he triumphed because Plato succeeded in making the atomists look mean fellows who strived after lust instead of the higher things. In sum: it seems to me that you do not tell history as what it is: a series of accidents combined with struggle for power etc. etc. You perceive some inherent reason in it. But in order to perceive inherent reason in a series of accidents one must distort reason itself. And this, I am afraid, is what you are doing. **[15|16]**
page 78 –[47] character of the anomaly that evokes the crisis. This is what I have just been talking about. How does a crisis come about? Of course there must be some reason to worry. But do not forget that a crisis very much depends on the character of those participating in it. If they have great influence, and if their voice is heard, and if they are conservative, then the crisis may be postponed simply on their authority. Assume Einstein had not invented the photon hypothesis. Planck was conservative. Nobody could have really said what would have happened. There have been attempts of something like a photon hypothesis before. Not much changed between 1900 and 1905 – only Einstein was persuasive enough. Now if a conservative authority dies, then the crisis may come in full blast – simply because some personal influence, some very irrational thing, does not any longer exist. It is not simply a matter of the kind of anomaly arising. Some people may make a mountain out of a molehill and they may succeed in persuading their contemporaries that there is indeed a mountain. Hence, all you can say is that at some point people feel that something must be done. But this feeling of theirs can never be justified by reference to an objective crisis-situation alone (degree of misfit between theory and facts; number of counter instances). Conversely you can not lay down a rule to the effect: do not consider abandoning a theory when the counter evidence is slight – and try to justify this rule by reference to history. For what happens in history is that people sometimes exaggerate and create a crisis that way, especially when they are influential. Your attitude that "there is probably (again your probably) no fully general answer to the question as to what makes an anomaly seem worth concerted scrutiny"[48] is quite correct, but you do not seem to see the reason why it is correct: it is correct because people turn their attention to an anomaly sometimes not because of the anomaly, but because of somebody else who says that the anomaly is worth considering, that it is decisive. Hence there is no inherent property of an anomaly which makes it create a crisis. But you make it look as if there were. You make it

look as if the great men who consider the anomaly have first studied it and now have reason to consider it decisive.

I think I shall now finish*. I am looking forward to your reaction.

* tired

⟨UNSIGNED⟩

Editors' Notes

PF 5-3-1 (16 ss., T+H)

[1] German retranslation of the second part of Hegel's principle "What is rational is real; And what is real is rational [*Was vernünftig ist, das ist wirklich; und was wirklich ist, das ist vernünftig*]" (HW 7: 24); see also Heine's famous reformulation of the relevant part of Hegel's principle: "All real is rational [*Alles was ist, ist vernünftig*]" (Heinrich Heine, *Briefe über Deutschland*, Frgament 19844/45).
[2] Feyerabend is probably referring to Feyerabend 1962b, which was finalized at the end of January 1961. On the history of the development and publication of Feyerabend 1962a, see *above*, Vol. 1, Part II, 1961-1, n2.
[3] See Kuhn 1962^2: 1, top.
[4] Similarly in Kuhn 1962^2: 1, middle: "Its aim is a sketch of the quite different concept of science that …".
[5] See Kuhn 1962^2: 2, bottom.
[6] Similarly in Kuhn 1962^2: 2, bottom.
[7] See Kuhn 1962^2: 3, bottom.
[8] See Kuhn 1962^2: 4, bottom.
[9] Similarly in Kuhn 1962^2: 4, middle.
[10] See Kuhn 1962^2: 3-4.
[11] See Kuhn 1962^2: 5, bottom.
[12] See Kuhn 1962^2: 6.
[13] "If they [distinctions like that of the context of discovery and the context of justifications] are to have more than pure abstraction as their content, that content must be discovered by observing them in application to the data they are meant to elucidate." (see Kuhn 1962^2: 9, bottom).
[14] See Kuhn 1962^2: 12, line 7.
[15] See Kuhn 1962^2: 13, top.
[16] See Kuhn 1962^2: 15-16.
[17] See Kuhn 1962^2: 17.
[18] See Kuhn 1962^2: 19.
[19] See Kuhn 1962^2: 20, top.
[20] See Kuhn 1962^2: 21, top.
[21] See Kuhn 1962^2: 23.
[22] See Kuhn 1962^2: 24.
[23] Similarly in Kuhn 1962^2: 24, middle: "Perhaps these are defects."
[24] See Kuhn 1962^2: 24, bottom.
[25] See Kuhn 1962^2: 26, top.
[26] Similar in Kuhn 1962^2: 34, middle: "Work under the paradigm can be conducted in no other way, and to desert the paradigm is to cease practising the science it defines"
[27] See Kuhn 1962^2: 42, bottom.
[28] Similarly in Kuhn 1962^2: 42, bottom: "those rules may not by themselves specify all that the practice of those specialists has in common".
[29] See Kuhn 1962^2: 50-51.

[30] See Kuhn 1962^2: 52, top.
[31] See Kuhn 1962^2: 52, bottom.
[32] See Kuhn 1962^2: 55, middle.
[33] See Kuhn 1962^2: 57, bottom.
[34] See Kuhn 1962^2: 59.
[35] See Kuhn 1962^2: 62, middle.
[36] Similarly in Kuhn 1962^2: 65.
[37] See Kuhn 1962^2: 74, bottom.
[38] Similarly in Kuhn 1962^2: 76, middle.
[39] See Kuhn 1962^2: 77, top.
[40] See Kuhn 1962^2: 77, middle.
[41] See Kuhn 1962^2: 77, middle.
[42] See Kuhn 1962^2: 78, top.
[43] See Kuhn 1962^2: 79, top.
[44] See Kuhn 1962^2: 80, top.
[45] See Kuhn 1962^2: 80, middle.
[46] See Kuhn 1962^2: 80, middle.
[47] See Kuhn 1962^2: 82, top.
[48] Similarly in Kuhn 1962^2: 82, middle.

2. Feyerabend to Kuhn, [1961]

⟨UNDATED⟩

Dear Tom,

If you don't have any time – don't read this letter. I only want to make some comments on your reply. I shall not rest content until either I myself realise that I have been mistaken, or you realise that you have been mistaken or we realise that we agree. This attitude of mine is prompted by the fact that I regard the topic we are discussing as one of the most important topics, not only of philosophy, but quite in general. Perhaps you will not agree, but I judge the importance of a topic from the influence a specific solution of it may have upon the well being of mankind. And by 'well being of mankind' I do not mean physical well being, but the well being which derives, among other thing ⟨sic⟩, from the exercise of one's imagination, from the full development of human faculties, and from spiritual happiness. A topic is important if a certain decision with respect to it, or a certain solution of the problems connected with it implies an increase, or a decrease of that well being. That is I am totally opposed to any attitude which says: "I am out to find the truth, come what may". What truth? And why? would be my question. The world is today full of people who deal with all sorts of problems simply because they are part of a tradition which deals with these problems. But traditions do not develop in a rational manner – this you admit yourself. Therefore the fact that certain problems are part of certain traditions does not impress me at all. For perhaps the whole tradition is superfluous, or dangerous and must be stopped (not by force, oh no, but by the sweet smell of reason). More specifically: why should we discuss complementarity and quantum theory? Why should we be interested in the history of a subject? If this interest does not have any consequences, of the kind indicated with respect to the well being of mankind, it is useless and should perhaps be given up.

Why do I say all this? I say all this in order to indicate why I am interested in the problems we are discussing. I think that these problems are very important because their solution, one way or an other, will have a tremendous influence upon the way we think, the way we act, and thereby upon our well-being. This conviction is the motor which makes me go on with the discussion and which makes me think that the discussion is worth while, however long it may take. I also go on with the discussion because I think that some of your ideas, when published, may have a disadvantageous influence. Now I do not think this is a case of truth being dangerous (as you seem to think; by the way your attitude seems to be: if it is true <u>and</u> dangerous, then I cannot do anything about it; I am a defender of the truth and I shall say the truth, whatever the consequences; this seems to be your attitude; I think that this is a very dangerous attitude; for you are not only a defender of truth; first of all you are a member of humanity and have obligations towards humanity; these obligations come first; these obligations do have the character of <u>certainty</u>; but whatever truths you may unearth – they have the character of <u>hypotheses</u> and therefore ought to be treated with great care when confronted with the certain obligations you possess).

Ours is not such a case as I think that you are here not a defender of truth, but are simply mistaken.

Before I am going on let me say the following. You will forgive me, please, when I develop all the reasons for my interest in the topic. I am not doing this in order to preach to you about the obligations of a scientist (or historian), but in order to make clear to you why this topic makes me so exited ⟨sic⟩ and why I do not leave you in peace.

Now back to the main topic. Let me tell you only one thing – and having done so I shall at once return to your letter. You say that you are a historian who describes what happens. You will not describe it differently, for [1|2] this would mean being untruthful. Now the important thing with historical descriptions is that they are of a misleading ambiguity and that they very often contain normative and descriptive elements mixed up in one process. This process, this mixture it is which is then put in front of our eyes as a "fact" and we are not aware that in swallowing this alleged fact we at the same time swallow adherence to norms which are all but acceptable. Let me take a very crude example first which is not as unrealistic as it may seem at first sight. Consider a society which has been a society of warriors and on top of the warfaring societies. Assume also that a development sets in in this society to greater refinement, science starts (theoretical science, that is, and not engineering), the arts are developed, all the finer capabilities of man are developed, but at the same time the muscles degenerate and finally the society is overrun by a bunch of toughies. What can a historian do in this case? The obvious thing is to say that the society slowly degenerated, became weak and therefore finally was bound to collaps⟨e⟩[E]. This is the way in which, for example, the collapse of the Roman Empire has been described. Prima facie this description is a factual description. For it describes what really happened, the gradual decline of the military strength. But at the same time the following happens: it is a factual description, hence it is a correct description, no other description would be correct – this is assumed. But if it is assumed that no other description would be correct it is also assumed (silently) that military virtues are the only virtues worth mentioning. This apparently factual description therefore contains, wrapped up in it, and hidden from sight, an evaluation: military strength is the only important measure of a society. But I am not prepared to accept this evaluation. On the contrary I am inclined to give a completely different description of what happened. I am inclined to say that what happened was not a decline, but wonderful progress, progress however, which was rudely disturbed by uncultivated intruders. Now note that this description is compatible with the one given first and it is compatible because it is based upon a different evaluation. What I want to derive from this example is the following: many historical description⟨s⟩[E] contain a normative component. It is this normative component which allows for different, and yet compatible descriptions to be given of "the same process". Now if this is the case then the historian is obliged, if he wants to be honest, to do one of two thing⟨s⟩[E]. One: he has to eliminate the normative component. My hypothesis is that then he will not be able to write history as it is the normative component which guides him in making the selection of material he does make. Two: he uses the normative component. But if he is doing this, then he better use the norms to which he himself is prepared to agree, i.e. the norms which belong to his own value system and the norms, the ethics which he would defend as a responsible citizen. What you do is (seems to me to be) the following: you swallow norms with which, I trust, you would disagree violently when presented with them naked, but you swallow them because they are part of what you regard as something completely factual. So under the cloak of FACT you accept what, I trust, you would dislike when presented with it in its real form, i.e. in the normative form.

This brings me at once to "science". You enter here the stage as a factual historian; as the man who is not interested in evaluations, but who is interested in finding the structur⟨e⟩[E] of what is historically given. Like Galileo who looks at his inclined place ⟨sic⟩ and who waits what is going to happen you look at the historical material and discover in this material factual relations which you are then going to describe. These relations are there, and if they are undesirable, then we have to accept them as such. Now the important thing is that in describing them you do not bring out their undesirable character. You are the opposite of the man who says of the society in the above example that it degenerated, assuming silently [2|3] that military values are the only values of importance, or perhaps even the only values in existence. You look at history and find certain identities which constitute a tradition. You find relations between thought, instruments of research, publication, terminology etc. etc. and you concentrate on these relations (which include a paradigm). You concentrate on them and in this way discover an identity (of adherence to one and the same paradigm) and this identity it is which you regard as the characteristic of science. So your "factual" description consists in exhibiting this identity. However the important thing is that in this way you overlook that in other respects an identity may not at all exist. If we consider, for example, the way in which the paradigm is treated, we may discover a radical change from a more hypothetical to a more dogmatic attitude (to a certain extent this has already happened with special relativity; to start with it was regarded as a very peculiar, and therefore suspicious hypothesis; now relativistic invariance is a conditio sine qua non of a reasonable theory just as was causal character about a century ago). That is, if one does not put the paradigm into the center of one's considerations, but if one puts from the very beginning the methods for the handling of any possible paradigm used by a community into the center then the history will perhaps be not a history of continuous development, but a history of gradual change from a more hypothetical attitude to a more dogmatic attitude, or the other way round. Now concentrating on the paradigm and investigating its role in the history of a subject you discover an identity which you represent as a factual discovery. But just as in the above artificial example you have combined in your allegedly 'factual' discovery a factual element and a normative element. The normative element consists in the silent assumption that it is the paradigm that is important rather than the way in which it is handled (dogmatically, as the unshakeable basis of truth, or hypothetically, as a promising working hypothesis). Now, of course, you are under the impression that the central role of the paradigm is a factual discovery rather than the result of selection according to interest. But it is not. How do you say you discovered this central role? By looking under which circumstances a research community is established with definite problems as defined in terms of a paradigm. Without knowing it yourself you already start with an evaluation: what is good is research guided by a paradigm. Small wonder that this evaluation leads you to discovering the identities you discover. Look at the Presocratics. There is no paradigm which is retained all the way through from Thales to Democritos ⟨sic⟩ and Aristotle (unless we artificially construct one from the hindsight). And there is therefore no unity if by unity we understand reference to one paradigm. But if by unity we understand the way of handling paradigms, any paradigm, then there is a wonderful unity here: it is the unity of a tradition in which ideas are not accepted as infallible truth, but are regarded as fallible, and therefore in need of improvement.

Now considering all this (and, by God, I do hope I have made myself clear) I arrive at the conclusion that your allegedly "factual" discovery is a mixture of normative elements (attention directed towards unity of paradigm, rather than unity of method) and factual elements (how history looks if analysed from the point of view of a person who is interested in paradigms and their influence). The question which is still in need of being answered is why you direct your attention to paradigms rather than to method. Of course, the unity following from a paradigm is more concrete, and therefore may seem to be more "real" than the unity following from adherence to the same method. But is it therefore more valuable? This question still remains unanswered and it must be answered explicitly as you are already answering it, without knowing, implicitly. You may also point out that we are the followers of the paradigm-identity-development and that our science is a science which is the result of a series of developments in which always a paradigm stands in the center. But this is not essential either. For [3|4] as I have by now pointed out again and again alternative descriptions of this development are also possible, for example that considering method. Contemporary science is the result of a development towards dogmatism – and this description is as factually correct as yours, only it is based upon different implicit values and it brings out an aspect that is not brought out by your description, namely that not all may be well in the house of science.

To sum up: you say you describe facts. I say you don't as you introduce evaluations by the back door. And if you do, then you better introduce the values with which you agree rather than those with which you don't agree. And this you can as the alternative description which results from the adoption of a different value system will also be factually adequate. Also in presenting these hidden values and in pointing out that modern science as we know it is based upon a paradigm in the center you invite people to look for a paradigm and to stick to it as they otherwise would not be scientific. You suggest (implicitly in your essay, explicitly in discussion) that we could not even proceed in a different manner. This is only partly correct. Of course, we could not proceed in a different manner and still get the same results in terms of continuity of tradition, i.e. transference of a paradigm. But we would still obtain something – and as a matter of fact such a something did exist (hence a different procedure does lead to results), it was the Ionian cosmology which gave us the atomic theory. What more do you want? Also unity of paradigm existed long before modern science which you want to characterize by this unity: it existed when the Aristotelian philosophy dominated everything. This was unity of paradigm and it was against this dogmatic unity, I think, that Galileo fought and that all those fought who took the side in Democritos, ⟨sic⟩ both in physics, and in ethics and who were called 'slaves of matter and of the flesh' by their opponents. Thirdly modern science is possible with<u>out</u> unity of paradigm: see the development in the 20th century. See especially all the rival hypotheses which were considered at the time Einstein's theory first came up: Lorentz' first theory, Lorentz' second theory, the theory of Aethermitfuehrung (Planck), Ritz' theory and which were gradually eliminated in a time when no paradigm existed in the domain of motion with respect to coordinate frames.

I Repeat ⟨sic⟩: your reference to paradigms as essential for science is not purely factual, it contains a normative element; secondly, paradigms are not essential to what you call science only – Arist.[1] philosophy and physics; thirdly, what you call science is possible without paradigms. Of course, with respect to the last point you may point out that research concerning relativity did not occur in a vacuum: But neither did the research occur in a vacuum which preceded the Newtonian paradigm.

Now to your letter: Item (1) You say that you are not interested in a prescriptive methodology, but in a "more realistic notion of the practice actually used successfully in physical science". Successfully – according to what standard? Physicists may succeed to well entrench a certain theory such that an alternative account may in the end seem to be completely impossible. This would be success according to some dogmatic standard, but not success according to a standard which puts emphasis upon the fallibility of our knowledge. Conversly ⟨sic⟩ the fact that the Ionians did not arrive at a stable theory may induce some other people to say that they did not really succeed – i.e. succeed in establishing a final theory. As a matter of fact all the criticism which has been put forth against the Presocratics proceeds from this assumption: that success must lead to the establishment of the <u>one and true theory</u>. This criticism is the criticism of Plato, of the Baconians, of Descartes. According to the standard of the dogmatists the Presocratics were anything but successful. According to a standard in which progress and change rather than stability are to be valued, the Presocratics are extremely successful in the continuous improvement, elimination etc. of theories. [4|5] Again your "realistic notion of scientific practice" contains value judgements hidden in it; and my contention is that it is better to make these value judgements explicit than presenting them in "factual" covers and thereby putting over values in the masks of facts.

I agree with your point (2) – that from history one should not draw inferences concerning the validity of methodological rules – but, do <u>you</u> agree with yourself on this point (cf. our last discussion). And if you do not, what is your position? All this should be made explicit. Your essay contains so much that is hidden, because implicit. And it may impress for reasons which are different from the reasons for which people think they are impressed.

"But then … I don't know any other source from which one could hope to derive … conclusions about a "valid" prescriptive method" (your item 3). And you accuse me that I am doing what I forbid you to do, viz. I refer to consequences of social life, democracy etc. in arguments of <u>validity</u>. Now a <u>full</u> argument is of course not so simple. I start from a certain evaluation. For example I regard development of imaginative abilities of human beings, and, more general, development of as many abilities as possible as valuable and anything opposing such development (for example adherence to one system) as less valuable. Hence, such development should be encouraged. It can be encouraged within a democracy, within a hypothetical tradition, hence this tradition is to be encouraged. I therefore derive my evaluation of the hypothetical tradition from other evaluations plus factual knowledge, and not from factual knowledge alone.

Your description in (4) cannot be correct for the reasons indicated above. Galileo dealt with the relations between two objectively given objects – the inclined plane, and the rolling sphere on it. When you assert a certain unity of development you assert this unity with respect to certain situations, for example with respect to a paradigm. But you could have chosen a different point of reference; for example you could have chosen the amount of deity inherent in scientific theories as your criterion and then you would have found that what you call the scientific tradition is characterised by the gradual elimination of divine causation for the purpose of explaining natural events (for example restoration of the loss of momentum in non-elastic collisions). You do not mention the point of reference you choose. You take it for granted that methodological procedure is not the point of reference to be chosen. This is so because you are fascinated by a particular strand of development which stands out for you with your particular evaluations (this is like the duck-rabbit; only in this latter case evaluations do not enter). Looking at history you say – this is the real line of development and report it as if you had made a factual discovery. But if you had chosen, consciously a different point of reference, if, for example, you would have chosen methodological considerations as your point of reference, then you would have to report completely different facts. And a different factual strand again if your point of reference had been the amount of divine meddling allowed. And all these strands would be successful with respect to their peculiar aims. This is also why I cannot dig your item 5. It is not I who wants the facts to be different. It is you who reports mixtures of facts and values as facts and who thereby creates a muddle.

I think this is all I want to write today. I have to go to bed now. My next letter will deal with your remaining items. Be patient with me; that is, be patient with the letters you have and with those you do not yet have.

Paul

P.S. I hear our correspondence will be edited posthumously. So we ought to be careful.

Editors' Notes

TK MC 240-21/28 (4 ss., T+H)

[1] Abbreviation for "Aristotelian".

3. Feyerabend to Kuhn, [1961]

University of California, Berkeley, Department of Philosophy

⟨UNDATED⟩

Dear Tom,

eccola la lettera promessa – I am also enclosing one of my opera magna for your scrutiny and I would like you to return it to me together with the letter, as it is one of two copies which I have left and this, you will admit, is very little. The thing is going to be published by the University of Pittsburgh Press in January.[1] If you should discover some scandalous inaccuracy I shall still have an opportunity to correct it (I am expecting the proofs to arrive in November, or some such time).

Returning to your own opus I would first like to repeat, briefly, two points which to me seem to be important and which you conceded: First, that a revolution is brought about not only by the failure to fit nature into the categories of one paradigm, it is brought about also by the further development of the paradigm at hand which development may gather a momentum of its own and turn into an entirely different, and incompatible paradigm. There is not only revolution from 'below' (i.e. from confrontation with 'nature') but also from 'above' (i.e. by confrontation with ideas which have developed out of the original paradigm). By the way, it still seems very paradoxical to me that you allow for such deviations, i.e. that you allow for deviations which are brought about unintentionally (deviations, that is, from the original paradigm) whereas you frown upon the explicit development of alternatives. What is your reason for this position of yours (and a reason you must give!) i.e. that alternatives to the paradigm which are unintentional side effects of an attempt to develop the paradigm are to be welcomed whereas alternatives which are the result of an explicit effort to look for something different are not so good. The end effect will be exactly the same: there will be a new paradigm. However in the one case this new paradigm is the result of a mistake, as it were, whereas in the other case it is the result of conscious and rational planning. You prefer it to be the result of a mistake whereas it seems to me that of course the rational procedure is to be preferred by far – unless you provide some very weighty arguments in your favour. I am waiting for these arguments, also for my benefit. As an aside I may tell you that I believe that here again your historicism has got the better of you. You believe in blind historical development. Consciously inventing a new theory means disturbing this development, which is something very deep, by bringing in something very superficial, viz. human reasoning (so fragile when compared with the forces of history!) Anyway, please answer me this question.

The second repetition is this: three (and not only two) things are needed in order that a certain gestalt switch be brought about: (a) nature (b) a paradigm (c) a certain attitude with respect to paradigms in general or what I call a methodological attitude (dogmatism: the paradigm cannot possibly go; if trouble arises then this shows that mankind is rotten and perhaps needs some help from the inquisition; non dogmatic: a paradigm is something which, if need arises, very pressing need, has to be replaced by a different paradigm and can be replaced as it is not an absolute truth). This is important. For you suggest that there is only one procedure which will be adopted when the paradigm comes into trouble: to look for a new paradigm; and that this is the procedure which will be adopted by all those who are interested in the paradigm. Not so. Not so. An alternative would be to admit the sinfulness of human nature and to attempt to purge it from sin. Another alternative (which is accepted by some primitive tribes, for example the Zande) would be to point out that this was obviously a year where demons were very active and made understanding impossible. etc. etc. Only if it has first been admitted, either explicitly, or implicitly, that a paradigm is something that may be in need of replacement and not something that lasts forever, only then will the reaction which you describe be the normal one. [1][2] This is how historical development is guided by methodological beliefs and attitudes.

My third repetition (which is very brief, and which I did not foresee when I started the letter) is a reference to the problem of tautological degeneration of your thesis that "great trouble leads to a search for a new paradigm". From one point of view the problem of the moon equation in Newton's time was very troublesome. However it did not lead to the search for a new paradigm. Will you then call trouble "great" only if it leads to search for a new paradigm? But then your thesis that "a new paradigm is looked for only if great trouble has arisen" is a tautology.

Now for the details of the second part of your essay.

[1] Remember my reservations concerning your comparing political revolutions with scientific revolutions. The most fundamental revolution, to me, in the domain of knowledge, would be the transition from a stage of dogmatism to a stage where replacement of any paradigm is possible (although, as you have pointed out, one will be very careful not to give up the old paradigm too quickly and too lightheartedly). Now it seems to me that political revolutions are more closely related to this fundamental revolution than to changes of paradigms about nature. That is they are more closely related to attitudes concerning any possible paradigm (methodological attitudes, as I would call them) than to the change of attitude with respect to one particular paradigm. This is just a guess on my part, and it may be a mistake.

By the way, it is quite interesting to note what application of your attitude means in the field of politics. In the sciences you advocate considerations of alternatives only if the existing paradigm is in real trouble (again, you must explain what real trouble is. For the moon equation was real trouble – what difficulty could be greater than a factor of two! And yet you say that physicists were right not to give up Newton's theory, or at least, not to look for something better implying, of course, that this trouble was not real enough. Is this not wisdom from hindsight? And how can such wisdom from hindsight guide the scientist who does not yet possess the hindsight because he has not yet made up his mind as to how to proceed? That is what use can your historical evaluation be to him if this historical evaluation, based upon historical fact, can exist only after he has made up his own mind?) But back to politics: in the domain of politics your rule not to consider alternatives unless the existent paradigm has got into trouble would mean that nobody should consider alternatives to a tyranny until it has become very obvious that it cannot work; this means revisions, or democratic ideas should be considered only after quite a lot of

people have been killed and only after it has become apparent that the tyrant cannot keep his power a moment longer. Only then the democrats are allowed to come forth and to suggest different means of ruling the people. That is, they are supposed to be cowards, they are not supposed by conscious propagation of their ideas to contribute to the downfall of tyranny and the philosophy connected with it (like racialism, or the idea that there are people chosen to rule by destiny, and others who are chosen to obey); they are supposed to wait until others have made life difficult for the tyrant, or until economic circumstances have weakened the rule of the tyrant etc. etc. and only then may they come forth and preach the creed of democracy. Now let us apply this immediately to the domain of the sciences. You see: preaching the creed of democracy may just be one of the things which lead into difficulties for the tyrant. And suggesting an alternative paradigm may just be one of the things which leads into difficulties for the existing paradigm (for it may turn out to be much more attractive, much more simple, and much better able to cope with problems than the existing paradigm). You propagate ceaseless efforts to fit nature into categories of one paradigm because it is in this way the crises will be brought about. You therefore implicitly assume that everything that leads to a fullfledged crisis (and not only to a temporary inconvenience) is to be [2|3] welcomed. Now if that is so then consideration of paradigms which are alternatives to the paradigm in existence is to be welcomed too, for it too may lead to crises (this you have admitted yourself, however you have admitted it only in connection with paradigms which are the unintended results of an attempt to further elaborate the existing paradigm. The above argument is quite general and I do not see why it should not be made general in this way: after all the fact that Maxwell used all the little wheels is a historical accident. Electrodynamics might well have been invented on the basis of a completely different metaphysics, on the basis of a metaphysics of 'waving' spirits such as was the metaphysics of Henry More. I conclude, then, that if you welcome acceleration of the development towards crises you must also welcome consideration of alternative paradigms which, as you admit yourself, may lead to crises. Only, you admit this only for paradigms which have been invented in a very particular manner, i.e. on the basis of a mistake, as it were: they were intended to be further applications of the existing paradigm, and they turned out to be rivals of that paradigm. I do not see any reason for that restriction and I would be very glad if you could give me one, a single good reason. If you don't I take it you admit that consideration of alternatives is not the anathema you want it to be but, on your own premises, one of the means to make the crises (which you want) occur sooner, or perhaps to lead to a stage which is a crisis in permanence.
[2] page 90:[2] when the country is divided into different camps then political recourse fails only if being divided into camps is something quite out of the ordinary. If it is not, as is the case in a democracy (many-party-systems) then there exist political institutions which have been designed with the express purpose of dealing with such a situation of polarization. One may even say that the battlecry of a democracy is 'polarization in permanence'.

Now if it is correct, on the other hand, that alternative paradigms are needed in order to accelerate the development that in the end leads to scientific revolutions, then there must exist scientific institutions which are capable of dealing with polarizations that may arise in the scientific community and which prevent these polarizations to become “incompatible modes of community life”.[3] A theory of such institutions is as yet missing. All existing philosophies of science (yours included!) are monistic in that they deal with what happens when one paradigm reigns supreme.* They are in that sense also non democratic. (This even applies to Popper!) And that your point of view is not democratic, and possesses an element of dogmatism can be seen most clearly from the fact that failure of a paradigm, as you describe it (a whole world breaks down; the activity of the whole group undergoes a most acute crisis etc. etc.) is seen as a dogmatist would see it: when his absolute truth breaks down, this means, to him, the end of his world. Only for a dogmatist can breakdown of a paradigm have such tremendous consequences, as you say breakdown of a paradigm has. For only he is not accustomed to the view that anything he believes in may at some time fail and that he never possesses the absolute truth. Now it is very well possible that scientists working in a certain tradition do tend to become dogmatists. This may be seen, very clearly, from the fact, that they sometimes use methods such as transcendental deduction etc. for defending their point of view. But this is not necessary – and it is not even true of all scientists. Alternatives are considered, and should be considered. (By the way, another difficulty of Newton’s mechanics, besides the empirical one and those created by the existence of alternatives such as Maxwell’s electrodynamics and the phenomenological thermodynamics is the lack of conceptual clarity of, say, the notion of an inertial system. This lead ⟨sic⟩ to publications such as Lange’s, Neumann’s (the older Neumann who was no ‘von’) and Hertz’s as well as to Boltzmann’s defence of the traditional point of view. This was dissolution [3|4] from within comparable to the dissolution of the older set theory (Cantor) through the discovery of the paradoxes). We have, therefore, at least three clearly distinguishable causes for crisis concerning a given paradigm: (1) failure to fit nature into its categories, which I call, positivistically, ‘revolution from below’; (2) inconsistency with successful alternatives that have been developed, either with the intention of elaborating the main paradigm considered at a given time (this is the case you admit), or on the basis of a completely different metaphysics that has been sleeping for some time (this is the case you want to exclude – with no arguments to support you, not even history) – this I shall call ‘revolution from above’ (as it involves ideas which, positivistically, may be placed ‘above’ nature); (3) internal unclarities such as are discussed by Berkeley (with respect to Newton’s idea of absolute space and with respect to his calculus of fluxions), Mach, Lange, Neumann (all these about the role of inertial systems and the method to ascertain one; also the real meaning of the law of inertia was involved here), Hertz (eliminating forces), Boltzmann (trying to represent the “orthodox” point of view, as he calls it himself, as clearly as possible in order to show, thereby, that it is not in difficulty as Hertz has assumed), Voss (who in his Encyclopaedia article reports about all these difficulties without being able to resolve them), Ostwald (who thinks that the difficulties are due to the metaphysical character of mechanics) – and which may lead to a ‘revolution from within’ or a ‘crisis from within’. It seems to me that every crisis contains all these three elements: ill fitting

with nature, inconsistency with successful alternatives, and internal inconsistency or unclarity, though in different degrees. The purest case of a revolution from above is perhaps special relativity; the purest case of a revolution from within is the development of modem set theory; the purest case of a revolution from below – well I am completely unable to think of such a case which in your presentation seems to play such a great, and almost the only role.
[3] p. 91: "pure logic"[4] – I have explained to you my misgivings about that phrase. To me your intention is beyond doubt. But some readers may be misguided into thinking that you attack a theory much more primitive and ridiculous than the theory you intend to attack. Other places where reference to logic occurs in that misleading manner are: pp. 92,
[4] p. 91: "techniques of persuasive argumentation"[5] – it is here that reference to method comes in.
[5] p. 93: "cumulation of unanticipated novelties proves to be an almost non existent exception to the rule of scientific development"[6] – excellent! Popper says, in his lectures, that discovery is always intimately connected with refutation or, as you would say, with the changes of paradigm brought about by a crisis.

By the way, development by cumulation – (and without much revolution) – is this not what happens today in many parts of psychology and sociology where we have papers with plenty of data and with no apparent purpose. One is here almost reminded of the character of some of the early publications of the Royal Society where anything is taken up, checked empirically, whether it is not ⟨sic⟩ interesting, or whether it is not.

On p. 94 you are exactly Popper when you say "Unanticipated novelty, the new discovery, can emerge only to the extent that his anticipations about nature and his instruments prove wrong".[7] This I do not say in order to make priority claims, but in order to show to you who your friends are (or, to be more precise, who the friends are of your better self).

This applies to the whole of page 94[8] – and it is for exactly these reasons that alternative paradigms should be considered all the time, and not only when there is a mess. Their consideration will accelerate the appearance of the mess, but also of its resolution. If you start looking for alternatives only after a mess has arisen, then it will most likely take [4|5] you a long time to discover them and thereby to establish a new order. If you have them present all the time and use them as instruments for the downfall of one of the paradigms used, then they may be used at once for bringing about order. The more I think and the more I write, the more reasons I find why scientists should consider alternatives at any time, and not only in a crisis. I have even found now a reason why it seems to you to be otherwise. Let me explain it: you say that historical research has taught you two things: (1) that normal science is science which is carried through according to a paradigm, and to a single paradigm and which refuses, and to your mind correctly so, to consider alternatives; and (2) crises arise when this paradigm is found to be in trouble when confronted with nature and not only at one place, but at many places. It is then that alternatives are considered, and legitimately considered. This is your original thesis, the thesis of your paper. Now in the meantime you have admitted that crises may be created, and have been created, not only by absence of fit between nature and the paradigm, but by the existence of alternatives, which are very successful, and which are yet inconsistent with the main paradigm (if it can still be called the main paradigm). Alternatives are therefore desirable because they lead to crisis (you do say that crises are desirable as they lead to interesting discoveries; hence anything that leads to them is desirable). But what happens now to "normal science". Normal science (if that period exists at all and if the truth is not rather "crisis in permanence") is a period where nobody uses this particular means of precipitating crises and the reason for this may be lack of ideas, or the fact that the initial success of the main paradigm has dazzled people to such an extent that they cannot think of anything else. The latter thing, however, is far from desirable and so it may well be that the absence, from "normal science", of alternatives, is far from desirable, too. Hence, "normal science" is far from desirable, too. You say, it has inbuilt guarantees to the effect that it will be upset sooner or later. Let me first point out that one of them is the non-dogmatic attitude of the community of scientists, i.e. an attitude which is prepared to change the paradigm if very good reasons arise. This guarantee and the concentration upon the elaboration of puzzles is bound to lead to crises. I completely agree (although I do not like the historicistic flavor of the whole affair). But why should we not add another guarantee ("doppelt hält besser" as the Germans say – doubly bound is more secure) viz. alternatives which will work in the same direction, namely in the direction of crisis and therefore new discoveries? "Normal Science" as you describe it is therefore an activity where only one of the crisis-creating means is being used and this for reasons which are psychologically understandable, but not otherwise impressive: people are dazzled by the initial success of the main paradigm; they say they have no time etc. etc. It is as if a person whose only aim is to have as many children as possible sleeps with one woman only because this is what everybody does the reason being that nobody is bold enough to approach more than one woman in five years.

[6] p.96: "It is hard to see how new theories could arise without … destructive changes in beliefs of nature."[9] Read in one way this is a tautology: ideas don't change unless they change. But with you one has to read the quoted statement as follows: "It is hard to see how new theories could arise without a previous crisis due to ill fitting between one paradigm and nature." The answer to this is, as you yourself have admitted, that they do arise without such a prelude (thermodynamics etc. etc.). See also the above.

[7] p. 96 again: concerning an interpretation "closely associated with logical positivism"[10] which "would restrict the range and meaning of an accepted theory so that it could not possibly conflict with any later theory …"[11]. Let me comment, at this point, that the intention of the [5|6] Copenhagen people was to construct exactly such a theory and that they think they have succeeded in constructing it. One element of the Copenhagen thought may be represented as follows: classical physics ran into trouble because it was not purely factual but went beyond the factual evidence that was available at any time of its existence. Or, to express it in terms which are used by some members of the Copenhagen school: because it contained metaphysical elements. It was the aim of the quantum theory (again only for some members of the

Copenhagen school) to remedy this shortcoming of the classical physics and thereby to achieve, at one stroke, two different things: first, to obtain a theory that was correct from an empirical point of view; secondly, to obtain a theory that did not contain any metaphysical elements and all of whose basic postulates were a mere expression of fact, of empirical fact. According to Bohr the remedy lies in restricting the application of the classical concepts. The classical concepts, Bohr says, are abstractions. That means that they are partly related to experience (they are after all abstracted from experience), and go partly beyond it. It is the part that goes beyond experience that has caused the trouble. If this part is removed we shall obtain a theory which is wholly empirical and wholly correct. Now the question arises: which part of the classical concepts is the part that goes beyond experience? According to Bohr the answer is to be found by an analysis of the experimental fact of duality. Duality forces us to restrict the use of the classical concepts (the wave concepts, the particle concepts) and it also tells us to what extent their use must be restricted: it must be restricted exactly to the extent that is given by the uncertainty relationship. Classical physics was bad, incorrect, and it was incorrect because it went beyond experience. Duality teaches us how far it went beyond experience and it thereby teaches us what parts of the classical theory must be cut out, as it were, in order to obtain a theory that is correct and wholly empirical. The trouble of the classical paradigm therefore does not show that it must be replaced by a different paradigm; it only shows that this paradigm was too pretentious and that its pretense must be curtailed by cutting out what is not justified by experience. The quantum theory is therefore not a new paradigm; it is rather what remains of the classical paradigm when its metaphysical pretensions have been eliminated. To express it even more boldly: the quantum theory presents us with the empirical content of the classical physics, rather than with an entirely new paradigm. This, I think, is the basic idea of Copenhagen. Any theory that is obtained by treating some classical body of theory in this manner is called by Bohr a "rational generalization" of the corresponding classical theory. Now what does this theory of theories say about the progress of science? It says the following: many new discoveries will be made. These discoveries will lead to new laws. But all these laws will be compatible with the uncertainty relationships and it cannot be otherwise as these relations are repetition of fact. The basic assumptions of the present theory therefore "cannot possibly conflict with any later theory" as you express it – and of this the Copenhagen people are quite certain as they have constructed the quantum theory and the uncertainties in such a manner that they simply repeat facts without making idealizations. The quantum theory is not a paradigm in your sense. It is what remains of a former paradigm (the classical theories) when this has been freed from anything that goes beyond experience. All this must be known if one wants to understand the Copenhagen philosophy. They have not simply added another theory to the theories of the past which at some future time may be replaced by again another theory. They have taken what they think to be the quintessence of all past theories, viz. classical physics and reconstructed it in such a manner that it fits the facts closely and can therefore not any more run into any crisis. From now on we have [6|7] entered a new age of scientific activity. There will be no more revolutions, there will only be accumulation. This, as you will also see from my enclosed paper,[12] is the Copenhagen point of view. And this makes it understandable why they react so violently against Bohm. For them Bohm wants to do nothing less than turning back history. He wants to make undone the tremendous progress towards unmetaphysical theories made in Copenhagen and again invent theories that go beyond the facts thereby reverting to a history that will again be full of crises, revolutions etc. In this connection I beg you to read (and then to return) the enclosed review, by Rosenfeld, of Bohm's book (it was published 1958 in the Manchester Guardian;[13] I have written a very impolite counterreview. Only be aware what you will face in Copenhagen and how careful you will have to be.[14] You will face a metaphysics. And metaphysicians usually are very dogmatic; but they are even more dogmatic when they believe their metaphysics to be truly factual). You will now also see that your "today this remains a minority view" (of line 7 from bottom of page 96 of your essay)[15] is even more true. Not only are the philosophers reluctant to accept such a view; but the majority of physicists agrees that this view correctly describes the relation between the theories of the past, that this property of past scientific development was due to the partly metaphysical character of the theories employed; that this shortcoming has in the meantime been remedied by Bohr; and that the future of science is therefore bound to be very different from its past and is to be all "normal science", i.e. totally accumulative. You see that the situation is now even more difficult: the Copenhagen people admit what the positivists do not admit, viz. the past science has been as you have described it. They therefore seem to be on your side. But they then point out that this was due to metaphysics, and that this won't happen any more in the future as now we have finally arrived at truly empirical theories.

[7] p⟨.⟩[E] 97:[16] excellent! You could not have said it better.

[8] page 98, last paragraph:[17] I completely agree. I have to add two things. First, the absurd view derived by you from the point of view you criticise viz. that if it were correct all a theory would say would be a repetition of past experiment, this view has been held both by philosophers and physicists. Schrödinger, of all people, defended it in a long discussion we had in Alpbach. Secondly, however, I don't understand how you can say what you say here and at the same time assert, in your paper on measurement in ISIS (page 166, footnote 9) that "to the extent that a scientific theory must be accompanied by a statement of the evidence for it … the full theory … must be analytically true".[18]

pages 99, 100, 101 again excellent.[19]

[9] page 101, beginning of last paragraph:[20] you have not shown that the differences are necessary!

[10] page 107:[21] the fluctuations of the community of scientists described by you on this page, taken together with Planck's dictum (and the consequences it implies!) shows very clearly, at least to me, that history is the history of human idiosyncrasies, of folly, and therefore without reason. Anybody who wants to derive reason from history is therefore bound to cheat at one place or another!

[11] page 108, bottom:[22] but one thing must remain: the belief that there is no absolute paradigm. If this belief is given up as well, then there may still be historical

continuity in the sense that there will be the same institutions, the same journals. But science as a rational undertaking won't exist any longer (and this even if people continue to call the new dogmatism 'science'). [7|8]
[12] page 109: "they will inevitably talk through each other".[23] This is not true! If Bohm were to produce a prediction which is forbidden by the current theory, if the experiment were carried out and decided in Bohm's favor, then a discussion would ensure which would be to the point and where people would not talk through each other. An important historical example is Eddington's expedition in order to decide between Einstein and Newton.

paradigms "necessarily" self-justifying? (3rd line from bottom)[24]
[13] page 110, line 5 from bottom:[25] almost always irreversible? This does not seem to be correct.

page 111 "always slightly at cross-purposes"[26] – this seems to be a little exaggerated. bottom: things "are again seen right side up"[27] – this is not factually correct. Things are not seen right side up but the situation is rather like the situation of a person who has learned reading a paper that is upside down. He does not see now the paper right side up, but rather he sees it still upside down without this disturbing him in his attempt to read the print.
[14] page 114, middle "that light was a self-consistent entity"[28] – this is not quite what people think in Copenhagen. There is no such thing as "light", there are, however, phenomena such as interference, Doppler effect, photoelectric effect which in the classical theory were explained, all of them, by reference to a single entity. Now this can't be done. Nor can we introduce a new entity (remember that the quantum theory is not a new paradigm, but it is classical physics, disinfected from metaphysics, and therefore also disinfected from the idea of real entities which are independent of observation. This having been done we cannot any longer speak of light being an entity, except in a picturesque manner, but we can only speak of different experimental results, and we can use the formalism of the quantum theory in order to predict what happens under these different circumstances).
[15] pp. 115 ff:[29] quite excellent. Here, at last, is empirical evidence, over and above vague talk such as propounded by Hanson, that we see different things when we see things differently.
[16] p. 118, last line of second paragraph.[30] In discussion you told me that in order to see something, or to experience a change of vision there must be a nature, a paradigm, and then you admitted, some general 'methodological' attitude concerning any paradigm. Now what you call 'the world' or 'a world' here is nature as seen through the spectacles of a paradigm, and 'different worlds' are simply different spectacles; (i.e. what Kant would have called 'nature'). This still presupposes the existence of one nature that is unchanged, although the paradigms are changed. I do not see how this nature can be eliminated by some "principle of economy". This is an almost positivistic argument. It also makes nonsense of the attempts of the psychology of perception to explain how one and the same world can be mirrored in a different manner in people who have accepted a different paradigm.
[17] page 121:[31] you talk about the traditional epistemological paradigm, about its past successes, about the fact that it was developed at the time of Newton, about the

need for its revision. All this I do not understand. What are you referring to? Are you referring to the theory of sense data? Then your history is incorrect: this theory was discussed in antiquity (see its explicit discussion in THEAITETOS), it was defended by Antisthenes, known to Aristotle etc. etc. "Its exploitation" you say "has been fruitful of a fundamental understanding that perhaps could not have been achieved in another way".[32] What is the evidence for this? What were the successes? What did it explain? It seems to me that here you simply want to please the philosophers by saying that in some sense they were right without knowing yourself the sense in which they were right. I think you are not confident of your judgement here and you should either omit the matter altogether, or deal with it in a much more explicit manner. Even more so as you [8|9] seem to identify the theory of sense data with a rarely accepted particular version of it, i.e. the version that sense data are pictures on the retina. What I would like you to retain, however, is your excellent argument against the assumptions that a sense data language is the neutral starting point (rather than the already infected point of arrival) for a scientific theory. This is a very good argument that should remain. Another thing that puzzles me is your diffident tolerance towards the sense data theory and your strenuous attempt not to hurt the sense dataists too much ["None of these remarks is intended to indicate" you write, on page 122 "that scientists do not characteristically interpret observations and data"[33] – this sound to me like "do not worry, babes, I am not out to kill you, I might even become your friend" etc. etc.]

[18] page 123: "Paradigms are not corrigible by normal science at all"[34] – this is a tautology if I ever heard one. But the last lines of this page are quite excellent. By the way: look for all places when you speak of 'retinal impressions' and eliminate this expression for the reasons stated above.

[19] page 127: "In the absence of a developed alternative".[35] May I modestly remind you of my "Explanation, Reduction, and Empiricism"?

"it no longer functions effectively"[36] – did it ever function? And if so, when? You can tell me in a minute which problems were solved at what time by Newton's theory. Can you tell me the same thing about what you call the "epistemological paradigm that has dominated philosophy for three [twenty would be better] centuries"?[37] In the last paragraph you say that a sense datum language will perhaps some day be devised.[38] And this after you have shown that this could not be done! This whole sense datum paragraph is very unsatisfactory. It seems to me that you are not quite sure of yourself, at least much less sure than you are when you are talking about science, and that your prose is very often dictated by the wish to be polite to the philosophers, and perhaps to get their sympathy, and their interest by not being too rough to them. Carnap and Goodman? (p. 128). Ridiculous! Very far? Carnap did not even arrive at tables let alone at atoms. And he explicitly abandoned his "Aufbau" ideas in 1935 (Testability and Meaning). And the main objection against all these attempts, however far they might get in constructing physical objects is that their elements of construction, i.e. the elements of a sense data language, have not yet been shown to exist in a neutral manner. For all these reasons I emphatically agree with you when you say "that scientists are right in principle as well as in practice when they treat oxygen and penduli (and perhaps also atoms and electrons) as the

fundamental ingredients of their immediate experience" (p. 128f).[39] By the way, a theory which would allow for such procedure has been sketched, in 1932, by Carnap himself and I have discussed it under the title of "the pragmatic theory of experience" in my explanation essay. Now from this there is still a long step to the point of view which you at some time seem to want to adopt (p. 118,[40] and my item [16] above) viz. that in addition to these immediate experiences no further 'nature' should be assumed to exist. You want to identify the 'worlds' of the scientist with 'nature', or omit the latter altogether which would mean that you identify experiences with 'nature' (experiences of the more complicated kind you are defending such as atom-experiences, penduli-experiences etc. etc.) or omit the idea of nature altogether (of a nature that is independent of experiences) which would mean that you fall back upon the 'traditional paradigm of epistemology' the only difference being that your experiential language is more complicated than theirs. I must repeat that this chapter on epistemology makes a somewhat confused impression upon me.
page 130: again the 'retinal imprints'. The same on page 131⟨.⟩[E41]
page 131 bottom:[42] very important (same operations, different aspects) the **[9|10]** same applies, of course, also to words: same words obtain a very different meaning.
[20] page 112: your reference to Hanson.[43] Your defence, in conversation, of this reference, seems to me to amount to saying that you prefer being member of a community to searching after the truth. True, Hanson, in some vague sense, defends the same point of view which you also defend. But first of all he has added nothing at all to Wittgenstein whom he imitates (all you need is to read, in this connexion, and really should read, are Wittgenstein's investigations on seeing in the second part of the INVESTIGATIONS), and not at all well. Of course, Hanson is closer to you than, say, to Ayer, and his book is widely read. But as regards the former argument I must retort that in some sense he is very far away indeed from you and this is care in presentation of point of view. And as regards the second argument I really cannot accept it: when many people read Hanson, the worse for them; no obligation for you can be derived from the follies of others. Worse still, by referring to the book you as it were make legitimate their follies, which is a very bad thing indeed. Do not misunderstand me: I love Hanson, and this is the reason why my review was so tame (I should have referred to the incredible mess he makes of the purely formal parts of quantum theory in the footnotes). You really seem to practice what you preach: belonging to a community is the highest value for you as well as your criterion for the selection of quotations. You quote Hanson because he supports the epistemological paradigm you are prepared to defend (though neither you nor he has succeeded in formulating clearly what is involved in this paradigm) and you do not at all care about his sloppiness, inaccuracies on physical, philosophical (and perhaps even historical) points. You by your quotation you recommend Hanson to people despite the fact that they will acquire sham knowledge etc. etc. (misleading popularization) – but this does not matter for you as soon as, by hook or by crook, they are made to join the very varied community of those who believe in a new paradigm. Using a political example this seems to me like putting the party above the morals. If somebody is an influential party member, he is to be welcomed and to be drowned in honors, even if it is well known that all his influence derives from his contacts in

the underworld and that his money has been obtained by holding up banks. Tom, you are immoral, as immoral as a Hegelian who identifies the main forces of history with what is good. I take a completely different course, also in my classes. People ask me to talk about Wittgenstein, because he is an influential contemporary philosopher, and they also ask me to talk about Ayer, etc. etc. I refuse to do this. And if they ask "shall we read the PHILOSOPHICAL INVESTIGATIONS?" I reply "If you want to waste your time, yes". In reading Hanson, however, I cannot see the slightest merit. Wittgenstein, at least, is an original source for some craze, and so there may be a justification, for historical reasons, to study him. But for reading Hanson there is not the slightest justification. By the way it may interest you that Stanley seems to be on your side in the question of quoting Hansons ⟨sic⟩. This became clear to me when we discussed Wagner. "Wagner is an influence" he said, "therefore we must study him in order to overcome him". All this is historicism: Wagner is a historical influence, and will act, unless one counteracts. Has nobody ever hit upon the simple idea of simply forgetting about Wagner? This, surely is less time consuming, and just as effective. Why first teach something, and then spend all one's energy in order to refute it? Simply don't mention it. **[10|11]**

[21] page 138 "must be slightly at cross purposes"[44] – this may happen as regards this or that problem, but not as regards every problem. For if the two theories were at cross purposes as regards every problem there could not be a "conflict" (your page 94)[45] between the two paradigms. However as soon as there is a conflict, in the very same moment there exists also common ground and the possibility to decide, to the satisfaction of both parties, which paradigm has to go. It seems to me that you exaggerate, and generalise a situation that may occasionally occur (that people talk through each other) in such a manner that you cannot any longer say what you want to say, namely, that they hold conflicting points of view.

Let me mention in this connexion that the assertion, made by many ordinary language philosophers (including your clever friend Stanley in his attack upon poor Pole) that certain philosophical theories do not conflict with the ordinary point of view and are therefore nothing but misleading reformulations of that point of view, let me say that this assertion is incorrect for exactly the same reasons. Take the assertion which is sometimes made by philosophers (and which may also be subscribed by a Popperian) that we never can fully establish any statement. Now the ordinary language philosophers (OLP for short – I like this abbreviation, it sounds like GULP) will at once point out that the word 'fully establish' is used here in a very unusual sense; so, he will say, when the ordinary man says, having seen and touched a table, that he has 'fully established' the presence of a table in front of him, the attacked philosopher will say that this has not been fully established etc. etc. And as both use the words 'fully established' in different senses, they are not in conflict when the one says 'fully established' and the other 'not fully established'. This is quite correct – in this single instance. Still there is a world of difference between the two as regards questions of test. The ordinary man is satisfied with the testimony of his eyes and he therefore stops after a finite amount of steps. The attacked philosopher is not so satisfied and he will regard any stopping as only preliminary and due to lack of time. He will not ascribe any logical preference to the process of looking, assuming, for example, that it is very convincing evidence that settles the matter. For him there is nothing that settles the matter except lack of time. And this point of view is very different from the corresponding point of view of the everydayman (or from the point of view which people such as Cavell ascribe to somebody they call the everydayman and who does not exist).

[22] page 140 good reasons why textbooks "should be systematically misleading"[46] – this sounds like there being good reasons for stealing. Also I hope I have shown above that alternatives to any present paradigm should be considered, such consideration leading to the more speedy arrival of crises. Hence it is imperative that textbooks be not written in the manner indicated. I therefore cannot accept the "as pedagogy this technique ... is unexceptionable" of p. 144 either.[47] You are too good to too many people at once. This can only lead to anarchy.

[23] Your chapter XI does not at all take into consideration the creation of crises 'from above' and from 'within' (cf. my item [2] on page 4 of my letter). Adding this will, I think, lead to a radically different summing up! For example you will not be able to say that testing occurs only "after the sense of crisis has evoked an alternative candidate for paradigm" (page 150, line 8 from below)[48] as you have admitted that a crisis may be brought about by the existence of an alternative paradigm from whence it follows that alternatives will not be invented only after the arrival of a crisis.

[24] page 151: "probabilistic verification theories demand the comparison [11|12] of all theories that could be possibly constructed for the explanation of a given body of sensory experience" Not so! Carnap's theory, for example, does not need reference to all alternatives (Reichenbach's does), nor does it refer to sensory experience as evidential basis: it refers to experimental results.[49]

[25] page 153: "to the historian, at least etc. etc."[50] But the historian can only show that people are sloppy, for example, that they do not all obey the 10 commandments. He cannot show, that the 10 commandments are wrong. Same about scientific method.

page 154: end of first paragraph: resolved by proof:[51] this suggests that you attack the very nonsensical theory that logic alone ⟨sic⟩ (see my above reference to other occurrences of this expression)⟨.⟩[E]

[26] page 155: "The men who scoffed at Einstein etc."[52] –You say that their space could not be curved[53] and you thereby seem to indicate that they had some reason to smile when Einstein said that space was curved. You ought also mention, if you give the defenders of this argument such an advantage (you seem here to lean towards Heisenberg's idea of a 'closed theory' and also a little towards Hanson), that space in their sense does not exist, or at least that it does not exist, if Einstein's theory is correct and this quite irrespective of whether the latter theory is now expressed by saying that 'space' is curved, or in a different, and to the people you mention, less offensive manner. Same about 'earth'. ARGUMENTS FROM MEANING NEVER LEAD ANYWHERE, AT THE VERY MOST, IF YOU GRANT THEM, YOU HAVE TO ADD THAT THE THING WITH THIS PARTICULAR MEANING DOES NOT EXIST.

page 156:[54] again the different worlds with ideas which are at cross purpose: how can such different worlds conflict? And if they conflict, on the other hand, is it then not possible to force the transition from one paradigm to the next?

Let me repeat this argument. You say (1) that succeeding paradigms will conflict; (2) that people holding them will talk through each other; and that (3) they live therefore in two different worlds such that the transition from the one to the other (4) cannot be forced, but is something like a conversion. Now (2), (3), and (4) seem to represent too radical a view of the matter (crucial experiments do exist), and moreover they are inconsistent with (1): if paradigms do conflict then there must be at least two statements, the one from the first paradigm, the other from the second, which are inconsistent, and therefore do not "talk through each other". Hence, if the holder of the first paradigm can, by experiment, show that his statement represents the facts, then the second chap must give up and this without any conversion being involved. For as both share a statement we must also assume that they share a set of ideas concerning the circumstances under which this statement is established by experiment (otherwise they would not share the statement, but at most a formulation of a statement, or a sentence!)
line 5 from bottom of page 157[55] is therefore much too general. Conversion may be necessary, it is not necessary in the case of conflicting paradigms. That is, it is never necessary (for you assert that all succeeding paradigms conflict – by the way, you assert this without giving reasons for it!)

I think I stop here. I shall perhaps send you another letter.

Paul

---

[3] * You only say that if there are more paradigms, then there will be a mess.

Editors' Notes

PF 5-3-2 (12 ss., T+H)

1 See FEYERABEND 1962a. On the history of the development and publication of Feyerabend 1962a, see *above*, Part I, Ch. 4, 1961-1, n14 and Vol. 1, Part II, 1957-12, n1.
2 See KUHN 1962²: 93-94.
3 See KUHN 1962²: 94, top.
4 Similarly in KUHN 1962²: 94, but without "pure".
5 See KUHN 1962²: 94.
6 Very similarly in KUHN 1962²: 96.
7 See KUHN 1962²: 96, bottom.
8 See KUHN 1962²: 96-97.
9 See KUHN 1962²: 98.
10 Similarly in KUHN 1962²: 98, but with 'early' inserted: "early logical positivism".
11 See KUHN 1962²: 98.
12 See *above*, n1.
13 The attached review (see ROSENFELD 1957a) is here reproduced:

1983

1. Feyerabend to Kuhn, 4 April 1983

APR 14 1983[1]
4.4.83

Dear Tom,

first, let me congratulate you on the Sarton Medal – I just read about it in ISIS.[2]

Then let me say how pleased I was to meet you after such a long time and to find you in such good spirits. There were many things I would have liked to discuss with you but, alas, there was no time. Maybe the next time I come through Boston.

You said you were returning to the subject of incommensurability including a criticism of those who made a mess of it, myself among them. Now I think I have written a very good case study of a transition between incommensurable frameworks in chapter 17 of Against Method;[3] it deals with the transition from the Homeric world view to the 'substance universe' of some of the Presocratics and with problems such as: under what circumstances can we infer a change of perception from a change of style (in painting and sculpture), concepts and grammar i.e. when can we say that a group of people sees the world as it paints it (as it talks about it etc.) and when we cannot make such an inference. I also try to define incommensurability in this special case. I chose Homer in order to get away from the temptation to use philosophy-of-science terms and to develop more general terminology. Unfortunately none of the many people who cursed me ever looked at this chapter – but it might interest you.

Before that I came to incommensurability (one narrow aspect of it) via a critique of the double language model. In Minnesota where I spent two years (in installments) the situation was presented thus: there is a theoretical language, there is an observation language and meaning flows from the observation language to the theoretical language by "upward seepage" (this is how Paul Meehl, the psychologist expressed the matter). In my thesis (1951)[4] I tried to show that observations taken by themselves cannot determine meanings. I therefore suggested a "downward seepage" theory i.e. I suggested that all there is are theoretical meanings, all terms are, semantically speaking, theoretical terms and the difference between observation and theory has nothing to do with meaning. To this Feigl objected that basing meanings on theories makes nonsense of crucial experiments between vastly different theories – such theories will have no common semantic elements. I replied that crucial experiments do not need common semantic elements, all that is needed are certain behavioral overlaps. Semantically, I said, the theories are incommensurable but that does not matter, for scientists do not argue in accordance with philosophical semantics. All of this I explained in a paper that appeared in 1958 in the Proceeding ⟨sic⟩ of the Aristotelian Society[5] (Feigls ⟨sic⟩ objection and my reply in footnote 21 of the reprint in Vol. i of my Philosophical Papers)[6] though without the word 'incommensurable'. Your notion of incommensurability is of course much wider. It is not tied to the statement view, as Stegmueller calls the idea that science is the same as a collection of statements and it takes non-articulated tendencies into account. But I don't think that you will find faults with either the Homer case or the earlier and more narrow version.

A few weeks ago I was asked if I was interested in an Albert Schweitzer Chair at NYU. $ 100 000 per year and lots of prestige. I thought about it for a few days and then said no: here in Berkeley nobody pays much attention to me, I can leave a semester each year and go to Zuerich. In New York I would be in a much more prominent position and, contrary to what you seem to think about me I like anonymity. As a matter of fact I very much yearn for my student days when nobody knew me and I could go to any meeting and open my mouth on any matter without people saying: look, this is PKF making a fool of himself.

Well, and now I'm overdue for watching Dallas – all the best and maybe we meet again, soon.

Paul

TK MC 240-21/28 (2 ss., T)

[1] The date, impressed with a stamp, probably indicates either the date Kuhn received the letter or the date he read it.

[2] See *Isis*, 74: 1 (March, 1983), p. 153.

[3] See FEYERABEND 1975: Ch. 17.

[4] See FEYERABEND 1951.

[5] See FEYERABEND 1958a.

[6] See FEYERABEND 1981a: 31n21 = Feyerabend 1958a: 163n22.

2. Kuhn to Feyerabend, 20 April 1983

April 20, 1983

Professor Paul Feyerabend
Department of Philosophy
University of California
Berkeley, CA 94720

Dear Paul,

Many thanks for your congratulations and even more for your warm letter of April 4th which included them. I too missed the chance to have some real talk when we saw each other so briefly at the Cohens',[1] and it pleases me greatly to think that the chance may come again, for I have often remembered our earlier talks with pleasure and wish that there had been some way to continue them. Please do let me know if you are going to be anywhere near here, for we should definitely give things a try once again.

I say all this even though I fear that your letter may have been evoked by a misunderstanding. I did not mean to imply that the paper I was working on would be directed to partisans of the notion of incommensurability but only to its critics. In short, I did not at all think of you as a target, something you will see if you look at the copy of the final version that I'm sending under separate cover. There are, as you will see, a number of significant respects in which I think you were more nearly right than I the first time around.

I have not read the part of Against Method[2] that you refer me to, but shall try to do so in near future. As you say, it is very likely to interest me. More generally, I shall be surprised if we are very far apart on this particular subject, though there is one thing in your letter that does seem to me a poor move. I think, as you do, that a language embodies cognition, and that it may therefore be necessary to change language here and there when one makes the transition to a new theory. On the other hand, I am reluctant to speak of language as embodying a theory. It seems to me better to reserve the term "theory" for an explicitly articulated structure. (it ⟨sic⟩ may be a structure of statements or one more like Stegmüller's, indifferently), ⟨sic⟩ and to acknowledge that there are other ways of embodying cognition. One advantage of this mode of speech is that it permits one to acknowledge that, given a language, there are an infinity of possible theories that might be stated in it, even though it may also be the case that that infinity does not include the theory that is going to emerge next in science.

Many congratulations on the Schweitzer Chair. I do, however, entirely understand your reasons for declining (what you mean when you say that "contrary to what you [Tom] seem to think about me [Paul] I like anonymity") it ⟨sic⟩ would have been fun to have you nearer by, and I think we would have managed more talk, but you are right in saying it is hard to be private in New York. You will be amused to know that I also was offered a Schweitzer Chair at NYU just over four years ago, and the difficulty of finding intellectual seclusion in New York was among my reasons for declining.[1|2]

Yes, ⟨sic⟩ let us by all means find a way to talk soon. Will you be in Berkeley when I visit briefly next September?[3]

As ever,
⟨UNSIGNED⟩
Thomas S. Kuhn

TK MC 240-21/28 (2 ss., T, c)

[1] At the turn of 1982, or more likely of 1983, Feyerabend and Kuhn met after about 17 years for the first time at a dinner with Robert. S. Cohen (1923-2017) and Marx W. Wartofsky (1928-1997) at the Cohens' place (see Feyerabend to Albert, [4 January 1982], in FEYERABEND-ALBERT 2008: 249; the letter was possibly misdated by the editors, who do not specify based on what evidence they assigned that date).
[2] See FEYERABEND 1975.
[3] Feyerabend and Kuhn met again briefly in Berkeley at Feyerabend's place in mid-September 1983 (see Feyerabend to Hoyningen-Huene, 8 October 1983 in FEYERABEND-HOYNINGEN-HUENE 2010: 57; and *below*, Kuhn to Feyerabend, 7 February 1988, 1988-1 and Feyerabend to Kuhn, 12 February 1988, 1988-2).

1988

1. Kuhn to Feyerabend, 7 February 1988

7 February 1988

Professor Paul Feyerabend
1168 Miller
Berkeley, CA 94708

Dear Paul:

You've known for a long time that I'm a dreadful letter writer. But having Paul Hoyningen as a go-between makes a difference.[1] He's shown me your answer to Putnam[2] and also your letter about your route to incommensurability.[3] I want to return the favor, especially since Paul (the other one) indicated you had expressed an interest.

Under separate cover I'm sending you three manuscripts. The first, "Commensurability, Comparability, Communicability" appeared in PSA 1982.[4] It's already somewhat out of date (as the other manuscripts will indicate), but it's got my answer to Putnam as well as a partial statement about our relationship. It's remarkable how close we are on the former, and I sense only one disagreement on the latter. (I think I've never read or even heard of the Parry article you mention.)[5] But is there perhaps an item missing in your letter? I've sometimes wondered whether dissatisfaction with quantum theory didn't play a role for both of us. In neither case do I know the answer but my wondering continues.

The second piece, "Possible Worlds in History of Science," was prepared for a Nobel Symposium the summer before last and is still working its way through the publication mill.[6] You'll see that the bit about structure at the end of the previous paper has become a great deal more central in the interim. And that development continues in the last item I'm sending, a series of three lectures I delivered last November in London. They will not be published in their present form.[7] I mean to develop the ideas they expressed further in a book (there are pieces missing in the present form, and the last lecture needs vast expansion). Probably that's going to take a few more years, so I'd be glad if you'd be careful with the document and its contents for the time being.[8] On all these matters, your reactions would interest me greatly.

Our reunion in Zurich continues memorable ⟨sic⟩ for both Jehane (who sends the very warmest greetings) and to me. I indeed was a great deal moved by the experience[9] as I had also been by my visit with you in Berkeley.[10] We must find some other opportunities. Are you by any chance going to be around during [1|2] July? Jahane and I will be in Santa Cruz for most of the month, and I'm sure we'll be getting to Berkeley from time to time.

As ever,
⟨UNSIGNED⟩
Thomas S. Kuhn

TK MC 240-21/28 (2 ss., T, c)

[1] Paul Hoyningen-Huene was born in Pfronten/Allgäu, West Germany in 1946. He studied Physics and Philosophy in Munich (1966-1971), Mathematical Physics at Imperial College, London (1971-72) and did graduate work in Theoretical Physics at the University of Zurich (1972-1975). He taught in Switzerland and Germany (1976-) and was a visiting scholar at the Massachusetts Institute of Technology with Thomas S. Kuhn (1984-85). Based on regular, extensive discussions with Kuhn about his views, Hoyningen-Huene advanced an original interpretation of Kuhn's ideas in HOYNINGEN-HUENE 1989. The book was enthusiastically endorsed by Kuhn in the 'Foreword' dated August 1988. Kuhn recounts that when he met him in the mid-1980s, "Hoyningen knew my work better than I and understood it very nearly as well," declaring: "No one, myself included, speaks with as much authority about the nature and development of my ideas" (KUHN 1993: p. xi). Hoyningen-Huene was also instrumental in reuniting Kuhn and Feyerabend in Zurich and Meilen in June 1985. Kuhn arrived in Zurich on 21 June and gave a talk "What are Scientific Revolutions?" at the ETH (Audimax ETH Zentrum, 5 p.m.). Feyerabend and Kuhn spent the following two days together: on 22 June, they took an 'intellectual cruise' around the lake; on 23 June, they had lunch together after a walk in Meilen. Kuhn returned to Paris on 24 June. (See *below*, main text at n9; Feyerabend to Kuhn, 2 February 1988, 1988-2; Kuhn to Feyerabend, 8 November 1988, 1988-3; FEYERABEND 1995:152-153 – the photograph in Meilen is from 23 June 1985; Paul Hoyningen-Huene, personal communication).

[2] See FEYERABEND 1987.

[3] See Feyerabend to Hoyningen-Huene, 23 October 1987, in FEYERABEND-HOYNINGEN-HUENE 2010: 151-152.

[4] See KUHN 1983.

[5] See Feyerabend to Hoyningen-Huene, 23 October 1987, in FEYERABEND-HOYNINGEN-HUENE 2010: 151; and *below*, Feyerabend to Kuhn, 12 February 1988, 1988-2 and n2.

[6] See KUHN 1988.

[7] From 23-25 November 1987, Kuhn delivered three Shearman Memorial Lectures at the University College London under the collective title "The Presence of Past Science"; see KUHN 2022a.

[8] Kuhn never completed the book project that he was working on in his final years. The edited draft was posthumously published as KUHN 2022b.

[9] See *above*, n. 1.

[10] Kuhn is probably referring here to his meeting with Feyerabend in Berkeley in September 1983 (see *above*, Kuhn to Feyerabend, 20 April 1983, 1983-2, n3).

2. Feyerabend to Kuhn, 12 February 1988

FEB 24 1988[1]
02/12/88

Dear Tom,

I was enormously pleased to get your letter. And the fact that you hardly ever write played a large role. So, valuing quality over quantity I think you should continue with your habit (which you will anyway).

I also was very pleased by our reunion in Switzerland. In Berkeley, unfortunately, I was full of pills because of pains and so I noticed you only dimly, but in Zuerich and Meilen I was fully there.[2] Unfortunately I won't be in Berkeley in July – this will be my time in Switzerland – but one of these days I shall stop on my way east or on my way back west and invade you in Boston.

I am looking forward to receiving your essays. I am now preparing a 'book', working title, 'Reality'.[3] It will take me at least another five years to finish and it will deal not only with science and philosophy but lots of it will be about painting, music, architecture and it will be a kind of farewell to relativism. I have great sympathy for that view and I think that most objections against it don't work, but it has its limits and I am trying to find out what the limits are. I understand you have never been a relativist but are also trying to find its limitations.

The Parry article has been reprinted in G.S. Kirk (ed.) The Language and Background of Homer, W. Heffer & Sons Ltd, Cambridge 1964, 48f,[4] it is short it is well worth reading. It deals with the problem of a person, Achilles, trying to transcend a certain paradigm which, for Perry, is fully defined by the Homeric language: 'Achilles has no language with which to express his disillusionment. Yet he expresses it, and in a remarkable way. He does it by misusing the language he disposes of. He asks questions that cannot be answered and makes demands that cannot be met. He uses conventional expressions where we least expect him to...

There are wonderful examples in painting. For example, the halo. First, when symbols abound it surrounds the holy heads in golden splendour. When perspective intrudes it becomes elliptical and some Saints are surrounded by veritable rings of Saturn. In a sense this is completely absurd: what has holiness yet to do with perspective? The socalled master of Flemalle 'The Virgin and Child before a Fire Screen' has an interesting way out: the firescreen behind the Virgin has exactly the shape of a halo, and it is in the right place, but it is a piece of matter.[5]

You say that quantum theory must have played a role in my way towards incommensurability. I'm not so sure, for I don't remember all the details. The idea that there exists no semantic distinction between observation statements and theoretical statements (observation statements are not merely theory-laden but fully theoretical) came from the protocol statement discussion of the Vienna Circle; just good old positivism. What I added was a discussion and refutation of what I called the 'stability thesis' viz. that some meanings are unaffected by theoretical change. In my 1958 paper[6] I discussed two ways of defending that thesis, one being connected with Bohr's version of complementarity. But this was only an example and I don't remember how much it meant to me at the time. The objection that crucial experiments would now be impossible because theories would be incomparable came from Feigl;[1|2] it did not bother me at all for I interpreted crucial experiments as practical decisions not as theoretical comparisons of statements with stable meanings.* In 1962 I gave some examples of the difficulties Feigl seemed to have in mind and called them 'incommensurable'.[7] Reading Anneliese Maier's Die Verlaeufer Galilei's ⟨sic⟩ im 14. Jahrhundert[8] where she emphasizes the difference between impetus and momentum played a great role here. This was a real revelation to me. But again I though ⟨sic⟩ that while the incommensurability might be a disaster for philosophical theories of explanation and reduction I did not regard it as a disaster for science and when I said that scientific progress involves 'irrational' elements I only meant: 'irrational' in the sense of certain philosophical theories. The criticism was of philosophy, not of science.+

All the best, best wishes to Jehane and I am sure we are soon going to meet again. We simply must!

Paul

**crucial experiments, therefore, had a 'subjective' component*
+All this was carried out within the anti- (or a-) historical framework of Neopositivism up to Quine. The historical approach I adopted only later – and there you were my teacher.

TK MC 240-21/28 (2 ss., T+H)

[1] The date, impressed with a stamp, probably indicates either the date Kuhn received the letter or the date he read it.
[2] See *above*, Kuhn to Feyerabend, 20 April 1983, 1983-2, n3 and 7 February 1988, 1988-1 and n1, respectively.
[3] Feyerabend may be referring here to the book project on which he would be working over his last years, whose edited draft was posthumously published in FEYERABEND 1999b.
[4] See PARRY 1956.
[5] See FEYERABEND 1984: 98.
[6] See FEYERABEND 1958a.
[7] See FEYERABEND 1962b.
[8] See MAIER 1949.

3. Kuhn to Feyerabend, 8 November 1988

8 November 1988

Dear Paul,

I know you are not a joiner, but I nevertheless welcome you to the club. As one who got there eighteen months before you, I can tell you that sixty-five is a good age, not at all the ancient state we thought it was in our teens, twenties, and thirties. There are of course changes, not all of them welcome. But I report with pleasure and surprise that peace of mind and depth of mind are with me as never before. I wish the same to you.

I'm working again on incommensurability, and it brings back fond memories of the days in Berkeley when we discovered we had selected the same term to cover much the same set of phenomena.[1] Did you know what we had let ourselves (and the word) in for? That's a question I ask myself, and I'm by no means sure of the answer.

The talks that followed on that discovery, I've greatly missed, though I do think they were too often devoted to the wrong question. But I've come to recognize that what we mainly deferred about was not substance, as it then seemed, but goals: I wanted a theory of the way things were; you, I think, wanted to change them. Though my primary goal remains what it was, I now see the need for change far more clearly.

One of the rewards of the last few years has been being in touch with you again. I was especially moved by our time together in Zurich.[2] Let us make sure there is more! Until then,

Happy birthday,
Tom

PF 3-9-1 ≈ TK MC 240-21/28 (1 s., T)

[1] See Feyerabend 1970a: 197-198, 219 = Feyerabend 1981a: 277-278, 300.

[2] See *above*, Kuhn to Feyerabend, 7 February 1988, 1988-1 and n1.

CAUSE IN PHYSICS

By L. Rosenfeld

Causality and Chance in Modern Physics. By D. Bohm. Foreword by Louis de Broglie. Routledge and Kegan Paul. Pp. xi. 170. 21s.

This is the most paradoxical book I have seen for many years. Its author is not only a distinguished physicist but a thinker with progressive views on most things; yet he is blind to one of the greatest of the advances in rational thinking which the development of physics has initiated. He purports to discuss fundamental aspects of modem science but, scorning the scientific method established since Newton's time, he revives the metaphysical attitude of system-makers like Descartes, whose self-assurance he wonderfully imitates. He holds out the expectation that his ideas will eventually elucidate the strange properties of matter in its states of highest energy, but there is not a word in the book about the possible relation of these properties to the themes treated at great length in it. The elementary style of exposition could easily mislead the unwary layman into the belief that he is here presented with a straightforward account of the problem stated in the title; but what the specialist sees in these dreary pages is a picture distorted beyond recognition by misinterpretation and misrepresentation.

The author's original endeavour was to vindicate the traditional conception of deterministic causality in atomic physics. Having failed in this, he now tries to argue that one ought to consider more general types of causality, partaking of both determinism and chance: the element of chance appearing at any "level" of description would have to be underpinned by the determinism of some hidden processes going on at a "deeper" level. This novel conception, however, is not "synthetic", as he thinks, but just confused; at any rate, it does not correspond to the actual situation with which we are confronted in atomic physics In fact, the infinite sequence of deeper and deeper levels of reality which our author imagines, unsupported as it is by any concrete evidence, appears to serve no other purpose than to provide him with a perpetual line of escape from embarrassing difficulties.

But why, the reader may well ask, all this fuss about determinism? The true character of the causal relations of atomic physics has been fully elucidated by Bohr and Heisenberg thirty years ago, and all that we have learned since about atoms and nuclei and other fundamental agencies of Nature has only strengthened the conclusions then reached and extended their scope. It turns out that the laws governing the behaviour of atomic systems are "statistical", in the same sense as those governing the behaviour of human communities: physicists thrive on chance just as insurance people do. This situation may look strange at first sight, and indeed its explanation would exceed the bounds of a review; but there is nothing "irrational" about it. If it still disturbs some physicists, it is because, in contradiction with the exigencies of sound scientific method, they give to some metaphysical prejudice, like determinism, greater weight than to the immediate lessons of experience. That such irrational dogmatists should hurl the very accusation of irrationality and dogmatism at the defenders of the common-sense, uncommitted attitude of the true scientist is the crowning paradox which gives a touch of comedy to a controversy so distressingly pointless and untimely.

Most likely, the underlining was added by Feyerabend's hand.

[14] Feyerabend is probably referring to the project "Sources for History of Quantum Physics" which was carried out in 1961-64 by a research committee based at the University of California, Berkeley and composed of Thomas S. Kuhn, John L. Heilbron, Paul Forman, and Lini Allen, with the aim of collecting both manuscript and oral records pertaining to the development of quantum physics from 1898 to 1933; their results constituted the original basis of the Archive for History of Quantum Physics (see *above*, Vol. 1, Part II, Feyerabend to Popper, 17 August 1961, 1961-9; and KUHN-HEILBRON-FORMAN-ALLEN 1967; HEILBRON 1968; <http://ohst.berkeley.edu/publications/quantum.html>).

[15] KUHN 1962^2: 98 bottom.

[16] Kuhn 1962^{2}: 99.
[17] Kuhn 1962^{2}: 100, middle paragraph.
[18] See Kuhn 1961: 166n9 [186n9].
[19] See Kuhn 1962^{2}: 101-103.
[20] See Kuhn 1962^{2}: 103, beginning of second paragraph.
[21] See Kuhn 1962^{2}: 108.
[22] Feyerabend is probably referring to Kuhn 1962^{2}: 109: "Therefore, when paradigms change, there are usually significant shifts in the criteria determining the legitimacy both of problems and of proposed solutions."
[23] Kuhn 1962^{2}: 109, bottom.
[24] No matching phrase is present in Kuhn 1962^{2} (see Kuhn 1962^{2}: 110).
[25] See Kuhn 1962^{2}: 111.
[26] See Kuhn 1962^{2}: 112.
[27] Similarly in Kuhn 1962^{2}: 112: "are again seen as they had been".
[28] See Kuhn 1962^{2}: 114, bottom.
[29] See Kuhn 1962^{2}: 115 ff.
[30] See Kuhn 1962^{2}: 118, last line of second paragraph.
[31] See Kuhn 1962^{2}: 121, middle.
[32] See Kuhn 1962^{2}: 121.
[33] See Kuhn 1962^{2}: 122.
[34] See Kuhn 1962^{2}: 122.
[35] See Kuhn 1962^{2}: 126.
[36] See Kuhn 1962^{2}: 126.
[37] Similarly in Kuhn 1962^{2}: 126: "epistemological viewpoint that has most often guided Western philosophy for the three centuries".
[38] See Kuhn 1962^{2}: 126, bottom.
[39] See Kuhn 1962^{2}: 127-128.
[40] See Kuhn 1962^{2}: 118.
[41] See Kuhn 1962^{2}: 128, 129.
[42] See Kuhn 1962^{2}: 130, middle.
[43] See Kuhn 1962^{2}: 113.
[44] Similarly in Kuhn 1962^{2}: 148: "are always at least slightly at cross-purposes".
[45] See Kuhn 1962^{2}: 96, middle.
[46] See Kuhn 1962^{2}: 137.
[47] See Kuhn 1962^{2}: 140.
[48] See Kuhn 1962^{2}: 145, top.
[49] The respective sentence is changed in Kuhn 1962^{2} to: "One probabilistic theory asks that we compare the given scientific theory with all others that might be imagined to fit the same collection of observed data" (see Kuhn 1962^{2}: 145).
[50] See Kuhn 1962^{2}: 147.
[51] See Kuhn 1962^{2}: 148.
[52] Similarly in Kuhn 1962^{2}: 149.
[53] The word "their" is not present in Kuhn 1962^{2}.
[54] See Kuhn 1962^{2}: 150.
[55] Feyerabend is probably referring to Kuhn 1962^{2}: 151: "The transfer of allegiance from paradigm to paradigm is a conversion experience that cannot be forced".

References

Unless otherwise specified, page references in the text are always to the most recent version of a book. Page references to the English translations are given within square brackets.

ANDRESEN, Jensine

1999 "Crisis and Kuhn", *Isis*, **90**: Supplement (1999): *Catching up with the Vision: Essays on the Occasion of the75th Anniversary of the Founding of the History of Science Society*, S43–S67.

ASTER, Ernst von

1925 *Platon: mit einem Titelbild*, Strecker und Schröder: Stuttgart 1925.

AUGUSTIN, Christian (ed.)

2010 *Aber ein Paul hilft doch dem anderen. Paul Feyerabend – Paul Hoyningen-Huene. Briefwechsel 1983–1994*, Passagen Verlag: Vienna 2010.

AUSTIN, John L.

1962 *Sense and sensibilia*, G. J. WARNOCK (ed.), Clarendon Press: Oxford.

AXTELL, Guy S.

1993 "In the Tracks of the Historicist Movement: Re-assessing the Carnap–Kuhn Connection", *Studies in History and Philosophy of Science Part A*, **24**: 1 (March 1993), 119–146.

AYER, Alfred Jules

1936 *Language, Truth and Logic*, Victor Gollancz: London 1936, 1946[2].

1940 *Foundations of Empirical Knowledge*, Macmillan: London 1940.

BACON, Francis

1620 *Novum Organon Scientiarum* in *The New Organon*, JARDINE–SILVERTHORNE (eds.), Cambridge University Press: Cambridge 2000.

M. Collodel, E. Oberheim (eds.), *Feyerabend's Formative Years. Volume 2. Feyerabend on Logical Empiricism, Bohm & Kuhn*, Vienna Circle Institute Library 10, https://doi.org/10.1007/978-3-031-57519-8

BAR-HILLEL, Yehoshua

1950 "Bolzano's Definition of Analytic Propositions", *Theoria*, **16**: 2 (August 1950), 91–117.

BARNETT, S. H. G.

1957 "Foreword" in *Observation and Interpretation: A Symposium of Philosophers and Physicists, Proceedings of the Ninth Symposium of the Colston Research Society*, KÖRNER–PRYCE (eds.), Butterworths Scientific Pub.: London and New York 1957, p. v.

BELL, John Stewart

1966 "On the Problem of Hidden Variables in Quantum Mechanics", *Reviews of Modern Physics*, **38**: 3 (July 1966), 447–452.

BEYER, Robert T. (ed.)

1949 *Foundations of Nuclear Physics*, Dover: New York 1949.

BLACK, Max

1949 *Language and philosophy: Studies in Method*, Cornell University Press: Ithaca (NY) 1949.

BOHM, David Joseph

1951 *Quantum Theory*, Prentice Hall: New York 1951.

1952a "A Suggested Interpretation of the Quantum Theory in Terms of "Hidden Variables" I", *Physical Review*, **85**: 2 (January 1952), 166–179.

1952b "A Suggested Interpretation of the Quantum Theory in Terms of "Hidden Variables" II", *Physical Review*, **85**: 2 (January 1952), 180–193.

1957a *Causality and Chance in Modern Physics*, Routledge & Kegan Paul: London and Van Nostrand: Princeton (NJ) 1957.

1957b "A Proposed Explanation of Quantum Theory in Terms of Hidden Variables at a Sub-quantum Mechanical Level" in *Observation and Interpretation: A Symposium of Philosophers and Physicists, Proceedings of the Ninth Symposium of the Colston Research Society*, S. KÖRNER – PRYCE (eds.), Butterworths Scientific Pub.: London and New York 1957, pp. 33–50.

[1959] "On Certain Epistemological and Cosmological Problems Raised in Dr. Feyerabend's Review of *Causality and Chance in Modern Physics*" in *Feyerabend's Fromative Years. Volume 1. Feyerabend and Popper*, COLLODEL–OBERHEIM (eds.), Springer Nature Switzerland AG: Cham 2020, Part 2, Ch. 7.

1961 "On the Relationship between Methodology in Scientific Research and the Content of Scientific Knowledge" *British Journal for the Philosophy of Science*, 12: **46** (August 1961), 103–116.

1962 "Classical and Non-Classical Concepts in the Quantum Theory. An Answer to Heisenberg's Physics and Philosophy", *The British Journal for the Philosophy of Science* **12**: 48 (February 1962), 265–280.

1964 "On the Problem of Truth and Understanding in Science" in *The Critical Approach to Science and Philosophy in Honor of Karl R. Popper*, M. BUNGE (ed.), Free Press: New York 1964, pp. 212–223.

1963a "Problems in the basic concept of physics. An inaugural lecture delivered at Birkbeck College 13th February 1963" in *Satyendranath Bose 70th Birthday Commemoration Volume, Part II*, Calcutta 1965, pp. 279–318.

1963b "Commentary" in *Scientific Change: Historical Studies in the Intellectual, Social, and Technical Conditions for Scientific Discovery and Technical Invention, from Antiquity to the Present* (Symposium on the History of Science, University of Oxford, 9–15 July 1961), A. C. Crombie (ed.), Heinesvann: London and Basic Books: New York 1963, pp. 477–486.

1966 *The Special Theory of Relativity*, W.A. Benjamin: New York 1966.

1968 "On the Relationship of Science and Art" in *DATA: Directions in Art, Theory and Aesthetics: An Anthology*, A. Hill (ed.), Graphic Society: New York 1968.

1969 "Science as Perception–Communication" in *The Structure of Scientific Theories*, P. Suppe (ed.), University of Illinois Pres: Illinois 1974, pp. 374–391.

1971a *An inquiry Into the Function of Language and Thought, Parts A and B*, Institute of Contemporary Arts: London 1971.

1971b "Quantum Theory as an Indication of a New Order in Physics, Part A: The Development of New Orders as shown Through the History of Physics", *Foundations of Physics*, **1**: 4 (December 1871), 359–71.

1972 "Indication of a New Order in Physics" in *The Rules of the Game*, T. Shanin (ed.), Tavistock: London 1972, pp. 249–275.

1974 "On the Subjectivity and Objectivity of Knowledge" in *Beyond Chance and Necessity*, J. Lewis (ed.), Humanities Press: Atlantic Highlands, pp. 121–130.

1976 *Fragmentation and Wholeness*. Van Leer Jerusalem Foundation: Jerusalem 1976.

1977 "On the Implicate or Enfolded Order: A New Order for Physics" in *Mind in Nature*, Griffin–Cobb (eds.), University Press of America: Washington, DC (1977), pp. 37–42.

1978 "The Implicate Order: A New Order for Physics", *Process Studies*, **8**: 2 (Summer 1978), 73–102.

1980 *Wholeness and the Implicate Order*. Routledge: London 1980.

1985 *Unfolding Meaning: A weekend of dialogue with David Bohm*, Factor (ed.), Routledge: London and New York 1985.

1986a "The Implicate Order: A New Approach to the Nature of Reality" in *Beyond Mechanism: The Universe in Recent Physics and Catholic Thought*, D. Shindler (ed.), University Press of America, Inc.: Lantham (MD) 1986, pp. 13–37.

1986b "The Implicate Order and Quantum Theory: A discussion with Professor P. C. W. Davies" in *The Ghost in the Atom*, Davies–Brown (eds.), Cambridge University Press: Cambridge, 1986, pp. 118–134.

1988 "A Realist View of Quantum Mechanics", in *Microphysical Reality and Quantum Formalism. Proceedings of the Conference 'Microphysical Reality and Quantum Formalism' Urbino, Italy, September 25th – October*

3rd, 1985 Volume 1, VAN DER MERWE–TAROZZI–SELLERI (eds.), Springer: Dordrecht, 1988, pp. 3–18.

1986 "A New Theory of the Relationship of Mind and Matter", *The Journal of the American Society for Psychical Research*, **80**: (1986), 113–135.

1987a "Bohm, David and the Mind-Matter Problem – Reply", *Journal of the American Society for Psychical Research*, **81**: 4 (1987), 405–407.

1987b "Fragmentierung und Ganzheit" in *Leben mit "Dden acht Todsünden der zivilisierten Menschheit"?*, FEYERABEND–THOMAS (eds.), Verlag der Fachvereine: Zurich 1987, pp. 331–352 [German version of *Fragmentation and Wholeness in Science and Society*, Science Council of Canada: Ottawa 1984.

1990 "A New Theory of the Relationship of Mind and Matter", *Philosophical Psychology* **3**: 2&3 (1990), 271–286.

1996 *On Dialogue*. L. NICHOL (ed.), Routledge: London 1996.

BOHM, David Joseph – BUB, Jeffery

1966 "A Proposed Solution of the Measurement Problem in Quantum Mechanics by a Hidden Variable Theory" *Rev. Mod. Phys.* **38**: 3 (1 July 1996), 435–469.

BOHM, David Joseph – FACTOR, Donald – GARRETT, Peter

1991 "Dialogue – A Proposal", http://www.david-bohm.net/dialogue/dialogue_proposal.html.

BOHM, David Joseph – HILEY, Basil J.

1985 "Unbroken Quantum Realism, from Microscopic to Macroscopic Levels", *Phys. Rev. Lett.* **55**: (2 December 1985), 2511.

1993 *The Undivided Universe: An Ontological Interpretation of Quantum Theory*, Routledge: London 1993.

BOHM, David Joseph – PEAT, Francis David

1987 *Science, Order, and Creativity*. Routledge: London 1987.

BOHM, David Joseph – VIGIER, Jean-Pierre

1954 "Model of the Causal Interpretation of Quantum Theory in Terms of a Fluid with Irregular Fluctuations", *Physical Review*, **96** 1 (1 October 1954), 208–216.

BOHR, Niels

1935 "Can Quantum-Mechanical Description of Physical Reality Be Considered Complete?", *Physical Review*, **48**: 8 (October 1935) 696–702.

BOLZANO, Bernard

1837 Wissenschaftslehre, 4 Voll., Seidel: Sulzbach 1837, reprinted in *Bernard Bolzano Gesamtausgabe*, I, Voll. 11–14, J. BERG (ed.), Frommann-Holzboog: Stuttgart-Bad Cannstatt 1985–2000.

BORN, Max

1971 *The Born-Einstein letters, correspondence between Albert Einstein and Hedwig Born from 1916 to 1955 with commentaries by Max Born*. Walker: New York 1971.

BORRADORI, Giovanna

1994 "Paradigms of Scientific Revolutions" [Interview with Thomas S. Kuhn] in *The American Philosopher: Conversations with Quine, Davidson, Putnam, Nozick, Danto, Rorty, Cavell, MacIntyre, Kuhn*, University of Chicago Press: Chicago 1994, pp. 153–167.

BRITT, Stewart H.

1941 *Social Psychology of Modern Life*, Farrar and Rinehart: New York 1941.

BUCHWALD, Jed Z. – SMITH, George E.

1997 "Thomas S. Kuhn, 1922–1996", *Philosophy of Science*, **64**: 2 (June 1997), 361–376.

CARNAP, Rudlof

1929 *Abriss der Logistik, mit besonderer Berücksichtigung der Relationstheorie und ihrer Anwendungen*, Springer: Vienna 1929.

1931 "Überwindung der Metaphysik durch logische Analyse der Sprache", *Erkenntnis*, **2** (1931), 219–241.

1934 *Logische Syntax der Sprache*, Springer: Vienna 1934.

1936 "Testability and Meaning", Part I, *Philosophy of Science*, **3**: 4 (October 1936), 419–471.

1937 "Testability and Meaning", Part II, *Philosophy of Science*, **4**: 1 (January 1937), 1–40.

1938 "Logical Foundations of the Unity of Science" in O. Neurath et al., *Encyclopedia and Unified Science*, The International Encyclopedia of Unified Science, Vol. 1, No. 1, University of Chicago Press: Chicago 1938, pp. 42–62.

1939 *Foundations of Logic and Mathematics*, The International Encyclopedia of Unified Science, Vol. 1, No. 3, University of Chicago Press: Chicago 1939.

1942 *Introduction to Semantics*, Studies in Semantics, Vol. I, Harvard University Press: Cambridge (MA) 1943.

1943 *Formalization of Logic*, Studies in Semantics, Vol. II, Harvard University Press: Cambridge (MA) 1943.

1950 *Logical Foundations of Probability*, University of Chicago Press: Chicago 1950.

1956 "The Methodological Character of Theoretical Concepts" in *Current Issues in the Philosophy of Science: Symposia of Scientists and Philosophers, Proceedings of Section L of the American Association for the Advancement of Science, 1959*, FEIGL–MAXWELL (eds.), Holt, Rinehart, and Winston: New York 1961, pp. 38–76.

1963 "Intellectual autobiography" in *The Philosophy of Rudolf Carnap* P. A. Schilpp (ed.), Open Court: LaSalle (IL) 1963, pp. 3–84.

1966 "The Aim of Inductive Logic", *Studies in Logic and the Foundations of Mathematics*, **44**: (1966), 303–318.

1977 *Two Essays on Entropy*, A. SHIMONY (ed.), University of California Press: Los Angeles, Berkeley and London 1977.

CARUANA, Louis

1995 "John von Neumann's 'Impossibility Proof' in a Historical Perspective", *Physis: Rivista Internazionale di Storia della Scienza*, **32**: (1995), 109–124.

COHEN, Robert S. – WARTOFSKY, Marx W. (eds.)

1965 *Proceedings of the Boston Colloquium for the Philosophy of Science, 1962–64: In Honor of Philipp Frank*, Boston Studies in the Philosophy of Science, Vol. II, Humanities Press: New York 1965.

COLLODEL, Matteo

2016 "Was Feyerabend a Popperian? Methodological issues in the History of the Philosophy of Science" *Studies in History and Philosophy of Science Part A*, **57**: (2016), 27–56.

2022 "Ehrenhaft's Experiments on Magnetic Monopoles: Reconsidering the Feyerabend-Ehrenhaft Connection", *International Studies in the Philosophy of Science*, **45**: 1 (2022), 69–94.

COLLODEL, Matteo – OBERHEIM, Eric

2020 "Introduction" in *Feyerabend's Formative Years. Volume 1. Feyerabend and Popper, Correspondence and Unpublished Papers*. COLLODEL–OBERHEIM (eds.), Springer Nature AG: Switzerland, Cham 2020, pp. 1–16.

COLLODEL, Matteo – OBERHEIM, Eric (eds.)

2020 *Feyerabend's Formative Years. Volume 1. Feyerabend and Popper, Correspondence and Unpublished Papers*. Springer Nature AG: Switzerland, Cham 2020.

Feyerabend's Formative Years. Volume 3. Feyerabend among Popperians, Correspondence and Unpublished Papers. Springer Nature AG: Switzerland, Cham 2020.

COLODNY, Robert G. (ed.)

1962 *Frontiers of Science and Philosophy*, University of Pittsburgh Series in the Philosophy of Science, Vol. 1, University of Pittsburgh Press: Pittsburgh 1962.

1970 *The Nature and Function of Scientific Theories: Essays in Contemporary Science and Philosophy*, University of Pittsburgh Series in the Philosophy of Science, Vol. 4, University of Pittsburgh Press: Pittsburgh 1970.

CORNFORD, Francis M.

1952 *Principium Sapientiae*, W. K. C. GUTHRIE (ed.), Cambridge University Press: Cambridge 1953.

CUSHING, James T.
1992 "Causal Quantum Theory: Why a Nonstarter?" in *Wave-particle duality*, F. Selleri (ed.), Plenum: London 1992, pp. 37–68.
1993 "What If Bell Had Come Before 'Copenhagen'?", in *Bell's Theorem and the Foundations of Modern Physics* (International Conference, Palazzo Del Ridotto, Cesena, Italy, 7–10 October 1991), VAN DER MERWE–SELLERI–TAROZZI (eds.), World Scientific: Singapore 1993, pp. 125–134.
1994 *Quantum Mechanics: Historical contingency and the Copenhagen Hegemony*, University of Chicago Press: Chicago 1994.

DE BROGLIE, Louis
1953 *La physique quantique restera-t-elle indeterministe? Comptes Rendus*, Paris 1953.

DEL SANTO, Flavio
2022 "Beyond Method: The Diatribe Between Feyerabend and Popper Over the Foundations of Quantum Mechanics", *International Studies in the Philosophy of Science*, **35**: (2022), 5–22.

DEL SANTO, Flavio – KRIZEK, Gerd Christian
2019 "Against the 'nightmare of a mechanically determined universe': Why Bohm was never a Bohmian" Retrieved from https://arxiv.org/pdf/2307.05611.pdf.

DESCARTES, René
1637 *Discours de la methode*, Ian Maire: Leyden 1637. English translation by J. Cottingham, R. Stoothoff, D. Murdoch, "Discourse on Method" in *The Philosophical Writings of Descartes*, Vol. I, Cambridge, Cambridge University Press, 1999.

DIEKS, Dennis
2017 "Von Neumann's impossibility proof: Mathematics in the service of rhetorics", *Studies in History and Philosophy of Science Part B. Studies in History and Philosophy of Modern Physics*, **60**: (2017), 136–148.

DIRAC, Paul A. M.
1930 *The Principles of Quantum Mechanics*, Clarendon Press: Oxford 1930, 1935^2, 1947^3, 1958^4, 1867^{4R}.

DRIVER-LINN, Erin
2003 "Where is psychology going? Structural fault lines revealed by psychologists' use of Kuhn", *American Psychologist*, **58**: 4 (April 2003), 269–278.

DUPREE, Anderson Hunter – KUHN, Thomas Samuel
1958 "Teaching the History of Science", *Isis*, **49**: 2 (June 1958), 172–173.

EDGLEY, Roy
1996 "Anarchy in Academia", review of FEYERABEND 1995, *New Left Review*, 217 (May-June 1996), 155–160.

EINSTEIN, Albert

1953 "Elementare Überlegungen zur Interpretation der Grundlagen der Quanten-Mechanik" in *Scientific Papers Presented to Max Born on His Retirement from the Tait Chair of Natural Philosophy in the University of Edinburgh*, Oliver and Boyd: London and Hafner: New York 1953, pp. 33–40.

EARMAN, John

1993 "Carnap, Kuhn, and the philosophy of scientific methodology" in *World Changes*, P. HORWICH (ed.), The MIT Press: Cambridge (MA) 1993, pp. 9–36.

FANO, Ugo

1957 "Description of States in Quantum Mechanics by Density Matrix and Operator Techniques", *Reviews of Modern Physics*, **29**: 1 (January-March 1957), 74–93.

FEIGL, Herbert – MAXWELL, Grover (eds.)

1956 *Foundations of Science & the Concepts of Psychology and Psychoanalysis*, Minnesota Studies in the Philosophy of Science, Vol. I, University of Minnesota Press: Minneapolis 1956.

1961 *Current Issues in the Philosophy of Science: Symposia of Scientists and Philosophers, Proceedings of Section L of the American Association for the Advancement of Science*, Holt, Rinehart and Winston: New York 1961.

1962 *Scientific Explanation, Space & Time*, Minnesota Studies in the Philosophy of Science, Vol. III, University of Minnesota Press: Minneapolis 1962.

FENYES, Imre

1952 "Eine wahrscheinlichkeitstheoretische Begründung und Interpretation der Quantenmechanik", *Zeitschrift für Physik*, **132**: (February 1952), 81–106.

FERRIANI, Maurizio

1995 "Ancora su Enriques e Carnap: le ragioni di un dialogo mancato" in *L'eredità di Rudolf Carnap*, A. PASQUINELLI (ed.), CLUEB: Bologna 1995, pp. 163–174.

FEYERABEND, Paul Karl

1951 *Zur Theorie der Basissätze* [Ph.D. thesis], Universität Wien: Vienna 1951.

1953 "Bemerkungen zu Interpretation and Preciseness" in *Semantiske problemer: inledningsforedrag og diskusjonsreferater fra Krets 5, Nordisk sommeruniversitet, 2. sesjon, Ustaoset 1952*, GJERLØW–BJERRUM–REGNÉLL (eds.), Akademisk forlag: Oslo 1953, pp. 77–84.

1954a "Wittgenstein und die Philosophie, I", *Wissenschaft und Weltbild: Monatsschrift für alle Gebiete der Forschung*, **7**: 5–6 (May–June 1954), 212–220.

1954b "Wittgenstein und die Philosophie, II", *Wissenschaft und Weltbild Monatsschrift für alle Gebiete der Forschung*, **7**: 7–8 (July–August 1954), 283–287.

1954c "Determinismus und Quantenmechanik", *Wiener Zeitschrift für Philosophie, Psychologie, Pädagogik, vol. 5*, no. 2 (1954), pp. 89–111, reprinted in FEYERABEND 2016, Ch. 3.

1954d "Physik und Ontologie", *Wissenschaft und Weltbild: Monatschrift für alle Gebiete der Forschung*, **7**: 11–12 (November–December 1954), 464–476. English translation by Wetterstein et al. in FEYERABEND 2016, Ch. 2.

1954e Review of [Rudolf Carnap, *Einführung in die symbolische Logik mit besonderer Berücksichtigung ihrer Anwendungen*, Springer: Vienna 1954], *Internationale Mathematische Nachrichten*, 35–36 (November 1954), 67.

1955a Review of [Ludwig WITTGENSTEIN, *Philosophical Investigations*, ANSCOMBE–RHEES (eds.), translation by G. E. M. ANSCOMBE, Blackwell: Oxford 1953], *The Philosophical Review*, **64**: 3 (July 1955), 449–483.

1955b "Carnaps Theorie der Interpretation theoretischer Systeme", *Theoria*, **21**: 1 (April 1955), 55–62.

1956 "Eine Bemerkung zum Neumannschen Beweis", *Zeitschrift für Physik*, **145**: 4 (August 1956), 421–423. English translation by Wettersten et al., reprinted in FEYERABEND 2016, Ch. 4.

1957a "On the Quantum-Theory of Measurement" in *Observation and Interpretation: A Symposium of Philosophers and Physicists, Proceedings of the Ninth Symposium of the Colston Research Society*, KÖRNER–PRYCE (eds.), Butterworths Scientific Pub.: London and New York 1957, pp. 121–130, reprinted in FEYERABEND 1981a, Ch. 13.

1957b "Zur Quantentheorie der Messung", *Zeitschrift für Physik*, **148**: 5 (October 1957), 551–559.

1958a "An Attempt at a Realistic Interpretation of Experience", *Proceedings of the Aristotelian Society*, **58** (1957–1958), 143–170, reprinted in FEYERABEND 1981a: Ch. 2.

1958b "Reichenbach's Interpretation of Quantum-Mechanics", *Philosophical Studies*, **9**: 4 (June 1958), 47–59.

1958c "Complementarity", *Proceedings of the Aristotelian Society*, Supplementary Volume **32**: (1958), 75–104, reprinted in FEYERABEND 2016, Ch. 5.

1958d Review of [John von Neumann, *Mathematical Foundations of Quantum Mechanics*, English translation by R. T. Beyer, Princeton University Press: Princeton (NJ) 1955], *The British Journal for the Philosophy of Science*, 8: 32 (February 1958), 343–347, reprinted in FEYERABEND 2016a, Ch. 17.

1960a "On the Interpretation of Scientific Theories" in *Proceedings of the 12th International Congress of Philosophy, Venice, 12–18 September 1958*, Vol. 5: Logic, theory of knowledge, philosophy of science, philosophy of language, Sansoni: Florence 1960, pp. 151–159, reprinted in FEYERABEND 1981a, Ch. 3.

1960b "Professor Bohm's Philosophy of Nature", Review of BOHM 1957a, *The British Journal for the Philosophy of Science*, **10**: 40 (February 1960), 321–338, reprinted in FEYERABEND 1981a, Ch. 14.

1960c "Das Problem der Existenz theoretischer Entitäten" in *Probleme der Wissenschaftstheorie: Festschrift für Viktor Kraft*, E. TOPITSCH (ed.), Springer: Vienna 1960, pp. 35–72. English translation by D. Sirtes and E. Oberheim in FEYERABEND 1999, Ch. 1.

1961a *Knowledge Without Foundations: Two Lectures Delivered on the Nellie Heldt Lecture Fund*, Oberlin College: Oberlin (OH) 1961, reprinted in FEYERABEND 1999, Ch. 2.

1961b "Comments on Hanson's 'Is There a Logic of Scientific Discover?'" in *Current Issues in the Philosophy of Science: Symposia of Scientists and Philosophers, Proceedings of Section L of the American Association for the Advancement of Science, 1959*, FEIGL–MAXWELL (eds.), Holt, Rinehart, and Winston: New York 1961, pp. 35–39.

1961c "Niels Bohr's Interpretation of the Quantum Theory" in *Current Issues in the Philosophy of Science: Symposia of Scientists and Philosophers, Proceedings of Section L of the American Association for the Advancement of Science, 1959*, FEIGL–MAXWELL (eds.), Holt, Rinehart, and Winston: New York 1961, pp. 371–390.

1962a "Problems of Microphysics" in *Frontiers of Science and Philosophy, University of Pittsburgh Series in the Philosophy of Science, Vol. 1*, R. G. COLODNY (ed.), University of Pittsburgh Press: Pittsburgh 1962, pp. 189–283, reprinted in FEYERABEND 2016, Ch. 7.

1962b "Explanation, Reduction, and Empiricism" in *Scientific Explanation, Space and Time*, FEIGL–MAXWELL (eds.), Minnesota Studies in the Philosophy of Science, Vol. III, 1962, pp. 28–97.

1963a "How to be a Good Empiricist: A Plea for Tolerance in Matters Epistemological" in *Philosophy of Science: The Delaware Seminar Vol. II*, B. BAUMRIN (ed.), Interscience: New York 1963, pp. 3–39, reprinted in FEYERABEND 1999, Ch. 3.

1963b "Materialism and the Mind-Body Problem", *The Review of Metaphysics*, **17**: 1 (September 1963), 49–66.

1963c "Professor Hartmann's Philosophy of Nature", *Ratio*, **5**: 1 (1963), 81–94.

1964a "Realism and Instrumentalism: Comments on the Logic of Factual Support" in *The Critical Approach to Science and Philosophy: In Honor of Karl R. Popper*, M. BUNGE (ed.), The Free Press of Glencoe: London, New York 1964, pp. 280–308, reprinted in FEYERABEND 1981a, Ch. 11.

1964b Review of *The Concept of the Positron: A Philosophical Analysis* by Norwood Russell Hanson, *The Philosophical Review*, **73**: 2 (April 1964), 264–266, reprinted in FEYERABEND 2016, Ch. 22.

1965a "Problems of Empiricism" in *Beyond the Edge of Certainty. Essays in Contemporary Science and Philosophy*, R. G. COLODNY (ed.), University of Pittsburgh Series in the Philosophy of Science, Vol. 2, Prentice Hall: Englewood Cliffs (NJ) 1965, pp. 145–260.

1965b "Reply to Criticism: Comments on Smart, Sellars and Putnam" in *Proceedings of the Boston Colloquium for the Philosophy of Science, 1962–64: In Honor of Philipp Frank*, COHEN–WARTOFSKY (eds.), Boston

Studies in the Philosophy of Science, Vol. II, Humanities Press: New York 1965, pp. 223–261, reprinted in FEYERABEND 1981a, Ch. 6.

1965c "On the 'Meaning' of Scientific Terms", *Journal of Philosophy* **62**: 266–274, reprinted in FEYERABEND 1981a, Ch. 5.

1966a "On the Possibility of a Perpetuum Mobile of the Second Kind" in *Mind, Matter, and Method: Essays in Philosophy and Science in Honor of Herbert Feigl*, FEYERABEND–MAXWELL (eds.), University of Minnesota Press: Minneapolis 1966, pp. 409–412, reprinted in FEYERABEND 2016, Ch. 13.

1966b "The Structure of Science", review of *The Structure of Science* by Ernest Nagel, *The British Journal for the Philosophy of Science*, **17**: 3 (November 1966), 237–249, reprinted in FEYERABEND 1981b, Ch. 3.

1966c "Dialectical Materialism and the Quantum Theory", *Slavic Review*, **25**: 3 (September 1966), 414–417.

1968 "On a Recent Critique of Complementarity: Part I", *Philosophy of Science*, **35**: 4 (December 1968), pp. 309–331, reprinted in FEYERABEND 1981a, Ch. 16.

1969 "On a Recent Critique of Complementarity: Part II", *Philosophy of Science*, **36**: 1 (March 1969), pp. 82–105, reprinted in FEYERABEND 1981a, Ch. 16.

1970a "Consolations for the Specialist" in *Criticism and the Growth of Knowledge, Proceedings of the International Colloquium in the Philosophy of Science, London, 1965, Vol. 4*, LAKATOS–MUSGRAVE (eds.), Cambridge University Press: London, New York 1970, pp. 197–230.

1970b "In Defence of Classical Physics", *Studies in History and Philosophy of Science Part A*, **1**: 1 (May 1970), 59–85, reprinted in FEYERABEND 2016, Ch. 14.

1970c "Problems of Empiricism, Part II" in *The Nature and Function of Scientific Theories: Essays in Contemporary Science and Philosophy*, R. G. COLODNY (ed.), University of Pittsburgh Press: Pittsburgh 1970, pp. 275–353.

1975 *Against Method: Outline of an Anarchistic Theory of Knowledge*, New Left Books: London 1975[1], Verso: London 1988[2], 1993[3].

1977 "Changing Patterns of Reconstruction", *The British Journal for the Philosophy of Science*, **28**: 4 (December 1977), 351–369.

1978 *Science in a Free Society*, New Left Books: London and New York 1978 [Verso: London 1982].

1979 *Erkenntnis für freie Menschen*, Suhrkamp: Frankfurt am Main 1979, 1980[2].

1981a *Realism, Rationalism and Scientific Method: Philosophical Papers, Volume 1*, Cambridge University Press: Cambridge 1981.

1981b *Problems of Empiricism: Philosophical Papers, Volume 2*, Cambridge University Press: Cambridge 1981.

1984 *Wissenschaft als Kunst*, Suhrkamp: Frankfurt am Main 1984 – 1987: "Putnam on Incommensurability", *The British Journal for the Philosophy of Science*, **38**: 1 (March 1987), 75–81; reprinted in *Farewell to Reason*, Verso: London, Ch. 10 (pp. 265–272).

1987 “Putnam on Incommensurability”, *The British Journal for the Philosophy of Science*, 38: 1 (March 1987), 75–81; reprinted in *Farewell to Reason*, Verso: London, Ch. 10 (pp. 265–272).

1988 *Against Method: Outline of an Anarchistic Theory of Knowledge*, Verso: London, 1975[1], 1993[3].

1993 *Against Method: Outline of an Anarchistic Theory of Knowledge*, Verso: London, 1975[1], 1988[2].

1995 *Killing Time: The Autobiography of Paul Feyerabend*, University of Chicago Press: Chicago 1995.

1999a *Knowledge, Science and Relativism: Philosophical Papers, Volume 3*, J. Preston (ed.), Cambridge University Press: Cambridge 1999.

1999b *The Conquest of Abundance: A Tale of Abstraction versus the Richness of Being*, B. Terpstra (ed.), University of Chicago Press: Chicago and London 1999.

2010 “Die Dogmen der Empirismus” in *Vertreibung, Transformation und Rückkehr der Wissenschaftstheorie: Am Beispiel von Rudolf Carnap und Wolfgang Stegmüller*, F. Stadler (ed.), LIT: Vienna and Berlin 2010 [1951], pp. 410–436.

2016a *Physics and Philosophy: Philosophical Paper, Volume 4*, Gattei–Agassi (eds.), Cambridge University Press: New York 2016.

2016b “The concept of intelligibility in modern physics (1948)” *Studies in History and Philosophy of Science Part A* **57**: (June 2016), pp. 64–66.

2022 “Single Magnetic Northpoles and Southpoles and Their Importance for Science”, International Studies in the Philosophy of Science, **35**: 1 (2022), pp. 95–117.

Feyerabend, Paul K. – Albert, Hans. – , Paul. Feyerabend, Paul Karl – Maxwell, Grover (eds.)

1966 *Mind, Matter and Method: Essays in Philosophy and Science in Honor of Herbert Feigl*, University of Minnesota Press: Minneapolis 1966.

2008 *Briefwechsel. Band 2 (1972–1986)*, W. Baum and M. Mühlmann (eds.). Kitab: Klagenfurt and Vienna 2008.

Feyerabend, Paul K. – Hoyningen-Huene

2010 *Aber ein Paul hilft doch dem anderen: Briefwechsel Paul Feyerabend - Paul Hoyningen-Huene 1983–1994*, Ch. Augustin (ed.), Passagen: Vienna 2010.

Fischer, Kurt Rudolf

1999 “Paul K. Feyerabend: A Personal Reminiscence” in *Aufsätze zum angloamerikanischen und österreichischen Philosophie*, Peter Lang: Frankfurt am Main, Berlin, Bern, New York, Paris, and Vienna 1999, pp. 77–189.

2006 “Feyerabends Weltanschauung” in *Paul Feyerabend. Ein Philosoph aus Wien*, Stadler–Fischer (eds.), Springer Gabler: Vienna 2006, pp. 1–13.

FLÜGGE, Siegfried (ed.)
1955–84 *Handbuch für Physik*, **55** Voll., Springer: Berlin 1955–1984.

FORMAN, Paul
1971 "Weimar Culture, Causality, and Quantum Theory: Adaptation by German Physicists and Mathematicians to a Hostile Environment", *Historical Studies in the Physical Sciences*, **3** (1971): 1–115.

FREIRE JUNIOR, Olivier
2015 *The Quantum Dissidents: Rebuilding the Foundations of Quantum Mechanics (1950–1990)*. Springer: Berlin, Heidelberg 2015.
2019 *David Bohm A Life Dedicated to Understanding the Quantum World.* Springer Nature AG: Switzerland 2019.

FRIEDMAN, Michael
1998 "On the Sociology of Scientific Knowedge and its Philosophical Agenda", *Studies in History and Philosophy of Science Part A*, **29**: 2 (June 1998), 239–271.
2003 "Kuhn and Logical Empiricism" in *Thomas Kuhn*, T. NICKLES (ed.), Cambridge University Press: Cambridge 2003, pp. 19–44.

FRÖHLICH, H. – GUTH, E.
1935 "The Fundamental Paradox of the Quantum Theory", *Nature*, **136**: 3431 (3 August 1935), 179–179.

GEIGER, Hans – SCHEEL, Karl (eds.)
1926–33 *Handbuch für Physik*, 24 Voll., Springer: Berlin, Vienna 1926–1933.

GOODMAN, Nelson
1949 "On Likeness of Meaning", *Analysis*, **10**: 1 (October 1949), 1–7.

GRÖNEWOLD, Hilbrand J.
1946 "On the Principles of Elementary Quantum Mechanics", *Physica*, **12**: 7 (October 1946), 405–460.

HACOHEN, Malachi
2000 *Karl Popper – The Formative Years, 1902–1945. Politics and Philosophy in Interwar Vienna*, Cambridge University Press: Cambridge 2000.

HANSON, Norwood Russel
1958a *Patterns of discovery: An inquiry into the conceptual foundations of science*, Cambridge University Press: Cambridge 1958.
1958b "The Logic of Discovery", *The Journal of Philosophy*, **55**: 25 (4 December 1958), 1073–1089.
1961 "Is There a Logic of Scientific Discovery?" in *Current Issues in the Philosophy of Science*, FEIGL-MAXWELL (eds.), Holt, Rinehart and Winston: New York 1961, pp. 20–35.

1966 "Equivalence: The Paradox of Theoretical Analysis" *Mind, Matter, and Method: Essays in Philosophy and Science in Honor of Herbert Feigl*, in FEYERABEND-MAXWELL (eds), University of Minnesota Press: Minneapolis 1966: 413–429.

HARTMANN, Nicolai
1926 *Ethik*, de Gruyter: Berlin and Leipzig 1926.

HASSE, Helmut – HECKE, Erich – DEURING, Max – SPERNER, Emanuel
1939–67 *Encyklopädie der Mathematischen Wissenschaften mit Einschluß ihrer Anwendungen*, **14** Voll., Teubner: Leipzig 1939–1967.

HEILBRON, John L.
1968 "Quantum Historiography and the Archive for History of Quantum Physics", *History of Science*, **7**: (1968), 90–111.
1998 "Thomas Samuel Kuhn, 18 July 1922–17 June 1996", *Isis*, **89**: 3 (September 1998), 505–515.

HEISENBERG, Werner
1955 "The Development of the Interpretation of Quantum Theory" in *Niels Bohr and the Development of Physics: Essays Dedicated to Niels Bohr On the Occasion of His Seventieth Birthday*, W. Pauli (ed.), McGraw-Hill: New York and Pergamon: London 1955, 12–29.
1958 *Physics and Philosophy*. Harper and Row: New York, 1958.

HELMHOLTZ, Hermann von
1847 *Über die Erhaltung der Kraft: Eine physikalische Abhandlung*, Vorgetragen in der Sitzung der physikalischen Gesellschaft zu Berlin am 23. Juli 1847, G. Reimer: Berlin 1847. English translation by J. Tyndall, "On the Conservation of Force: A Physical Memoir" in *Scientific Memoirs: Natural Philosophy*, TYNDALL–FRANCIS (eds.), London 1853, Vol. 1, Part II, pp. 114–162, "The Conservation of Force: a Physical Memoir" in *Selected Writings of Hermann von Helmholtz*, R. KAHL (ed.), Wesleyan University Press: Middletown 1971, pp. 3–55, "On the Conservation of Energy" in *Applications of Energy: Nineteenth Century*, R. LINDSAY (ed.), Dowden Hutchinson & Ross: Stroudsburg 1976, pp. 7–31 and 226–243.
1871 "Ueber die Erhaltung der Kraft" in *Populäre Wissenschaftliche Vorträge, Vol. 2*, Vieweg: Braunschweig 1871. English translation by E. Atkinson, "On the Conservation of Force: Introduction to a Series of Lectures Delivered at Carlsruhe in the Winter of 1862–1863" in *Popular Lectures on Scientific Subjects*, D. Appleton and Company: New York 1885, pp. 317–362.

HEMPEL, Carl G.
1945 "On the Nature of Mathematical Truth", *American Mathematical Monthly*, **52**: 10 (December 1945), 543–556, reprinted in *Readings in Philosophical Analysis*, FEIGL–SELLARS (eds.), Appleton-Century-Crofts: New York

1949, pp. 148–162, and in *The Philosophy of Carl G. Hempel: Studies in Science, Explanation, and Rationality*, J. H. FETZER (ed.), Oxford University Press: Oxford and New York 2001, pp. 3–17.

1958 "The Theoretician's Dilemma: A Study in the Logic of Theory Construction" in *Concepts, Theories and the Mind-Body Problem*, FEIGL–SCRIVEN–MAXWELL (eds.), Minnesota Studies in the Philosophy of Science, Vol. 2, University of Minnesota Press: Minneapolis 1958, pp. 173–226.

1962 "Explanation in Science and in History" in *Frontiers of Science and Philosophy*, R. G. COLODNY (ed.), University of Pittsburgh Series in the Philosophy of Science, Vol. 1, University of Pittsburgh Press: Pittsburgh, pp. 7–33.

1973 "The Meaning of Theoretical Terms: A Critique of the Standard Empiricist Construal" in *Logic, Methodology and Philosophy of Science IV: Proceedings of the Fourth International Congress for Logic, Methodology and Philosophy of Science*, NAGEL–SUPPES–TARSKI (eds.), North Holland: Amsterdam 1973, pp. 367–378, reprinted in *The Philosophy of Carl G. Hempel: Studies in Science, Explanation, and Rationality*, J. H. FETZER (ed.), Oxford University Press: Oxford and New York 2001, pp. 208–217.

HEMPEL, Carl Gustav – OPPENHEIM, Paul

1948 "Studies in the Logic of Explanation", *Philosophy of Science*, **15**: 2 (April 1948), 135–175.

HENKIN, Leon

1960 Review of PUTNAM 1957, FEYERABEND 1958b, LEVI 1959, *Journal of Symbolic Logic*, **25**: 3 (September 1960), 289–291.

HERMANN, Grete

1935 "Die Naturphilosophischen Grundlagen der Quantenmechanik", *Abhandlungen der Fries'schen Schule, n. F.*, **6**: 2 (1935), 69–152.

HERSHBERG, James

1993 *James B. Conant: Harvard to Hiroshima and the Making of the Nuclear Age*, Knopf: New York 1993.

HILL, Edward L.

1966 "Classical Mechanics as a Limiting Form of Quantum Mechanics" in *Mind, Matter, and Method: Essays in Philosophy and Science in Honor of Herbert Feigl*, FEYERABEND–MAXWELL (eds.), University of Minnesota Press: Minneapolis 1966, pp. 430–448.

HOYNINGEN-HUENE, Paul

1989 *Die Wissenschaftsphilosophie Thomas S. Kuhns: Rekonstruktion und Grundlagenprobleme*, Vieweg and Teubner Verlag: Wiesbaden 1989.

1993 *Reconstructing Scientific Revolutions. The Philosophy of Science of Thomas S. Kuhn*. University of Chicago Press: Chicago 1993. English translation of HOYNINGEN-HUENE 1989.

1995 "Two letters of Paul Feyerabend to Thomas S. Kuhn on a draft of The Structure of Scientific Revolutions" *Studies in History and Philosophy of Science Part A* **26**: 353–387.

2000 "Paul Feyerabend and Thomas Kuhn" in *The Worst Enemy of Science? Essays in Memory of Paul Feyerabend*. PRESTON–MUNÉVAR–LAMB (eds.), Oxford University Press: New York 2000, pp. 102–114.

2002 "Paul Feyerabend und Thomas Kuhn", *Journal for General Philosophy of Science*, **33**: 1 (2002), pp. 61–68.

2006 "More Letters by Paul Feyerabend to Thomas S. Kuhn on 'Proto-Structure'", *Studies in History and Philosophy of Science Part A* **37**: (2006), 610–632.

2013 *Systematicity: The Nature of Science*, Oxford University Press: New York 2013.

HOWARD, Don

2004 "Who invented the "Copenhagen Interpretation"? A Study in Mythology", *Philosophy of Science*, **71**: 5 (December 2004), 669–682.

HUFBAUER, Karl

2012 "From Student of Physics to Historian of Science: T. S. Kuhn's Education and Early Career, 1940–1958", *Physics in Perspective*, **14**: 4 (December 2012), 421–470.

HUME, David

1748 *An Enquiry Concerning Human Understanding*, London 1748, T. L. BEAUCHAMP (ed.), Oxford University Press: Oxford 1999. German translation by R. Richter, *Eine Untersuchung über den menschlichen Verstand*, Dürr: Leipzig 1907, Felix Meiner: Leipzig 1910.

IRZIK, Gürol – GRÜNBERG, Teo

1995 "Carnap and Kuhn: Arch enemies or close allies?", *The British Journal for the Philosophy of Science*, **46**: 3 (September 1995), 285–307.

JAMMER, Max

1974 *The Philosophy of Quantum Mechanics: The interpretations of quantum mechanics in historical perspective*, Wiley: New York 1974.

JASPERS, Karl

1950 *Einführung in die Philosophie: Zwölf Radiovorträge*, Artemis: Zurich 1950. English translation by R. Manheim, *Way to Wisdom: An Introduction to Philosophy*, Yale University Press: New Haven (CT) 1951.

JØRGENSEN, Jørgen

1931 *A Treatise of Formal Logic: Its Evolution and Main Branches, with Its Relation to Mathematics and Philosophy*. English translation by W. Worster, 3 Voll., Levin & Munksgaard: Copenhagen and Humphrey Milford, Oxford University Press: London 1931, Russell & Russell: New York 1962.

1948 *Den logiske Empirismes Udvikling*, Festschrift Edited by the University of Copenhagen in Honor of His Majesty the King's Birthday, Copenhagen 1948. English translation, "The Development of Logical Empiricism", *International Encyclopedia of Unified Science*, Vol. 2, No. 9, University of Chicago Press: Chicago 1951.

JUHOS, Béla von

1950 *Die Erkenntnis Und Ihre Leistung: Die Naturwissenschaftliche Methode*, Springer: Vienna 1950.

KAILA, Eino

1936 "Det främmande själslivets kunskapsteoretiska problem", *Theoria*, **2**: 1–2 (April 1936), 128–155.

1939 *Den mänskliga kunskapen: vad den är och vad den icke är*. English translation by G. H. von Wright, Natur och kultur: Stockholm and Söderström: Helsinki 1939.

KATZ, David

1944 *Gestaltpsychologie*, B. Schwab: Basel 1944.

KAUFMANN, Felix

1950 "Basic Issues in Logical Positivism" in *Philosophical Thought in France and the United States*, M. FARBER (ed.), University of Buffalo Publications in Philosophy: New York 1950, pp. 565–588.

KEUPINK, Alfons

2006 "Arthur Pap (1921–1959): A Brief Intellectual Biography" in *The Limits of Logical Empiricism: Selected Papers of Arthur Pap*, KEUPINK–SHIEH (eds.), Springer: Dordrecht 2006, pp. 365–368.

KEUPINK, Alfons – SHIEH, Sanford (eds.)

2006 *The Limits of Logical Empiricism: Selected Papers of Arthur Pap*, Springer: Dordrecht 2006.

KIRCHMANN, Julius H. von

1871 *Die Metaphysik des Aristoteles: übersetzt und mit einer Lebensbeschreibung des Aristoteles versehen*, L. Heimann: Berlin 1871.

KÖHLER, Wolfgang

1966 "A Task for Philosophers" in *Mind, Matter, and Method: Essays in Philosophy and Science in Honor of Herbert Feigl*, FEYERABEND–MAXWELL (eds.), University of Minnesota Press: Minneapolis 1966, pp. 70–91.

KÖRNER, Stephan

1957 "Preface" in *Observation and Interpretation: A Symposium of Philosophers and Physicists, Proceedings of the Ninth Symposium of the Colston Research Society*, KÖRNER–PRYCE (eds.), Butterworths Scientific Pub.: London and New York 1957, p. vii.

KÖRNER, Stephan – PRYCE, Maurice Henry Lecorney (eds.)
1957 *Observation and Interpretation: A Symposium of Philosophers and Physicists, Proceedings of the Ninth Symposium of the Colston Research Society*, Butterworths Scientific Pub.: London and New York 1957.

KOŽNJAK, Boris.
2018 "The Missing History of Bohm's Hidden Variables Theory: The Ninth Symposium of the Colston Research Society, Bristol, 1957", *Studies in History and Philosophy of Science Part B. Studies in History and Philosophy of Modern Physics* **62**: (2018), 85–97.

KLEIN, Felix – MEYER, Wilhelm Franz (eds.)
1904–35 *Encyklopädie der Mathematischen Wissenschaften mit Einschluß ihrer Anwendungen*, 6 Voll., Teubner: Leipzig 1904–1935, <http://gdz.sub.uni-goettingen.de/no_cache/en/dms/load/toc/?IDDOC=183743>.

KRAFT, Victor
1950 *Der Wiener Kreis: der ursprung des Neopositivismus*, Springer: Wien 1950. English translation by A. Pap, *The Vienna Circle: The Origin of Neopositivism, a Chapter in the History of Recent Philosophy*, Greenwood Press: New York 1953.
1966 "The Problem of Induction" in *Mind, Matter, and Method: Essays in Philosophy and Science in Honor of Herbert Feigl*, FEYERABEND–MAXWELL (eds.), University of Minnesota Press: Minneapolis 1966, pp. 306–318.

KRISHNAMURTI, Jiddu – BOHM, David Joseph
1985 *The Ending of Time*, Harper: San Francisco 1985.

KUBY, Daniel
2016 "Feyerabend's 'The concept of intelligibility in modern physics' (1948)" *Studies in History and Philosophy of Science Part A* **57**: 57–63.
2021 "Feyerabend's Reevaluation of Scientific Practice: Quantum Mechanics, Realism and Niels Bohr" in *Interpreting Feyerabend*, BSCHIR–SHAW (eds.), Cambridge University Press: Cambridge 2021, pp. 132–56.

KUHN, Thomas Samuel
1953 "Plans for Research" [Kuhn's application for a fellowship of the Guggenheim Foundation, 22 October 1953] in HUFBAUER 2012, pp. 456–460.
1959 "The essential tension: Tradition and innovation in scientific research" in *The Third (1959) University of Utah Research Conference on the Identification of Scientific Talent*, C. W. Taylor (ed.), University of Utah Press: Salt Lake City 1959, pp. 162–74, reprinted in KUHN 1977, pp. 225–239.
1961 "The Function of Measurement in Modern Physical Science", *Isis*, **52**: 2 (June 1961), 161–193, reprinted in KUHN 1977, pp. 178–224.

1962 *The Structure of Scientific Revolutions*, International Encyclopedia of Unified Science, Vol. 2, No. 2, University of Chicago Press: Chicago 1962, and [as an independent volume] University of Chicago Press: Chicago 1962, 1970[2].

1964 "A function for thought experiments" in *L'aventure de la science: Mélanges Alexandre Koyré*, Vol. 2, Herman: Paris 1964, pp. 307–334, reprinted in KUHN 1977, pp. 240–265.

1977 *The Essential Tension: Selected Studies in Scientific Tradition and Change*, University of Chicago Press: Chicago 1977.

1983 "Commensurability, Comparability, Communicability" in *PSA 1982: Proceedings of the 1982 Biennial Meeting of the Philosophy of Science Association*, P. Asquith. and T. Nickles (eds.), Philosophy of Science Association: East Lansing (MI), pp. 669–688; reprinted in KUHN 2000: 33–57.

1988 "Possible Worlds in History of Science" in *Possible Worlds in Humanities, Arts and Sciences: Proceedings of Nobel Symposium 65*, Allén Sture (ed.), De Gruyter: Berlin 1988, pp. 9–32; reprinted in KUHN 2000: 58–89.

1993 "Foreword" in HOYNINGEN-HUENE 1993, pp. xi–xiii – 1983: "Commensurability, Comparability, Communicability" in *PSA 1982: Proceedings of the 1982 Biennial Meeting of the Philosophy of Science Association*, P. Asquith. and T. Nickles (eds.), Philosophy of Science Association: East Lansing (MI), pp. 669–688; reprinted in KUHN 2000: 33–57. - 1988: "*Possible Worlds in History of Science" in Possible Worlds in Humanities, Arts and Sciences: Proceedings of Nobel Symposium 65*, Allén Sture (ed.), De Gruyter: Berlin 1988, pp. 9–32; reprinted in KUHN 2000: 58–89. – *The Road since Structure: Philosophical Essays, 1970–1993*, with an Autobiographical Interview, J. Conant and J. Haugeland (eds.), The University of Chicago Press: Chicago 2000. – 2022a: "The Presence of Past Science" in KUHN 2022c: 23–89. – 2022b: "The Plurality of Worlds. An Evolutionary Theory of Scientific Development in KUHN 2022c: 103–265. – 2022c: *The Last Writings of Thomas S. Kuhn: Incommensurability in Science*, B. Mladenović (ed.), University of Chicago Press: Chicago 2022.

2000 *The Road since Structure: Philosophical Essays, 1970–1993, with an Autobiographical Interview*, J. Conant and J. Haugeland (eds.), The University of Chicago Press: Chicago 2000.

2022a "The Presence of Past Science" in KUHN 2022c: 23–89.

2022b "The Plurality of Worlds. An Evolutionary Theory of Scientific Development in KUHN 2022c: 103–265.

2022c *The Last Writings of Thomas S. Kuhn: Incommensurability in Science*, B. Mladenović (ed.), University of Chicago Press: Chicago 2022.

KUHN, Thomas S. – HEILBRONN John L – FORMAN, Paul – ALLEN, Lili
1967 *Sources for History of Quantum Physics: An Inventory and Report*, American Philosophical Society: Philadelphia 1967.

LAKATOS, Imre – MUSGRAVE, Alan (eds.)
1970 *Criticism and the Growth of Knowledge, Proceedings of the International Colloquium in the Philosophy of Science, London 1965, Vol. 4*, Cambridge University Press: Cambridge 1970.

LEWIN, Kurt Z.
1931 "Der Übergang von der aristotelischen zur galileischen Denkweise in Biologie und Psychologie", *Erkenntnis*, **1** (1930/1931), 421–466.

LEVI, Isaac
1959 "Putnam's three truth values", *Philosophical Studies*, **10**: 5 (1959), 65–69.

LORENZ, Konrad
1973 *Die acht Todsünden der zivilisierten Menschheit*, R. Piper: Munich 1973. English translation, *Civilized Man's Eight Deadly Sins*, Harcourt Brace Jovanovich: New York 1974.

LUDWIG, Günther
1954 *Die Grundlagen der Quantenmechanik*, Springer: Berlin 1954 – MAIER, Anneliese 1949: *Die Vorläufer Galileis im 14. Jahrhundert, Studien zur Naturphilosophie der Spätscholastik*, Edizioni di "Storia e letteratura": Rome 1949.

MAIER, Anneliese
1949 *Die Vorläufer Galileis im 14. Jahrhundert, Studien zur Naturphilosophie der Spätscholastik*, Edizioni di "Storia e letteratura": Rome 1949.

MALCOLM, Norman

1942 "Moore and Ordinary Language" in *The Philosophy of G. E. Moore*, Library of Living Philosophers, Vol. 4, P. A. SCHILPP (ed.), Northwestern University: Evanston and Chicago 1942, pp. 345–368.
1951 Review of REICHENBACH 1951, *The Philosophical Review*, **60**: 4 (October 1951), 582–586.

MANNINEN, Juha – STADLER Friedrich (eds.)
2010 *The Vienna Circle in the Nordic Countries: Networks and Transformations of Logical Empiricism*, Dordrecht: Springer 2010.

MATES, Benson
1951 "Analytical Sentences", *The Philosophical Review*, **60**: 4 (October 1951), 525–534.

MATSON, Wallace I.
1955 "Cornford on the Birth of Metaphysics", review of CORNFORD 1952, *Review of Metaphysics*, **8**: 3 (March 1955), 443–454.

MEIER, Carl A. (ed.)

1992 *Wolfgang Pauli und C. G. Jung: Ein Briefwechsel 1932–1958*, Springer: Berlin and Heidelberg 1992.

MOORE, George E.

1903 *Principia Ethica*, Cambridge University Press: Cambridge 1903, T. Baldwin (ed.), Cambridge University Press: Cambridge 1993².

MORRIS, Charles

1960 "On the History of the International Encyclopedia of Unified Science", *Synthese*, **12**: 4 (December 1960), 517–521.

MYRVOLD, Wayne, C.

2003 "On some early objections to Bohm's theory", *International Studies in the Philosophy of Science*, **17**: 1 (July 2020) 7–24.

NAGEL, Ernest – SUPPES, Patrick – TARSKI, Alfred (eds.)

1949 "The Meaning of Reduction in the Natural Sciences" in *Science and Cillilization*, R. STAUFFER (ed.), University of Wisconsin Press: Madison 1949, pp. 99–145.

1961 *The Structure of Science. Problems in the Logic of Scientific Explanation*, Routledge & Kegan Paul: London 1966².

1962 *Logic, Methodology and Philosophy of Science: Proceedings of the 1960 International Congress*, Stanford University Press: Stanford (CA) 1962.

NEUMANN, John von

1932 *Mathematische Grundlagen der Quantenmechanik*, Springer: Berlin 1932, Springer: Berlin, Heidelberg, New York 1996.² English translation by R. T. Beyer, *Mathematical Foundations of Quantum Mechanics*, Princeton University Press: Princeton (NJ) 1955.

NEURATH, Otto – CARNAP, Rudolf – HAHN, Hans

1929 *Wissenschaftliche Weltauffassung: Der Wiener Kreis*, Veröffentlichungen des Vereines Ernst Mach, Verein Ernst Mach, Artur Wolf Verlag: Vienna 1929. English translation by M. Neurath and R. S. Cohen, "The Scientific Conception of the World: The Vienna Circle" in Otto Neurath, *Empiricism and Sociology*, NEURATH–COHEN (eds.), Kluwer: Dordrecht 1973, pp. 299–318.

OBERDAN, Thomas

1990 "Positivism and the Pragmatic Theory of Observation", *PSA 1990,* Fine–Forbes–Wessels (eds.), Proceedings of the Biennial Meeting of the Philosophy of Science Association, 1990, pp. 25–37.

OBERHEIM, Eric

2006 *Feyerabend's Philosophy*, de Gruyter: Berlin and New York 2006 2006 2024 "The Limited Validity of Falsificationism. Feyerabend's theoretical pluralism and its relation to Popper, Wittgenstein and Bohm", *Philosophical Aspects of*

Origin. https://fag.ifil.uz.zgora.pl/index.php/fag. Oberheim, Eric – Hoyningen-Huene, Paul 2024 "The Incommensurability of Scientific Theories" <https://plato.stanford.edu/archives/sum2024/entries/incommensurability/>.

PAP, Arthur
1949a *Elements of Analytic Philosophy*, MacMillan: New York 1949.
1949b "Synonymity and Logical Equivalence", *Analysis*, **9**: 4 (March 1949), 51–57.
1950 "Logic and the Concept of Entailment," *The Journal of Philosophy*, **47**: 13 (22 June 1950), 378–387.
1953 "Reduction Sentences and Open Concepts", *Methodos*, **5**: 17 (1953), 3–30 in *The Limits of Logical Empiricism: Selected Papers of Arthur Pap*, KEUPINK–SHIEH (eds.), Springer: Dordrecht 2006, Ch. 19.

PAP, Pauline
2006 "Arthur Pap: Biographical Notes" in *The Limits of Logical Empiricism: Selected Papers of Arthur Pap*, KEUPINK–SHIEH (eds.), Springer: Dordrecht 2006, pp. 369–374. PARRY, Adam 1956 "The Language of Achilles", *Transactions and Proceedings of the American Philological Association*, **87** (1956), 1–7; reprinted in *The Language and Background of Homer. Some Recent Studies and Controversies*, G. S. Kirk (ed.), pp. 48–54.

PARRY, Adam
1956 "The Language of Achilles", *Transactions and Proceedings of the American Philological Association*, 87 (1956), 1–7; reprinted in *The Language and Background of Homer. Some Recent Studies and Controversies*, G. S. Kirk (ed.), pp. 48–54.

PAULI, Wolfgang
1926 "Quantentheorie" in *Handbuch für Physik*, 24 Voll., GEIGER–SCHEEL (eds.), 1926–33, Vol. 23, Springer: Berlin, Vienna 1926–1933, pp. 1–278.
1933 "Die allgemeinen Prinzipien der Wellenmechanik" in *Handbuch für Physik*, 24 Voll., GEIGER–SCHEEL (eds.) 1926–33, Vol. 23, Springer: Berlin, Vienna 1926–1933, pp. 83–269.

PEAT, David F.
1996 *Infinite Potential: The Life and Times of David Bohm*, Addison-Wesley/Perseus Books/Helix Books/Basic Books: 1996.

PEIERLS, Rudolf Ernst
1935 "The Fundamental Paradox of the Quantum Theory", *Nature*, **136**: 3436 (7 September 1935), 395–395.

PINTO DE OLIVERIA, José Carlos
2007 "Carnap, Kuhn, and revisionism: on the publication of Structure in Encyclopedia", *Journal for General Philosophy of Science*, **38**: 1 (April 2007), pp. 147–157.

POPPER, Karl R.

1935 *Logik der Forschung. Zur Erkenntnistheorie der modernen Naturwissenschaft*, Springer: Vienna 1935. English translation by K. R. Popper, J. Freed and L. Freed in POPPER 1959a.

1950a "Indeterminism in Quantum Physics and in Classical Physics, Part I", *The British Journal for the Philosophy of Science*, **1**: 2 (August 1950), 117–133.

1950b "Indeterminism in Quantum Physics and in Classical Physics, Part II", *The British Journal for the Philosophy of Science*, **1**: 3 (November 1950), 173–195.

1957 "The Propensity Interpretation of the Calculus of Probability, and the Quantum Theory" in *Observation and Interpretation: A Symposium of Philosophers and Physicists, Proceedings of the Ninth Symposium of the Colston Research Society*, KÖRNER–PRYCE (eds.), Butterworths Scientific Pub.: London and New York 1957, pp. 65–67 and 88–89, correction in *The British Journal for the Philosophy of Science*, **8**: 32 (February 1958), 301n1.

1959a *The Logic of Scientific Discovery*. English translation by Karl R. Popper, Julius Freed and Lan Freed, Hutchinson: London (January) and Basic Books: New York (March) 1959, Routledge: London 2002.

1959b "Back to the Presocratics", *Proceedings of the Aristotelian Society*, N.S. Vol. **59** (1958–59), 1–24, reprinted in *Conjectures and Refutations: The Growth of Scientific Knowledge*, Routledge and Kegan Paul 1963, Ch. 5.

1963a "The Demarcation Between Science and Metaphysics" in *The Philosophy of Rudolf Carnap*, A. SCHILPP (ed.), Open Court: La Salle (IL) 1963, pp. 183–226, reprinted in POPPER 1963b, Ch. 11.

1963b *Conjectures and Refutations. The Growth of Scientific Knowledge*, Routledge and Kegan Paul 1963.

1966 "A Theorem on Truth-Content" in *Mind, Matter, and Method: Essays in Philosophy and Science in Honor of Herbert Feigl*, FEYERABEND–MAXWELL (eds.), University of Minnesota Press: Minneapolis 1966, pp. 343–363.

1974 "Autobiography of Karl Popper" in *The Philosophy of Karl Popper*, P. A. SCHILPP (ed.), Open Court: La Salle (IL) 1974, pp. 1–181, reprinted as an independent volume in a revised edition as POPPER 1976.

1974b "Replies to My Critics" in *The Philosophy of Karl Popper*, P. A. SCHILPP (ed.), Open Court: La Salle IL 1974, pp. 961–1197.

1976 *Unended Quest: An Intellectual Autobiography*, revised edition of POPPER 1974a, 1974b, Fontana/Collins: London 1976, 1978^{2R}, 19823RE, 19864^{E} Routledge: London and New York 1992, 2002^{2R}.

PUTNAM, Hilary

1957 "Three-valued Logic", *Philosophical Studies*, **8**: 5 (October 1957), 73–80.

1965 "How Not to Talk About Meaning" in *Proceedings of the Boston Colloquium for the Philosophy of Science, 1962–64: In Honor of Philipp Frank*, COHEN–WARTOFSKY (eds.), Boston Studies in the Philosophy of Science, Vol. II, Humanities Press: New York 1965, pp. 205–222.

QUINE, Willard van Orman
1951a "Two Dogmas of Empiricism", *The Philosophical Review*, **60**: 1 (January 1951), 20–43, reprinted in *From a Logical Point of View*, Harvard University Press: Cambridge (MA) 1953, 1961², 1980³, pp. 20–46.
1951b *Mathematical Logic*, Harvard University Press: Cambridge (MA) 1951, 1962².

RADNER, Michael – WINOKUR, Stephen (eds.)
1970 "Discussion at the Conference on Correspondence Rules" in *Analyses of Theories and Methods of Physics and Psychology*, Minnesota Studies in the Philosophy of Science, Vol. IV, University of Minnesota Press: Minneapolis 1970, pp. 220–259.

REICHENBACH, Hans
1944 *Philosophic Foundations of Quantum Mechanics*, University of California Press: Los Angeles and Berkeley 1944, Dover: New York 1988².
1947 *Elements of Symbolic Logic*, Macmillan: New York 1947, The Free Press: New York 1966, Dover: New York 1980.
1951 *The Rise of Scientific Philosophy*, University of California Press: Berkeley and Los Angeles 1951.

REISCH, George A.
1991 "Did Kuhn Kill Logical Empiricism?", *Philosophy of Science* **58**: 2 (1991), 264–277.
1994 "Planning science: Otto Neurath and the International Encyclopedia of Unified Science", *The British Journal for the History of Science*, **27**: 2 (June 1994), 153–175.
2004 *How the Cold War Transformed Philosophy of Science: To the Icy Slopes of Logic*, Cambridge University Press: Cambridge 2004.

RICHARDSON, Alan
2007 "'That Sort of Everyday Image of Positivism': Thomas Kuhn and the Decline of Logical Empiricist Philosophy of Science" in *The Cambridge Companion to Logical Empiricism*, RICHARDSON–UEBEL (eds.), Cambridge University Press: Cambridge 2007, pp. 346–369.

RONCHI, Vasco
1956 *Histoire de la lumière*, Librairie Armand Colin: Paris 1956. French translation by J. Taton, *Storia della luce*, Zanichelli: Bologna 1939.

ROSENFELD, Léon
1957a "Cause in Physics", review of BOHM 1957a, *The Manchester Guardian*, 28 June 1957, p. 10.
1957b "Misunderstandings about the foundations of quantum theory" in KÖRNER–PRYCE (eds.), Butterworths Scientific Pub.: London and New York 1957, pp. 41–46.

ROZEBOOM, William W.

1962a "Intentionality and Existence", *Mind*, N. S., **71**: 281 (January 1962), 15–32.

1962b "The Factual Content of Theoretical Concepts" in *Scientific Explanation, Space and Time*, FEIGL–MAXWELL (eds.), Minnesota Studies in the Philosophy of Science, Vol. III, 1962, pp. 273–357.

RUDNER, Richard

1950 "A Note on Likeness of Meaning", *Analysis*, **10**: 5 (April 1950), 115–118.

RUSSELL, Bertrand

1914 *Our Knowledge of the External World: As a Field for Scientific Method in Philosophy*, Open Court: Chicago and London 1914, Routledge, 1993

RYLE, Gilbert

1949 *The Concept of Mind*, Hutchinson: London 1949.

1954 *Dilemmas*, Cambridge University Press: Cambridge 1954.

SALMON, Wesley C.

1966 "Verifiability and Logic" in *Scientific Explanation, Space and Time*, FEIGL–MAXWELL (eds.), Minnesota Studies in the Philosophy of Science, Vol. III, 1962, pp. 354–376.

SCHILPP, Paul A. (ed.)

1963 *The Philosophy of Rudolf Carnap*, The Library of Living Philosophers, Vol. XI, Open Court: La Salle (IL) 1963.

SCHELER, Max

1916 *Der Formalismus in der Ethik und die materiale Wertethik: Neuer Versuch der Grundlegung eines ethischen Personalismus*, 2 Parts, Max Niemeyer: Halle an der Saale 1913–1916, 1921^2, 1927^3, reprinted in *Gesammelte Werke*, Vol. 2, M. SCHELER (ed.), Francke-Verlag: Bern and Munich 1954, M. S. FRINGS (ed.), 1980^2, English translation by M. S. Frings and R. L. Funk, *Formalism in Ethics and Non-Formal Ethics of Values: A new attempt toward the foundation of an ethical personalism*, Northwestern University Press: Evanston (IL) 1973.

SCHLICK, Moritz

1931 "The Future of Philosophy" in *Proceedings of the Seventh International Congress of Philosophy: held at Oxford, England, September 1–6, 1930*, G. RYLE (ed.), H. Milford: London 1931, pp. 112–116, reprinted in *Philosophical Papers, Vol. 2* (1925–1936), MULDER–VAN DE VELDE–SCHLICK (eds.), Reidel: Dordrecht 1979, pp. 171–175.

1934 "Über das Fundament der Erkenntnis", *Erkenntnis*, **4** (1934), 79–99. English translation by P. Heath, "On the Foundation of Knowledge", in *Philosophical Papers, Volume 2*, MULDER–VAN DE VELDE–SCHLICK (eds.), Reidel: Dordrecht, Boston, London 1979, pp. 370–387.

SCHORNER, Michael

2010 "Comeback auf Umwegen. Die Rückkehr der Wissenschaftstheorie in Österreich" in *Vertreibung, Transformation und Rückkehr der Wissenschaftstheorie: Am Beispiel von Rudolf Carnap und Wolfgang Stegmüller*, F. STADLER (ed.), LIT: Vienna and Berlin 2010: pp. 189–252.

SCHRÖDINGER, Erwin

1935 "On the Present Status of Quantum Mechanics", *Die Naturwissenschaften*, **23**: 48 (November 1935), 807–812, **23**: 49 (December 1935): 823–828, **23**: 50 (December 1935), 844–849.

1952a "Are There Quantum Jumps? Part I", *The British Journal for the Philosophy of Science*, **3**: 10 (August 1952), 109–123.

1952b "Are There Quantum Jumps? Part II", *The British Journal for the Philosophy of Science*, **3**: 11 (November 1952), 233–242.

SCHWINGER, Julian S. (ed.)

1958 *Selected Papers on Quantum Electrodynamics*, Dover: New York 1958.

SCRIVEN, Michael

1958 "Definitions, Explanations and Theories" in Minnesota Studies in the Philosophy of Science, Vol. 2, FEIGL–MAXWELL (eds.), 1958, University of Minnesota: Minneapolis 1958, pp. 99–195.

1966 "The Limitations of the Identity Theory" in *Mind, Matter, and Method: Essays in Philosophy and Science in Honor of Herbert Feigl*, FEYERABEND–MAXWELL (eds.), University of Minnesota Press: Minneapolis 1966, pp. 191–197.

SELLARS, Wilfrid S.

1965 "Scientific Realism or Irenic Instrumentalism: A Critique of Nagel and Feyerabend on Theoretical Explanation" in *Proceedings of the Boston Colloquium for the Philosophy of Science, 1962–64: In Honor of Philipp Frank*, COHEN–WARTOFSKY (eds.), Boston Studies in the Philosophy of Science, Vol. II, Humanities Press: New York 1965, pp. 171–204.

1966 "The Refutation of Phenomenalism: Prolegomena to a Defense of Scientific Realism" in *Mind, Matter, and Method: Essays in Philosophy and Science in Honor of Herbert Feigl*, FEYERABEND–MAXWELL (eds.), University of Minnesota Press: Minneapolis 1966, pp. 198–214.

SMART, John J. C.

1965 "Conflicting Views about Explanation" in *Proceedings of the Boston Colloquium for the Philosophy of Science, 1962–64: In Honor of Philipp Frank*, COHEN–WARTOFSKY (eds.), Boston Studies in the Philosophy of Science, Vol. II, Humanities Press: New York 1965, pp. 157–169.

SOMMERFELD, Arnold (ed.)
1923 *The Principle of Relativity: A Collection of Original Memoirs on the Special and General Theory of Relativity*, Methuen & Co.: London 1923, Dover: New York 1952².

STADLER, Friedrich (ed.)
2010 *Vertreibung, Transformation und Rückkehr der Wissenschaftstheorie: Am Beispiel von Rudolf Carnap und Wolfgang Stegmüller*, LIT: Vienna and Berlin 2010.

STADLER, Friedrich – FISCHER, Kurt R. (eds.)
2006 *Paul Feyerabend: Ein Philosoph aus Wien*, Springer: Vienna and New York 2006.

STEVENS, Stanley S.
1966 "Quantifying the Sensory Experience" in *Mind, Matter, and Method: Essays in Philosophy and Science in Honor of Herbert Feigl*, FEYERABEND–MAXWELL (eds.), University of Minnesota Press: Minneapolis 1966, pp. 215–233.

TARSKI, Alfred
1936 "Der Wahrheitsbegriff in den formalisierten Sprachen", *Studia Philosophica*, **1** (1936), 261–405. English translation by J. H. Woodger, "The Concept of Truth in Formalized Languages" in *Logic, Semantics, Metamathematics: Papers from 1923 to 1938*, Oxford University Press: Oxford 1956, J. CORCORAN (ed.), Hackett: Indianapolis 1983², pp. 152–278.

TEMPLE, George Frederick James
1935a "The Fundamental Paradox of the Quantum Theory", *Nature*, **135**: 3423 (8 June 1935), 957–957.
1935b "The Fundamental Paradox of the Quantum Theory", *Nature*, **136**: 3431 (3 August 1935), 179–180.

TER HAAR, Dirk
1955 "Foundations of Statistical Mechanics", *Reviews of Modern Physics*, **27**: 3 (July-September 1955), 289–338.

TOWNSEND, Burke
1971 "Feyerabend's Pragmatic Theory of Observation and the Comparability of Alternative Theories" in *PSA 1970*, BUCK–COHEN (eds.), Boston Studies in the Philosophy of Science, Vol. 8. Springer: Dordrecht 1971.

TRANEKJÆR RASMUSSEN, Edgar
1949 "Psychologie und Gesetzeserkenntnis" in *Gesetz und Wirklichkeit*, 4. Internationale Hochschulwochen des Österreichischen College, Alpbach (Tyrol), 21. August – 9. September, 1948, S. MOSER (ed.), Tyrolia Verlag: Innsbruck, Vienna 1949, pp. 111–???

UEBEL, Thomas

1991 "Neurath's Programme for Naturalistic Epistemology", *Studies in the History and Philosophy of Science Part A*, **22**: 4 (December 1991), 623–646.

2007 *Empiricism at the Crossroads: The Vienna Circle's Protocol-sentence Debate*, Open Court: Chicago 2007.

VAN STRIEN, Marij

2019 "Pluralism and Anarchism in Quantum Physics: Paul Feyerabend's writings on quantum physics in relation to his general philosophy of science", *Studies in History and Philosophy of Science Part A*: **80**: (2020), 1–18.

2024 "Why Bohm was never a determinist" in *Guiding Waves in Quantum Mechanics: 100 Years of de Broglie-Bohm Pilot-Wave Theory*, OLDOFREDI (ed.), Oxford University Press: Oxford 2024.

WAISMANN, Friedrich

1938 "Symposium: The Relevance of Psychology to Logic III", *Proceedings of the Aristotelian Society, Supplementary Volumes*, **17** (1938), 54–68.

1950 "Analytic-Synthetic II", *Analysis*, **11**: 2 (December 1950), 25–38.

WATKINS, John W. N.

1970 "Against 'Normal Science'" in *Criticism and the Growth of Knowledge, Proceedings of the International Colloquium in the Philosophy of Science, London, 1965, Vol. 4*, LAKATOS–MUSGRAVE (eds.), Cambridge University Press: London and New York 1970, pp. 25–37.

WAX, Nelson (ed.)

1954 *Selected Papers on Noise and Stochastic Processes*, Dover: New York 1954.

WEIZSÄCKER, Carl Friedrich von

1955 "Komplementarität und Logik. Niels Bohr zum 70. Geburtstag am 7. 10. 1955 gewidmet", *Die Naturwissenschaften*, **42**: 19 (October 1955), 521–529 and **42**: 20 (October 1955), 545–555.

WHITE, Morton G.

1950 "The analytic and the synthetic: an untenable dualism" in *John Dewey: Philosopher of Science and Freedom*, S. HOOK (ed.), Dial Press: New York 1950, pp. 316–330.

WILKINS, Maurice

1987 Interview of David Bohm by Maurice Wilkins on 1987 January 30, Niels Bohr Library & Archives, American Institute of Physics: College Park (MD), www.aip.org/history-programs/niels-bohr-library/oral-histories/32977-7

WITTGENSTEIN, Ludwig

1921 *Tractatus Logico-Philosophicus*, English translation by C. K. Ogden, Routledge & Kegan Paul: London 1922. English translation by D. F. Pears and B. F. McGuinness, Humanities Press: New York 1961.

1953 *Philosophical Investigations*, ANSCOMBE–RHEES (eds.). English translation by G. E. M. Anscombe, Blackwell: Oxford 1953, 1958[2], 2001[3], revised by P. M. S. Hacker and J. Schulte, Blackwell: Oxford 2009[4].

Index

M. Collodel, E. Oberheim (eds.), *Feyerabend's Formative Years. Volume 2. Feyerabend on Logical Empiricism, Bohm & Kuhn*, Vienna Circle Institute Library 10, https://doi.org/10.1007/978-3-031-57519-8

The manufacturer's authorised representative in the EU is Springer Nature Customer Service Centre GmbH, Europaplatz 3, 69115 Heidelberg, Germany. If you have any concerns regarding our products, please contact ProductSafety@springernature.com

Printed and bound by CPI Group (UK) Ltd, Croydon, CR0 4YY
30/09/2026
02225599-0017